AELIAN

ON THE CHARACTERISTICS
OF ANIMALS

I

BOOKS I—V

TO

A. S. F. G.

PREFACE

NINETY-THREE years have elapsed since Aelian's *De natura animalium* was edited for the Teubner series by Rudolf Hercher. His text was a revision of that which he had published six years earlier, in 1858, in the Didot series. Both these books have long been out of print and almost unobtainable. In one respect the Teubner edition is inferior to its predecessor, since the editor gives no more than a bare ' Index mutationum praeter codices factarum ' without specifying which ' codices ' he has used, and those who are concerned to know how he explains or defends some of his frequent desertions of the manuscripts must still turn to the preface and the ' Adnotatio critica ' of the Didot edition. It was Hercher's service to have detected the prevalence of glosses and interpolations, although in expelling them he is conscious that some will think that he has exceeded all bounds (Didot ed., *Praef.* p. ii). The text here printed is substantially that of Hercher's edition of 1864, and divergences from it are shewn in the critical notes, which lay no claim to be exhaustive. In 1902 E. L. De Stefani made a survey of the manuscripts in Continental libraries [1] and

---

[1] The British Museum Burney MS 80 contains only excerpts in a 16th-cent. hand; there is no MS of the *NA* in Bodley or in the Cambridge University Library, and I have not sought farther afield.

established their relations. It is to be regretted that no one has yet come forward to complete the task which he began and to provide a fresh text, with an adequate critical apparatus.

In rendering the names of Aelian's birds and fishes I have taken as my guides the two *Glossaries* of Sir D'Arcy Thompson, one of *Greek Birds* (2nd ed., 1936), the other of *Greek Fishes* (1947.) Botanical terms are those given on the authority of Sir William Thiselton-Dyer in the ninth edition of Liddell and Scott's *Greek Lexicon*. In identifying Aelian's reptiles and insects the various articles contributed, jointly or separately, by H. Gossen and A. Steier, by M. Wellmann and others to Pauly-Wissowa's *Real-Encyklopädie* have been of service. In 1935 Gossen published a systematic catalogue of all Aelian's animals, and perhaps I shall be blamed for not following him more often than I have done. In determining the modern equivalents and the scientific nomenclature of the fauna and flora of Ancient Greece the oracles do not always speak with one voice, and the best that a layman can hope for is that, when two or more interpretations have presented themselves, the result of his choice may be judged, if not correct, at any rate excusable.

My thanks are due to those who have kindly solved for me various problems that arose in the course of my work: to Professor H. W. Bailey, Professor W. I. B. Beveridge, Professor F. E. Fritsch (†), Dr. D. A. Parry, Dr. M. G. M. Pryor, Dr. G. Salt, Mr. A. F. Huxley, Mr. J. E. Raven. But my heaviest obligations are to Mr. A. S. F. Gow, who read considerable portions of my translation in typescript, saved me from more blunders than I care

PREFACE

to think of, and besides improving my English offered a number of corrections to the Greek text which I have gladly and gratefully adopted. The Syndics of the Cambridge University Press have courteously allowed me to reproduce two passages from an edition of Nicander published by them in 1953.

A. F. S.

Cambridge,
  1957.

INTRODUCTION

Life

THE life of Aelian has been sketched by his contemporary Flavius Philostratus (2nd–3rd cent. A.D.) in his *Lives of the Sophists* (2. 31), and he is the subject of a brief notice in 'Suidas.' Claudius Aelianus was born at Praeneste about the year A.D. 170. He came of *libertus* stock and assumed the name of Claudius. At Rome he studied under Pausanias of Caesarea, a noted rhetorician and pupil of Herodes Atticus for whom Aelian reserved his chief admiration. Although a Roman, as he himself is proud to assert (*VH* 12. 25; 14. 45), he obtained such a mastery of the Attic idiom that he came to be known as 'the honey-tongued or honey-voiced,' while his success as a declaimer was rewarded by the bestowal of the title of *Sophist*. (By the end of the second century the term had ceased to bear any philosophical implications and had come to denote one who taught or practised rhetoric.) Nevertheless, mistrusting, it may be, his ability to maintain his hold over pupils and audiences—for the demands on a successful rhetorician were heavy—he devoted himself to the writing of 'history' (τῷ ξυγγράφειν ἐπέθετο, Phil.). He held the office of ἀρχιερεύς presumably at Praeneste, but the greater part of his time must have been spent in Rome, where he had access to libraries and enjoyed the patronage of the empress Julia Domna, who had gathered around her on the Palatine a circle of learned men that included Oppian, Serenus Sammonicus, Galen, Philostratus, and others who figure in the *Deipnosophists* of Athenaeus.[1] It was his boast that he had never been outside Italy, had never been aboard a ship, and knew nothing of the sea—statements which most readers will find no difficulty in accepting.[2] He was over sixty years of age when he died, unmarried.

Works

Besides the *De natura animalium* (to give it the name by which it is commonly referred to) two other works by Aelian have survived—Ἐπιστολαὶ ἀγροικίαι, a literary exercise in the form of twenty-four letters, vignettes of life in the country, some with an erotic motive; and Ποικίλη ἱστορία (*Varia historia*) in fourteen books, beginning with some chapters on natural history, but consisting in the main of anecdotes historical and biographical, with excursions into mythology, and a variety of other topics. The greater part as we have it seems to be from the hand of an epitomator. It resembles the *De natura animalium* in its deliberate avoidance of any systematic order. Fragments of two treatises, Περὶ προνοίας and Περὶ θείων ἐναργειῶν have been

---

[1] J. Bidez in *Camb. Anc. Hist.* 12. 613; see also Wellmann in *Hermes* 51. 1.

[2] The words ἐθεασάμην ἐν τῇ πόλει τῇ Ἀλεξανδρέων (*NA* 11. 40) occur in a chapter borrowed wholly from Apion, and Wellmann (*RE* 1. 486) considers that Aelian is simply transcribing his authority. M. Croiset (*Hist. de la lit. gr.* 5. 774) demurs to this view; his explanation seems to me unconvincing.

preserved, most of them in 'Suidas.' So far as we can judge they were collections of stories illustrating heaven's retribution on unbelievers. Aelian has some bitter words for the scepticism of the Epicureans. A bare mention is enough for two sets of epigrams inscribed ἔνεκ' Αἰλιανοῦ, on 'herms' of Homer and Menander which are supposed to have stood in Aelian's house at Rome.[1]

The *De natura animalium* is a miscellany of facts, genuine or supposed, gleaned by Aelian from earlier and contemporary Greek writers (no Latin writer is once named) and to a limited extent from his own observation to illustrate the habits of the animal world. We are of course prepared to encounter much that modern science rejects, but the general tone with its search after the picturesque, the startling, even the miraculous, would justify us in ranking Aelian with the Paradoxographers rather than with the sober exponents of Natural History. Mythology, mariners' yarns, vulgar superstitions, the ascertained facts of nature—all serve to adorn a tale and, on occasion, to point a moral. His religion is the popular Stoicism of the age: Aelian repeatedly affirms his belief in the gods and in divine Providence; the wisdom and beneficence of Nature are held up to veneration; the folly and selfishness of man are contrasted with the untaught virtues of the animal world. Some animals, to be sure, have their failings, but he chooses rather to dwell upon their good qualities, devotion, courage, self-sacrifice, gratitude. Again, animals are guided by Reason, and from them we may learn contentment, control

[1] See G. Kaibel, ed., *Epigrammata Graeca ex lapidibus conlecta* (Berol. 1878), nos. 1084–5.

of the passions, and calm in the face of death. Suicide is commended as an escape from the ills of life, and riches are to be despised. Aelian's Stoicism hardly goes below the surface. His primary object is to entertain and while so doing to convey instruction in the most agreeable form. He was among the first to break away from the age-long tradition of the periodic structure of sentences, at least for works of a serious nature, and to affect a simpler prose of short, co-ordinated, sometimes paratactic, clauses. In this and in the rich variety of topics and in a certain fondness for piquant, not to say earthy, stories from the life of men and of animals one may trace the influence of the Milesian Tales. Unfettered by any canons of style or language, picaresque, and sometimes gross, they pandered to popular taste. To adopt their technique while refining the style and imparting a moral flavour to his narratives may well have seemed to Aelian a sure way of gaining a like popularity with educated readers.[1] Some might find fault with his random and piece-meal handling of his theme—of that he is well aware, and in the *Epilogue* he defends himself with the plea that a frequent change of topic helps to maintain the reader's interest and saves him from boredom, But as to the permanent value of his work he has no misgivings, and since Philostratus informs us that his writings were much admired, we may assume that they appealed to cultivated circles in a way that the voluminous and possibly arid compilations of grammarians did not.

[1] See W. Schmid, *Der Atticismus*, 3. 7 ff.

INTRODUCTION

Sources

The principal sources of the *De natura animalium* have been investigated by Max Wellmann and Rudolf Keydell in a series of articles which appeared in the journal *Hermes* between the years 1891 and 1937. Here it will be enough to state their conclusions and to indicate some of the reasons for them.

That the name of Aristotle should occur over fifty times in a work professing to deal with animals will surprise no one. Yet it is certain that Aelian knew Aristotle only at second hand through the epitome of his zoological works made by Aristophanes of Byzantium (3rd/2nd cent. B.C.). Even so there is little enough of genuine descriptive zoology, and it was not in any purely zoological work that Aelian found his chief inspiration and guide. It is noticeable how often his statements regarding the names, habits, and characteristics of animals reflect in their manner of presentation, their content and style, the comments of scholiasts and writers like Athenaeus, Clement of Alexandria, and Pollux, who took their materials from grammarians. It became a mannerism with the scholars of Alexandria to cite Homer whenever it was possible, and Aelian follows the fashion, less (so it would seem) with an aim to establishing some fact of natural history than to proving Homer's knowledge of the science. Specimens of grammarian's lore meet us in the excursions into etymology and lexicography, in the myths and proverbs relating to animals, with their illustrations from dramatists and poets, and in a wealth of other matter which a professed zoologist would disregard as being irrelevant. Aelian is not, like Athenaeus, scrupulous in always naming his authorities, as we shall see later, but from parallel passages in other writers ranging from Plutarch and Athenaeus down to the *Geoponica* (5th cent. A.D.) in which Pamphilus is expressly named as being the source, Wellmann concludes that the pattern and the chief source for Aelian was Pamphilus of Alexandria. He in his turn had based his work upon that most voluminous of grammarians Didymus, nicknamed Χαλκέντερος, excerpting and abridging into one work a number of separate treatises by his forerunner.[1] The title of the work is given by 'Suidas' as Λειμών, and he adds ἔστι δὲ ποικίλων περιοχή. It must have been a miscellany of ample scope embracing mythology, natural history, and *paradoxa* or 'tales of wonder,' historical and biographical notices, all derived from earlier Greek literature. In a number of places Aelian has grouped together, more or less closely, chapters derived from one and the same authority: thus, 12. 16–20 come from Democritus; 4. 19, 21, 26–7, 32, 36, 41, 46, 52 from Ctesias; 16. 2–22 from Megasthenes; 17. 31–4 from Amyntas. From this it would seem that his exemplar was arranged partly by animals and partly by authors.

Aelian has given us accounts of over one hundred birds. Many of his accounts correspond with those which we find in Athenaeus ((9. 387F–397c), but since Aelian is generally more detailed, the resemblances are to be traced to the use of a common source. For Athenaeus the principal authority on birds was ' that best of all ancient ornithologists, Alexander the

---

[1] Wellmann detects a hidden allusion to its title in some words of Aelian's *Epilogue*, οἱονεὶ λειμῶνά τινα ἢ στέφανον . . . ᾠήθην δεῖν τήνδε . . . διαπλέξαι τὴν συγγραφήν.

INTRODUCTION

Myndian,'[1] whom he cites more often than any other writer on natural history, Aristotle alone excepted, viz. thirteen times in Book IX and four times elsewhere. Photius describes him as having collected ' a multitude of marvellous, even incredible, tales from earlier writers touching animals, trees, places, rivers, plants, and the like.'[2] Aelian names him five times, and in a chapter (3. 23) relating to storks and their transformation into human beings takes occasion to praise his knowledge and to express his own belief in the story. It is not stretching probability to see in Alexander the source for Aelian's accounts of similar transformations (*e.g.* 1. 1; 5. 1; 15. 29), and for much besides, whether of fact or fable, regarding birds, their assignment to special gods (1. 48; 2. 32; 4. 29; 10. 34–5; 12. 4; and cp. Ath. 9.388A), their significance as omens (3. 9; 10. 34, 37; and cp. Plut. *Marius* 17, Artem. *Oneir.* 2. 66). Nevertheless since Athenaeus and Aelian concur in misrepresenting him on the spelling of σκώψ, it may be questioned whether they had direct access to his writings and whether their common error is not due to Pamphilus; see note on Ael. 15. 28. In his description of the κατώβλεπον (7. 5) Aelian differs from the account given by Alexander in Ath. 5. 221B.

Among ancient writers who treated of poisons and their antidotes the principal authority was Apollodorus (3rd cent. B.C.). Two of his works, or the essence of them, survive in the poems of Nicander. But though Aelian on seven occasions adduces Nicander as witness, there are discrepancies which

[1] D. W. Thompson, *Glossary of Greek birds*, p. vi.
[2] Fragments collected by Wellmann in *Hermes* 26. 546–55.

preclude the idea of a direct use of the poet. There are however indications that Aelian and the scholia to Nicander drew from a common source. Aelian states (9. 26) that the Agnus-castus, an antidote to snakebites, was used at the Thesmophoria to ensure chastity : the same note occurs in Σ Nic. *Th.* 71. In 9. 20 Aelian states on the authority of ' Aristotle ' (*Mirab.* 841 a 27) that the ' Pontic stone ' if burnt expels snakes : Σ Nic. *Th.* 45 cite the same passage. In 6. 51 ' Sostratus,' we are told, ' describes the Dipsas as *white.*' Here Aelian has forsaken Apollodorus-Nicander, who had written (*Th.* 337) ὑποζοφόεσσα μελαίνεται, and he then proceeds to tell the myth of the Dipsas and the Ass, adding that it has been treated by Sophocles (and other poets): Σ Nic. *Th.* 343 state specifically ' Sophocles ἐν Κωφοῖς.' (Clearly Σ did not borrow from Aelian.) The story of the Beaver and its self-mutilation is told by Aelian (6. 34); it is mentioned in Σ Nic. *Th.* 565, and Sostratus is named as the authority for it. From Ael. 4. 51 and 6. 37 we learn the difference between οἶστρος and μύωψ : according to Σ Ap. Rh. 1. 1265 and Σ Theoc. 6. 28 the distinction was first noted by Sostratus, though Aelian is the first to mention it. It seems then that Sostratus in his two works Περὶ βλητῶν καὶ δακετῶν and Περὶ ζῴων treated of insects as well as the lower animals and snakes. As a zoologist his reputation stood next to Aristotle, and we are justified in assuming that both for Aelian and for the scholiast on Nicander he was the source for more than they have openly acknowledged, in the case of Aelian for 1. 20–22; 6. 36–8; 9. 39; 10. 44; 12. 8.

Aelian has much to tell us of elephants, both

those of Libya and of India. Like Pliny (*HN* 8. 1–34) before him and like Plutarch in his *De sollertia animalium*, Aelian has drawn extensively upon Juba II, King of Mauretania (*c.* 50 B.C.–*c.* A.D. 23). He was the first to maintain that the elephant's tusks are *horns* and not *teeth*, and Aelian follows him (8. 10 ; 11. 15 ; 14. 5). And since we learn from Pliny (*HN* 5. 16) that he wrote about the Atlas mountains and their forests, he is a likely source for all that Aelian relates touching Mauretania, its people, and its animals. The chapters on pearls (15. 8) and on Indian ants (16. 15) are to be traced to Juba's work *De expeditione Arabica*.

The knowledge which Aelian displays of Egypt and its topography, its local traditions, customs, and religious beliefs, especially those relating to birds and animals, can come only from a writer well acquainted with the land and its people. We are given mystical and mythological reasons for the reverence or detestation in which certain creatures are held (10. 19, 21, 46); there are tales of wonder ranging from the merely curious to the impossible; quotations from Homer are introduced into chapters on Egyptian religion. The pattern fits Apion (1st cent. A.D.). Born in the Great Oasis, he became head of the Alexandrian school, was a Homeric scholar and a pretender to omniscience. His *Aegyptiaca* was a compilation dealing with the history and the marvels of Egypt and was based upon earlier writers with additions from his own experience. One such there is which ' every schoolboy knows,' the story of Androcles and the Lion (Ael. 7. 48).[1] Chapters on

---

[1] A. Gellius 5. 14 [Apion] *Hoc . . . ipsum sese in urbe Roma vidisse oculis suis confirmat.*

the ibis, on Apis, on the hawk, the bird sacred to
Apollo, and the mystical explanations of the nature
and behaviour of animals (10. 15, 16, 18, 28, 45) may
with fair certainty be traced to Apion, and perhaps
the same writer is to be detected beneath the phrases
λέγουσιν Αἰγύπτιοι (12. 3) and Αἰγυπτίων λόγοι
(16. 39).

All this is not to claim for Aelian a first-hand
knowledge of the authors from whom he has borrowed. Even his knowledge of Herodotus is not
above suspicion.[1] Nor need it surprise us that a
writer at the end of the second century should take
full advantage of such encyclopaedic compilations as
lay within easy reach rather than be at the pains of
reading through the long array of authors listed in
Index IV. Great as his debt to Pamphilus may
have been, there are however writers later than
Pamphilus whom he must have studied for himself,
Leonidas of Byzantium, Demostratus, Telephus of
Pergamum; and there is good ground for believing
that he had read the *Halieutica* of Oppian and Plutarch's *De sollertia animalium*.

Of the many successors of Aristotle who made a
special study of fishes Oppian alone has survived in
his entirety. Plutarch devotes some chapters of
his *De sollertia animalium* to them, and Aelian has
named some 115 fresh- and salt-water fishes and
animals. In four places he cites as his authority

---

[1] Herodotus is cited thirteen times in the *NA*, but (strange
as it may seem) only once in the *VH*, of which Macan has
written (*Herodotus vii–ix*, vol. 2. 113), 'Even where an
anecdote agrees with or reproduces an Herodotean incident,
A. seems to have found it in some other source. More frequently the items in A. show little or no sign of Herodotean
colouring and are plainly drawn from independent sources.'

INTRODUCTION

Leonidas of Byzantium, and also in four places Demostratus. Although Aelian names neither Plutarch nor Oppian there are some striking parallels with both which cannot be due to chance and which the assumption of a common source does not sufficiently explain.

Plutarch (*De soll. anim.* 24. 976D) gives a list of four creatures that fall victims to others: Aelian (8. 6) gives the same list in the same order, almost in the same words. The two preceding chapters (8. 4 and 5) repeat the substance of Plut. 23. 976C. In excerpting from his forerunners Aelian tends to preserve the order of their narratives: thus,

Ael. 8. 16	=	Plut. 980 B, C
25	=	980 E
28	=	981 D, E
9. 3, part	=	982 D, pt. 1
9	=	982 D, pt. 2
13	=	982 E
17	=	983 B–D

Not the order alone but the choice of the same words points to Aelian having copied Plutarch, though the rhetorician has here and there added some flowers of his own; compare Ael. 8. 4 with Plut. 23. 976 A, B, Crassus and his moray; Ael. 8. 5 with Plut. 23. 976 C, divination by means of fish.

There are twenty-six chapters in Aelian which bear a marked resemblance to passages in the first book of Oppian's *Halieutica*, and of these more than half occur within the narrow compass of Aelian 9. 34 to 10. 8. Once again the order of Aelian's descriptions is with two exceptions the same as in Oppian.

INTRODUCTION

In three of the above passages there can be little doubt that Aelian has paraphrased Oppian: compare

In both we find the same fishes in the same order, and, what is most significant, since a prose-writer is not bound by the exigences of metre, the same use now of the singular, now of the plural. These three chapters cannot be separated from the other four-teen, so that it is at least likely that they too are paraphrases of Oppian. Of the remaining nine

INTRODUCTION

passages some may have been derived from Oppian, others more probably from a common source.

One such source was Leonidas of Byzantium.[1] From him Aelian derived the story of the friendship between a boy and a dolphin at Poroselene (2. 6), which recurs in Oppian (5. 448–518). In 2. 8 Aelian tells how dolphins help men in the catching of other fish, and a similar account is given by Oppian (5. 425–47): it is probable that both drew upon Leonidas. A comparison of Aelian's two chapters on poisonous fishes, 2. 44 and 50 (where Leonidas is named), with Opp. 2. 422–505 points certainly to him as their common source. Other passages indicate despite differences that both made use of the same authority, whether Leonidas or some other: compare

Ael. 1. 4	with Opp. 3. 323–6
5 (τρώκτης)	„ 144–8 (ἀμία)
19	„ 2. 141–66
27	„ 241–6
30	„ 128–40.

The researches of Leonidas extended as far as the Red Sea (Ael. 3. 18). For information on fishes in western waters Aelian relied upon one Demostratus, who differs from Leonidas in being independent of any Aristotelian tradition and in concentrating upon paradoxa. To him Wellmann would attribute the accounts contained in Ael. 13. 23; 15. 9, 12; per-

---

[1] Keydell (*Hermes* 72. 430 ff.) puts the date of Leonidas in the 2nd cent. A.D. Leonidas is reported as having himself seen the boy and dolphin; Pausanias (3. 25. 7) also was a witness, and Oppian says that the memory of the event is still fresh, for it happened 'not long ago but in our own generation,' the last quarter of the 2nd century. Granting that it is incredible that the boy rode upon the dolphin, the rest of the tale may well be true.

haps also 13. 27; 14. 4, 15, 20, 21, 24–7. Since he was a Roman,[1] some acquaintance with the fishlore of the west might be expected of him, and he may also have been the source of 13. 5, 16, 17; 14. 8, 22–3; 15. 2, 3.

It must be confessed that Aelian often tries our patience : we are irritated by his lack of any kind of system, by his repetitions, his inconsistencies, his servile credulity, his failure to verify statements where the facts were within reach, his style ' mit dem öden Schlamm der sophistischen Diction übergossen.' He is at times a careless copyist, and his claims as set forth in the *Epilogue* to have contributed original discoveries to the subject are questionable. Enough has been said to show that he was for the most part a retailer of other men's wares. Yet he has shown judgment in choosing his authorities, Aristophanes of Byzantium, Sostratus, Juba, Alexander of Myndus, whether in fact he consulted their writings at first hand or simply relied upon the excerpts which he found in Pamphilus.

For subsequent ages Aelian was an authority not to be neglected. There are echoes of his ' learning ' in that remarkable compound of animal lore and pious allegory known as ' the Bestiary.' [2] His name is linked with that of Aristotle in a compendium of

---

[1] See Ael. 15. 19. Wellmann would place him in the 2nd cent. B.C. That a Greek of that date should have been a Roman Senator, Keydell regards as impossible, and from the manner in which A. speaks of him believes him to have been a personal acquaintance of A.

[2] See *The Book of Beasts . . . transl. . . . and ed. by T. H. White* (Lond. 1954).

natural history based upon the Epitome of Aristophanes of Byzantium which was prepared at the instigation of the emperor Constantine Porphyrogenitus (A.D. 905–59); Manuel Philes (1275–1345) in his poem Περὶ ζῴων ἰδιότητος owes much to Aelian, and in the following century or perhaps earlier it was thought worth while to rearrange the contents of the *De natura animalium* according to subjects in 225 chapters, a few of the original chapters being omitted.[1] But for us today the chief interest and value of Aelian consist in the relics which he has preserved of writers whose works are no longer extant.

Manuscripts

of the De Natura Animalium

A	Monacensis Augustanus 564	s. xiv/xv p.C.
B	Berolinensis Phillippsianus 1522	s. xvi
C	Parisiensis gr. 1695	s. xvi
D	Vaticanus Palatinus gr. 65	s. xvi
E	Parisiensis gr. 1694	s. xvi
F	Laurentianus 86. 8	s. xv
G	Barberinus II. 92	s. xvi
H	Vaticanus Palatinus gr. 260	s. xiv
L	Laurentianus 86. 7	s. xiii
M	Monacensis 518	s. xv
N	Neapolitanus III D 8	s. xv
O	Neapolitanus III D 9	s. xv
P	Parisiensis gr. 1756	s. xiv
Q	Vaticanus Palatinus gr. 267	s. xv
R	Marcianus 518	s. xv

[1] This rearrangement is to be found in cod. Laurentianus 86. 8, De Stefani's F.

S	Vindobonensis med. gr. 7	s. xv
V	Parisiensis suppl. gr. 352	
	[formerly Vat. gr. 997]	s. xiii
W	Vindobonensis med. gr. 51	s. xiv

From these De Stefani selected seven only as possessing value for the constitution of the text, viz. A, F, H, L, P, V, and W, the remainder being copies of one or other of those seven.

Editions

1556	C. Gesner (Zurich, F°). Ed. pr.
1611	P. Gillius and C. Gesner (Geneva, 16°).
1744	Abraham Gronovius (London, 4°).
1784	J. E. G. Schneider (Leipzig, 8°).
1832	C. F. W. Jacobs (Jena, 8°).
1858	R. Hercher (Didot, Paris, la. 8°).
1864	R. Hercher (Teubner, Leipzig, 8°).

Gesner provided a parallel Latin translation which was later revised by A. Gronovius and was reprinted in all editions down to 1858. The only translation into a modern language that I know of (but have not seen) is the German version by Jacobs (Stuttgart, 1839–42). Gossen in 1935 announced that he had ready for press a fresh translation equipped with full notes, indexes, etc., but I have not been able to trace it.

Criticism

Cobet (C. G.). Novae lectiones (p. 780). Leyden, 1858.

Variae lectiones, ed. 2 (pp. 131, 209, 341). *Ib.*, 1873.

INTRODUCTION

Cobet (C. G.). Aeliani locus [*NA* 1. 30] correctus.
    Mnemos. 7 (1858) 340.

  De locis nonnullis apud A. *Ib.* N.S. 12 (1884) 433.

Baehrens (W. A.). Vermischte Bemerkungen zur gr.
    u. lat. Sprache [*NA* 7. 8]. *Glotta* 9 (1918) 171.

De Stefani (E. L.). I manoscritti della *Hist. Animal.*
    di Eliano. *Studi ital. di filol. class.* 10 (1902) 175.

  Per l'*Epitome Aristotelis De animalibus* di Aristofane
    di Bizanzio. *Ib.* 12 (1904) 421.

Goossens (R.). L' ὀδοντοτύραννος, animal de l' Inde,
    chez Palladius. [See *NA* 5. 3.] *Byzantion* 4
    (1927–8) 34.

Gossen (H.). Die Tiernamen in A's . . . Π. ζ.
    *Quellen u. Stud. z. Gesch. d. Naturwissenschaften
    u. d. Medizin* 4 (1935) 280.

Grasberger (L.). Zur Kritik des A. *Jb. f. class.
    Philol.* 95 (1867) 185.

Haupt (M.). Conjectanea [*NA* 2. 22]. *Hermes*
    5 (1871) 321.

  Varia. *Ib.* 4 (1870) 342.

Hercher (R.). Zu A.'s Thiergeschichte. *Philol.* 9
    (1854) 748.

  Zu A.'s Thiergeschichte. *Jb. f. class. Philol.* 71
    (1855) 450.

  Aelian, *etc. Philol.* 10 (1855) 344.

  Interpolationen bei A. *Jb. f. class. Philol.* 72
    (1856) 177.

  Zu griech. Prosaikern. *Hermes* 11 (1876) 223.

Kaibel (G.). [A. and Callimachus.] *Hermes* 28
    (1893) 54.

Keydell (R.). Oppians Gedicht von der Fischerei u.
    A.'s Tiergeschichte. *Hermes* 72 (1937) 411.

Klein (J.). Zu A. [*NA* 6. 21, 46; 12. 33]. *Rhein.
    Mus.* N.F. 22 (1867) 308.

INTRODUCTION

Meineke (A.).   Zu griech. Schriftstellern [*NA* 4. 12].
   Hermes 3 (1869) 162.

Mentz (F.).   Die klass. Hundenamen.   *Philol.* 88
   (1933) 104, 181, 415.

Morel (W.).   Iologica.   *Philol.* 83 (1928) 345.

Radermacher (L.).   Varia.   *Rhein. Mus.* 51 (1896)
   463.

   Zu Isyllos von Epidauros.   [*NA* 13. 15.]   *Philol.*
   58 (1899) 314.

Roehl (H.).   Zu A.   [*NA* 11. 10].   *Jb. f. class.*
   Philol. 121 (1880) 378.

Scott (J. A.).   Misc. notes from A.   *Class. Jl.* 24
   (1929) 374.

Schmid (W.).   Der Atticismus . . . Vol. 3 : Älian.
   [A detailed examination of A.'s vocabulary,
   syntax, and style.]   Stuttgart, 1893.

Shorey (P.).   An emendation of A. Π. ζ.   [*NA* 8. 1.]
   Class. Philol. 3 (1908) 101.

Thouvenin (P.).   Untersuchungen über den Modus-
   gebrauch bei A.   *Philol.* 54 (1895) 599.

Venmans (L. A. W. C.).   Σέρφος.   *Mnemos.* N.S. 58
   (1930) 58.

   Λευκοὶ μύρμηκες.   *Ib.* 318.

Wellmann (M.).   Sostratos, ein Beitrag z. Quellen-
   analyse des A.   *Hermes* 26 (1891) 321.

   Alexander von Myndos.   *Ib.* 481.

   Juba, eine Quelle des A.   *Ib.* 27 (1892) 389.

   Leonidas von Byzanz und Demostratos.   *Ib.* 30
   (1895) 161.

   Aegyptisches.   *Ib.* 31 (1896) 221.

   Pamphilos.   *Ib.* 51 (1916) 1.

In addition to the works named in the Preface, I
should mention :

INTRODUCTION

Aristotle. *Historia animalium* [trans.] *by D. W. Thompson.* Oxf. 1910.

Keller (O.). *Die antike Tierwelt.* 2 vols. Leipz. 1909–13.

Oppian . . . with an Engl. transl. by A. W. Mair. (Loeb Cl. Lib.) Lond. 1928.

Radcliffe (W.). *Fishing from the earliest times.* Lond. 1921.

Saint-Denis (E. de). *Vocabulaire des animaux marins en Latin classique.* (Études et commentaires, 11.) Paris, 1947.

Abbreviations used in the critical notes.

Cas[aubon, I.]   *Oud*[endorp, F. van]
Ges[ner, C.]   *Schn*[eider, J. G.]
Gill[ius, P.]   *OSchn*[eider, Otto]
Gron[ovius, A.]   *Valck*[enaer, L. K.]
H[ercher, R.]   *Wytt*[enbach, D.]
Hemst[erhusius, T.]   add[ed by].
Jac[obs, C. F. W.]   conj[ectured by].
Mein[eke, A.]   del[eted by].
                    om[itted by].

AELIAN

ON THE CHARACTERISTICS
OF ANIMALS

SUMMARY

3

SUMMARY

SUMMARY

SUMMARY

SUMMARY

ΑΙΛΙΑΝΟΥ

ΠΕΡΙ ΖΩΩΝ ΙΔΙΟΤΗΤΟΣ

ΠΡΟΟΙΜΙΟΝ

Ἄνθρωπον μὲν εἶναι σοφὸν καὶ δίκαιον καὶ τῶν
οἰκείων παίδων προμηθέστατον, καὶ τῶν γειναμένων
ποιεῖσθαι τὴν προσήκουσαν φροντίδα, καὶ τροφὴν
ἑαυτῷ μαστεύειν καὶ ἐπιβουλὰς φυλάττεσθαι καὶ τὰ
λοιπὰ ὅσα αὐτῷ σύνεστι δῶρα φύσεως, παράδοξον
ἴσως οὐδέν· καὶ γὰρ λόγου μετείληχεν ἄνθρωπος
τοῦ πάντων τιμιωτάτου, καὶ λογισμοῦ ἠξίωται,
ὅσπερ οὖν ἐστι πολιαρκέστατός τε καὶ πολυω-
φελέστατος· ἀλλὰ καὶ θεοὺς αἰδεῖσθαι οἶδε καὶ
σέβειν. τὸ δὲ καὶ τοῖς ἀλόγοις μετεῖναί τινος
ἀρετῆς κατὰ φύσιν,[1] καὶ πολλὰ τῶν ἀνθρωπίνων
πλεονεκτημάτων καὶ θαυμαστὰ ἔχειν συγκεκληρω-
μένα, τοῦτο ἤδη μέγα. καὶ εἰδέναι γε μὴ ῥᾳθύμως
τὰ προσόντα αὐτῶν ἰδίᾳ ἑκάστῳ, καὶ ὅπως
ἐσπουδάσθη οὐ μεῖον τῶν ἀνθρώπων καὶ ⟨τὰ⟩[2] τῶν
ἄλλων ζῴων, εἴη ἄν τινος πεπαιδευμένης φρενὸς
καὶ μαθούσης πολλά. ὡς μὲν οὖν καὶ ἑτέροις
ὑπὲρ τούτων ἐσπούδασται, καλῶς οἶδα · ἐγὼ δὲ
[ἐμαυτῷ][3] ταῦτα ὅσα οἷόν τε ἦν ἀθροίσας καὶ
περιβαλὼν αὐτοῖς τὴν συνήθη λέξιν, κειμήλιον οὐκ

---

[1] φύσιν καὶ εἰ μὴ κατὰ τὴν οἰκείαν κρίσιν.

AELIAN

ON THE CHARACTERISTICS
OF ANIMALS

PROLOGUE

THERE is perhaps nothing extraordinary in the fact that man is wise and just, takes great care to provide for his own children, shows due consideration for his parents, seeks sustenance for himself, protects himself against plots, and possesses all the other gifts of nature which are his. For man has been endowed with speech, of all things the most precious, and has been granted reason, which is of the greatest help and use. Moreover, he knows how to reverence and worship the gods. But that dumb animals should by nature possess some good quality and should have many of man's amazing excellences assigned to them along with man, is indeed a remarkable fact. And to know accurately the special characteristics of each, and how living creatures also have been a source of interest no less than man, demands a trained intelligence and much learning. Now I am well aware of the labour that others have expended on this subject, yet I have collected all the materials that I could; I have clothed them in untechnical language, and am persuaded that my achievement is a treasure

---

² ⟨τά⟩ add. Jac.   ³ [ἐμαυτῷ] del. H.

ἀσπούδαστον ἐκπονῆσαι πεπίστευκα. εἰ δέ τῳ καὶ
ἄλλῳ φανεῖται ταῦτα λυσιτελῆ, χρήσθω αὐτοῖς ·
ὅτῳ δὲ οὐ φανεῖται, ἐάτω τῷ πατρὶ θάλπειν τε καὶ
περιέπειν · οὐ γὰρ πάντα πᾶσι καλά, οὐδὲ ἄξια
δοκεῖ σπουδάσαι πᾶσι πάντα. εἰ δὲ ἐπὶ πολλοῖς
τοῖς πρώτοις καὶ σοφοῖς γεγόναμεν, μὴ ἔστω
ζημίωμα ἐς [1] ἔπαινον ἡ τοῦ χρόνου λῆξις, εἴ τι καὶ
αὐτοὶ σπουδῆς ἄξιον μάθημα παρεχοίμεθα καὶ τῇ
εὑρέσει τῇ περιττοτέρᾳ καὶ τῇ φωνῇ.

[1] εἰς MSS always.

far from negligible. So if anyone considers them profitable, let him make use of them; anyone who does not consider them so may give them to his father to keep and attend to. For not all things give pleasure to all men, nor do all men consider all subjects worthy of study. Although I was born later than many accomplished writers of an earlier day, the accident of date ought not to mulct me of praise, if I too produce a learned work whose ampler research and whose choice of language make it deserving of serious attention.

BOOK I

A

1. Καλεῖταί τις Διομήδεια νῆσος, καὶ ἐρῳδιοὺς
ἔχει πολλούς. οὗτοι, φασί, τοὺς βαρβάρους οὔτε
ἀδικοῦσιν οὔτε αὐτοῖς προσίασιν · ἐὰν δὲ Ἕλλην
κατάρῃ ξένος, οἱ δὲ θείᾳ τινὶ δωρεᾷ προσίασι
πτέρυγας ἁπλώσαντες οἱονεὶ χεῖράς τινας ἐς
δεξίωσίν τε καὶ περιπλοκάς. καὶ ἁπτομένων τῶν
Ἑλλήνων οὐχ ὑποφεύγουσιν, ἀλλ' ἀτρεμοῦσι καὶ
ἀνέχονται, καὶ καθημένων ἐς τοὺς κόλπους κατα-
πέτονται, ὥσπερ οὖν ἐπὶ ξένια [1] κληθέντες.
λέγονται οὖν οὗτοι Διομήδους ἑταῖροι εἶναι καὶ
σὺν αὐτῷ τῶν ὅπλων τῶν ἐπὶ τὴν Ἴλιον μετ-
εσχηκέναι, εἶτα τὴν προτέραν φύσιν ἐς τὸ τῶν ὀρνίθων μεταβαλόντες εἶδος, ὅμως ἔτι καὶ νῦν διαφυ-
λάττειν τὸ εἶναι Ἕλληνές τε καὶ Φιλέλληνες.

2. Ὁ σκάρος πόας μὲν θαλαττίας σιτεῖται καὶ
βρύα · λαγνίστατος δὲ ἄρα ἰχθύων ἁπάντων ἦν,
καὶ ἥ γε πρὸς τὸ θῆλυ ἀκόρεστος ἐπιθυμία αὐτῷ
ἁλώσεως αἰτία γίνεται. ταῦτα οὖν αὐτῷ συνεγνωκότες οἱ σοφοὶ τῶν ἁλιέων, ἐπιτίθενταί οἱ τὸν
τρόπον τοῦτον. ὅταν θῆλυν συλλάβωσιν, ἐνέδησαν [2]
ὁρμιᾷ σπάρτου πεποιημένῃ λεπτῇ τοῦ στόματος
ἄκρου, καὶ ἐπισύρουσι διὰ τῆς θαλάττης τὸν ἰχθὺν
ζῶντα · ἴσασι δὲ εὐνάς τε αὐτῶν καὶ διατριβὰς καὶ

---

[1] Gron: ξενίᾳ.　　　　　　[2] ἔδησαν.

14

BOOK I

1. There is a certain island called Diomedea,[a] and The Birds of Diomede it is the home of many Shearwaters. These, it is said, neither harm the barbarians nor go near them. If however a stranger from Greece puts in to port, the birds by some divine dispensation approach, extending their wings as though they were hands, to welcome and embrace the strangers. And if the Greeks stroke them, they do not fly away, but stay still and allow themselves to be touched; and if the men sit down, the birds fly on to their lap as though they had been invited to a meal. They are said to be the companions of Diomedes[b] and to have taken part with him in the war against Ilium; though their original form was afterwards changed into that of birds, they nevertheless still preserve their Greek nature and their love of Greece.

2. The Parrot Wrasse feeds upon seaweed and The Parrot Wrasse wrack, and is of all fishes the most lustful, and its insatiable desire for the female is the reason why it gets caught. Now skilful anglers are aware of this, and they set upon it in this way. Whenever they capture a female, they fasten a fine line of esparto to its lip and trail the fish alive through the sea, knowing as they do where the fish lie, their haunts, and where

[a] Mod. San Domenico, one of the three ' Isole di Tremiti,' about 15 mi. N of the ' spur ' of Italy.

[b] King of Argos; settled later in Daunia, where he died and was buried in Diomedea.

15

ὅπου συναγελάζονται. μόλυβδος δὲ αὐτοῖς πεποίη-
ται βαρὺς τὴν ὁλκήν, περιφερὴς τὸ σχῆμα, καὶ
ἔχει μῆκος τριῶν δακτύλων, καὶ διείληπται ἐξ
ἄκρων σχοίνῳ, καὶ ἐπισύρει ¹ τὸν τεθηραμένον.
καὶ κύρτον τις τῶν ἐν τῇ πορθμίδι παραρτήσας
ἐπάγεται εὐρὺν τὸ στόμα, καὶ ἐς τὸν ἑαλωκότα
τέτραπται σκάρον ὁ κύρτος · βαρεῖται δὲ ἡσυχῇ
οὗτος λίθῳ μεμετρημένῳ. οὐκοῦν οἱ ἄρρενες,
ὥσπερ οὖν νύμφην ὡρικὴν ² νεανίαι θεασάμενοι,
οἰστροῦνταί τε καὶ μεταθέουσι, καὶ ἐπείγονται
φθάσαι ἄλλος ἄλλον καὶ γενέσθαι πλησίον καὶ
παραψαῦσαι, ὥσπερ οὖν δυσέρωτες ἄνθρωποι
φίλημα ἢ κνίσμα θηρώμενοι ἤ τι ἄλλο κλέμμα
ἐρωτικόν. ὁ τοίνυν ἄγων τὸν θῆλυν ἡσυχῇ καὶ
πεφεισμένως, λοχῶν τε καὶ ἐπιβουλεύων εὐθὺ τοῦ
κύρτου σὺν τῇ ἐρωμένῃ, φαίης ἄν, τοὺς ἐραστὰς
ἄγει. γενομένων δὲ ὁμοῦ τῷ κύρτῳ, τὸν μὲν
μόλυβδον μεθῆκεν ὁ θηρατὴς ἐς τὸ ἔσω³· ὁ δὲ
ἄρα ἐμπίπτων σὺν τῇ ὁρμιᾷ κατασπᾷ καὶ τὸν
θῆλυν. οὐκοῦν συνεσρεύσαντες ἑαλώκασι, καὶ διδό-
ασι δίκην ὁρμῆς ἀφροδισίου ταύτην οἱ σκάροι.

3. Ὁ ἰχθὺς ὁ κέφαλος τῶν ἐν τοῖς ἕλεσι
βιούντων ἐστί, καὶ πεπίστευται τῆς γαστρὸς
κρατεῖν καὶ διαιτᾶσθαι πάνυ σωφρόνως. ζῴῳ ⁴
μὲν γὰρ οὐκ ἐπιτίθεται, ἀλλὰ πρὸς πάντας τοὺς
ἰχθῦς ἔνσπονδος εἶναι πέφυκεν · ὅτῳ δ᾽ ἂν ἐντύχῃ
κειμένῳ, τοῦτό οἱ δεῖπνόν ἐστιν. οὐ πρότερον δὲ
αὐτοῦ προσάπτεται, πρὶν ἢ τῇ οὐρᾷ κινῆσαι. καὶ
ἀτρεμοῦντος μὲν ἔχει τὴν ἄγραν, κινηθέντος δὲ
ἀνεχώρησεν.

¹ ἐπισύρεται.   ² Jac : ἐρωτικήν.

they assemble. They prepare a heavy leaden sinker round in shape and three fingers in length; a cord is passed through both ends, and it trails the captured fish after it. One of the men in the boat attaches to the side a weel with a wide mouth; the weel is then turned towards the captured Wrasse and slightly weighted with a stone of appropriate size. Whereupon the male Wrasses, like young men who have caught sight of a pretty girl, go in pursuit, mad with desire, each trying to outstrip the other and to reach her side and rub against her, just as love-sick men strive to kiss or tickle ⟨a girl⟩ or to play some other amorous trick. So then the man who is towing the female gently and slowly and planning to entrap ⟨his fish⟩, draws the lovers (as you might call them) with the loved one straight towards the weel. As soon as they come level with the weel, the angler lets the lead weight drop into it, and as it falls in it drags the female down with it by the line. And as the male Wrasses swim in with her, they are captured and pay the penalty for their erotic impulse.

3. The Mullet is one of those fishes that live in pools and is believed to control its appetite and to lead a most temperate existence. For it never sets upon a living creature, but is naturally inclined to peaceful relations with all fish. If it comes across any dead fish, it makes its meal off that, but will not lay hold upon it until it has moved it with its tail: if the fish does not stir, it becomes the Mullet's prey; but if it moves, the Mullet withdraws.

The Mullet

---

³ εἴσω MSS *always*.　　　⁴ *Cobet*: ζῴῳ H.

AELIAN

4. Τιμωροῦσιν ἀλλήλοις ὡς ἄνθρωποι πιστοὶ καὶ
συστρατιῶται δίκαιοι οἱ ἰχθύες, οὕσπερ οὖν ἀνθίας
οἱ τῆς θήρας ἐπιστήμονες τῆς θαλαττίας φιλοῦσιν
ὀνομάζειν, ὄντας τὰ ἤθη πελαγίους. τούτων γοῦν
ἕκαστοι, ὅταν νοήσωσι τεθηρᾶσθαι τὸν σύννομον,
προσνέουσιν ὤκιστα, εἶτα ἐς αὐτὸν τὰ νῶτα
ἀπερείδουσι, καὶ ἐμπίπτοντες καὶ ὠθούμενοι τῇ
δυνάμει κωλύουσιν ἕλκεσθαι.

Καὶ οἱ σκάροι δὲ ἐς τὴν οἰκείαν ἀγέλην εἰσὶν
ἀγαθοὶ τιμωροί. προσίασι γοῦν, καὶ τὴν ὁρμιὰν
ἀποτραγεῖν σπεύδουσιν, ἵνα σώσωσι τὸν ᾑρημένον·
καὶ πολλάκις μὲν ἀποκόψαντες ἔσωσαν καὶ ἀφῆκαν
ἐλεύθερον, καὶ οὐκ αἰτοῦσι ζωάγρια· πολλάκις δὲ
οὐκ ἔτυχον, ἀλλ᾽ ἥμαρτον μέν, τὸ δ᾽ οὖν ἑαυτῶν
πεποιήκασιν εὖ μάλα προθύμως. ἤδη δὲ καὶ ἐς
τὸν κύρτον τὸν σκάρον ἐμπεσεῖν φασι καὶ τὸ
οὐραῖον μέρος ἐκβαλεῖν, τοὺς δὲ ἀθηράτους καὶ
περινέοντας ἐνδακεῖν καὶ ἐς τὸ ἔξω τὸν ἑταῖρον
προαγαγεῖν. εἰ δὲ ἐξείη [1] τὸ στόμα, τῶν τίς οἱ [2]
ἔξω τὴν οὐρὰν παρώρεξεν, ὁ δὲ περιχανὼν ἠκολού-
θησεν. οὗτοι μὲν δὴ ταῦτα δρῶσιν, ὦ [3] ἄνθρωποι,
φιλεῖν οὐ μαθόντες, ἀλλὰ πεφυκότες.

5. Ὁ ἰχθὺς ὁ τρώκτης, τούτου μὲν κατηγορεῖ
τὴν φύσιν καὶ τὸ ὄνομα, ἤδη δὲ καὶ τὸ στόμα·
ὀδόντες δὲ αὐτῷ συνεχεῖς τε ἐμπεφύκασι καὶ
πολλοί, καὶ πᾶν τὸ ἐμπεσὸν διατεμεῖν εὖ μάλα
καρτεροί. οὐκοῦν ἁλοὺς ἀγκίστρῳ μόνος ἰχθύων ἐς

[1] ἐξίοι κατά.   [2] Jac : τίς ὁ.   [3] Jac : ὡς.

18

4. As loyal men and true fellow-soldiers come to one another's aid, so do the fish which men skilled in sea-fishing call *Anthias*;[a] and their haunts are the sea. For instance, directly they are aware that a mate has been hooked, they swim up with all possible speed; then they set their back against him and by falling upon him and pushing with all their might try to stop him from being hauled in.

The 'Anthias'

Parrot Wrasses too are doughty champions of their own kin. At any rate they rush forward and make haste to bite through the line in order to rescue the one that has been caught. And many a time have they cut the line and set him free, and they ask for no reward for life-saving. Many a time however they have not contrived to do this, but have failed in spite of having done all they could with the utmost zeal. And it has even happened, they say, that, when a Parrot Wrasse has fallen into the weel and has left his tail-part projecting, the others that are swimming around uncaught have fixed their teeth in him and have dragged their comrade out. If however his head was projecting, one of those outside offered his tail, which the captive grasped and followed. This, my fellow-men, is what these creatures do: their love is not taught, it is inborn.

The Parrot Wrasse

5. Of the fish known as the ' Gnawer '[b] its name and, what is more, its mouth declare its nature. Its teeth grow in an unbroken line and are numerous and so strong as to bite through anything that comes their way. Therefore, when taken with a

The Gnawer

---

[a] Unidentified.

[b] Perhaps the fox-shark; see Thompson, *Gk. fishes*, s.vv. ἀλώπηξ, τρώκτης.

AELIAN

τὸ ἔμπαλιν ἑαυτὸν οὐκ ἐπανάγει, ἀλλὰ ὠθεῖται
τὴν ὁρμιὰν ἀποθρίσαι [1] διψῶν. οἱ δὲ ἁλιεῖς σοφίζον-
ται τἀναντία· τὰς γάρ τοι τῶν ἀγκίστρων λαβὰς
χαλκεύονται μακράς. ὁ δὲ (καὶ γὰρ πῶς ἐστι καὶ
ἁλτικὸς) καὶ ὑπὲρ ταύτας ἀνέθορε πολλάκις καὶ
τὴν τρίχα τὴν ἄγουσαν τεμὼν ἐς ἤθη τὰ τῶν
ἰχθύων αὖθις ἀπονήχεται. οὗτός τοι καὶ τὴν ἀγέλην τὴν σύννομον παραλαβὼν σὺν αὐτοῖς ἐκείνοις
χωρεῖ καὶ τοῖς δελφῖσιν ὁμόσε· καὶ ἕνα ἀποκρι-
θέντα πως περιελθόντες εἶτα ἐπιτίθενται τῷ θηρίῳ
καρτερῶς· ἴσασι γὰρ ὅτι τῶν ἐξ αὐτῶν δηγμάτων
οὐ ῥαθύμως ἐπαίει. οἱ μὲν γὰρ ἔχονται αὐτοῦ καὶ
μάλα ἐγκρατῶς, ὁ δὲ ἀναπηδᾷ καὶ κυβιστᾷ,[2] καὶ
ὡς ὑπὸ τῆς ὀδύνης στρεβλοῦται διελέγχεται·
ἀπρὶξ γὰρ ἐμφύντες συνεξαίρονται πηδῶντος. καὶ
ὁ μὲν ἀποσείσασθαι καὶ ἀποκροῦσαι σπεύδει
αὐτούς, οἱ δὲ οὐκ ἀνιᾶσιν, ἀλλὰ ἐσθίουσι ζῶντα.
εἶτα μέντοι ὅ τι ἂν ἕκαστος μέρος ἐκτράγῃ, τοῦτο [3]
ἔχων ἀπαλλάττεται· καὶ ὁ δελφὶς ἀσμένως
ἀπονήχεται, δαιτυμόνας, ὡς ἂν εἴποις, ἀκλήτους
ἑστιάσας σὺν τῇ ἑαυτοῦ ὀδύνῃ ἐκείνους.

6. Γλαύκης ἀκούω τῆς κιθαρῳδοῦ ἐρασθῆναι
κύνα· οἱ δὲ οὐ κύνα, ἀλλὰ κριόν· ἄλλοι δὲ χῆνα.
καὶ ἐν Σόλοις δὲ τῆς Κιλικίας [4] παιδός, ᾧ ὄνομα
ἦν Ξενοφῶν, κύων ἠράσθη· ἄλλου δὲ [5] ὡραίου
μειρακίου ἐν Σπάρτῃ κολοιὸς ἐπὶ τῷ εἴδει ἐνόσησεν.

---

[1] ἀποθερίσαι.
[2] κυβιστῶν δῆλός ἐστιν.
[3] Jac : εἶτα μέντοι τοῦτο ὅ τι . . . ἔχων.
[4] τοῖς Κιλικίοις.
[5] καὶ ἄλλου.

hook, it is the only fish that does not attempt to withdraw, but presses on in its eagerness to cut the line. Fishermen however counter this by a device: they have their hooks forged with a long shank. But the Gnawer, being a powerful jumper in its way, often leaps above the shank, and cutting the hair-line that is drawing it, swims away again to the places where fish haunt.

It also gathers round it a shoal of its fellows and with them also makes an attack upon the Dolphins. And if one chance to get separated from the rest, the Gnawers surround it and then set upon the creature furiously, knowing as they do that the Dolphin is by no means insensible to their bites. For the Gnawers cling most tenaciously to it, while the Dolphin leaps upwards and plunges; and it shows how it is being tormented by the pain, for the Gnawers that have fastened upon it are lifted out of the water with it as it leaps. And while the Dolphin struggles to shake them loose and beat them off, they never relax their hold, but would eat it alive. Then however when each Gnawer has bitten away a piece, they go off with their mouthful, and the Dolphin is thankful to swim away after having fed its uninvited guests (if one may so call them) to its own pain. *and Dolphins*

6. I am told that a dog fell in love with Glauce the harpist. Some however assert that it was not a dog but a ram, while others say it was a goose. And at Soli in Cilicia a dog loved a boy of the name of Xenophon; at Sparta another boy in the prime of life by reason of his beauty caused a jackdaw to fall sick of love. *Animals in love with human beings*

21

AELIAN

7. Λέγουσι τὸν θῶα τὸ ζῷον φιλανθρωπότατον εἶναι. καὶ ὅταν μέν που περιτύχῃ ἀνθρώπῳ, ἐκτρέπεται αὐτόν, οἷον αἰδούμενος· ὅταν δὲ ἀδικούμενον θεάσηται ὑπ' ἄλλου θηρίου, τὸ τηνικαῦτα ἐπαμύνει αὐτῷ.

8. Νικίας τις τῶν συγκυνηγετούντων [1] ἀπροόπτως παραφερόμενος [2] ἐς ἀνθρακευτῶν κάμινον κατηνέχθη, οἱ δὲ κύνες οἱ σὺν αὐτῷ τοῦτο ἰδόντες οὐκ ἀπέστησαν, ἀλλὰ τὰ μὲν πρῶτα κνυζώμενοι περὶ τὴν κάμινον καὶ ὠρυόμενοι διέτριβον, τὰ δὲ τελευταῖα μονονουχὶ τοὺς παριόντας ἠρέμα καὶ πεφεισμένως κατὰ τῶν ἱματίων δάκνοντες εἶτα εἷλκον ἐπὶ τὸ πάθος, οἷον ἐπικούρους τῷ δεσπότῃ παρακαλοῦντες τοὺς ἀνθρώπους οἱ κύνες. καὶ γοῦν εἷς ὁρῶν τὸ γινόμενον ὑπώπτευσε τὸ συμβάν, καὶ ἠκολούθησε καὶ εὗρε τὸν Νικίαν ἐν τῇ καμίνῳ καταφλεχθέντα, ἐκ τῶν λειψάνων συμβαλὼν τὸ γενόμενον.

9. Ὁ κηφὴν ὁ ἐν μελίτταις γεννώμενος μεθ' ἡμέραν μὲν ἐν τοῖς ἀνθρηνίοις κατακέκρυπται, νύκτωρ δέ, ἡνίκα ἂν παραφυλάξῃ καθευδούσας τὰς μελίττας, ἐπιφοιτᾷ τοῖς ἔργοις αὐτῶν καὶ λυμαίνεται τοῖς σίμβλοις. τοῦτο ἐκεῖναι καταμαθοῦσαι, αἱ μὲν πλεῖσται τῶν μελιττῶν καθεύδουσιν ἅτε πεπονηκυῖαι, ὀλίγαι δὲ αὐτῶν ἐλλοχῶσιν. εἶτα ὅταν ἕλωσι τὸν φῶρα, παίουσιν αὐτὸν πεφεισμένως καὶ ἐξωθοῦσι,[3] καὶ ἐκβάλλουσι φυγάδα εἶναι. ὁ δὲ οὐδ' οὕτω πεπαίδευται· πέφυκε γὰρ καὶ ἀργὸς καὶ λίχνος, δύο κακώ. ἔξω τοίνυν τῶν κηρίων ἑαυτὸν ἀποκρύπτει, εἶτα ὅταν ἐπὶ τὰς νομὰς ἐξορμήσωσιν

7. Men say that the Jackal is most friendly dis- The Jackal
posed to man, and whenever it happens to encounter a
man, it gets out of his way as though from deference;
but when it sees a man being injured by some other
animal, it at once comes to his help.

8. One Nicias unwittingly outdistanced his fellow Nicias and
his hounds
huntsmen[1] and fell into a charcoal-burners' furnace.
But his hounds, which saw this happen, did not leave
the spot, but at first remained whining and baying
about the furnace, until at length, by just daring to
bite the clothes of passers-by gently and cautiously,
they tried to draw them to the scene of the mishap,
as though the hounds were imploring the men to
come to their master's help. One man at any rate
seeing this, suspected what had occurred and followed. He found Nicias burned to death in the
furnace, and from the remains he guessed the truth.

9. The Drone, which is born among bees, hides The Drone
itself among the combs during the day, but at night,
when it observes that the bees are asleep, it invades
their work and makes havoc in the hives. When the
bees realise this (most of them are asleep, being
thoroughly tired, though a few are lying in wait for
the thief), directly they catch him they beat him,
not violently[2], and thrust him out[3] and cast him forth
into exile. Yet even so the Drone has not learnt his
lesson, for he is naturally slothful and greedy—two
bad qualities! So he secretes himself outside the
combs and later, when the bees fly forth to their

---

[1] κυνηγετούντων.　　　　[2] φερόμενος.
[3] ἐξωθοῦσι τοῖς πτεροῖς.

αἱ μέλιτται, ὁ δὲ ὠσάμενος ἔσω τὸ ἑαυτοῦ δρᾷ,
ἐμφορούμενος καὶ κεραΐζων ἐκεῖνος τὸν θησαυρὸν
τῶν μελιττῶν τὸν γλυκύν. καὶ ἐκεῖναι ἐκ τῆς
νομῆς ὑποστρέψασαι, ὅταν αὐτῷ περιτύχωσιν, ἐνταῦθα μὲν οὐκέτι πεφεισμένως αὐτὸν παίουσιν,
οὐδ' ὅσον ἐς φυγὴν τρέψαι, ἀλλὰ εὖ μάλα [1] βιαίως
ἐμπεσοῦσαι διαλοῶσι τὸν λῃστήν· καὶ οὐ μεμπτὴν
ὑπομείνας τὴν τιμωρίαν, ὑπὲρ τῆς γαστριμαργίας
καὶ ἀδηφαγίας τῇ ψυχῇ [2] ἔτισεν. μελιττουργοὶ
λέγουσι ταῦτα, καὶ ἐμὲ πείθουσιν.

10. Εἰσὶ δέ τινες καὶ ἐν ταῖς μελίτταις ἀργοὶ
μέλιτται, οὐ μὴν κηφηνώδεις τὸν τρόπον· οὐ γὰρ
λυμαίνονται τοῖς κηρίοις οὐδ' ἐπιβουλεύουσι τῷ
μέλιτι αὗται, ἀλλὰ τρέφονται [3] ἐκ τῶν ἀνθέων καὶ
αὗται πετόμεναι καὶ σύννομοι ταῖς ἄλλαις οὖσαι.
εἰ δὲ καί εἰσιν ἄτεχνοι περὶ τὴν ἐργασίαν καὶ τὴν
κομιδὴν τὴν τοῦ μέλιτος, ἀλλὰ γοῦν οὐκ εἰσὶν
ἄπρακτοι πάντῃ. αἱ μὲν γὰρ αὐτῶν ὕδωρ τῷ
βασιλεῖ κομίζουσι καὶ ταῖς πρεσβυτέραις δέ, αἵπερ
οὖν [4] τῷ βασιλεῖ παραμένουσι καὶ ἐς τὴν δορυφο-
ρίαν ἀπεκρίθησαν τὴν αὐτοῦ· ἕτεραι δὲ αὐτῶν [5]
ἔχουσιν ἐκεῖνο ἔργον, τὰς ἀποθνησκούσας τῶν
μελιττῶν ἔξω φέρουσι· δεῖ γὰρ αὐταῖς καθαρὰ
εἶναι τὰ κηρία, καὶ οὐκ ἀνέχονται νεκρὰν ἔσω
μέλιτταν· ἄλλαι δὲ [6] νύκτωρ φρουροῦσιν, ὥσπερ
οὖν πόλιν μικρὰν φυλάττουσαι τὴν τῶν κηρίων
οἰκοδομίαν ἐκεῖναί γε.

11. Μελιττῶν δὲ ἡλικίαν διαγνοίη τις ἂν τὸν
τρόπον τοῦτον. αἱ μὲν αὐτοετεῖς στιλπναί τέ εἰσι

---

[1] εὖ μάλα τοῖς κέντροις.　　　　[2] τὴν ψυχήν.

feeding-grounds, pushes his way in and does what is natural to him, cramming himself and plundering the bees' treasure of honey. But they on returning from their pasturage, directly they encounter him, no longer beat him with moderation nor merely put him to flight, but fall upon him vigorously and make an end of the thief. The punishment which he suffers none can censure: he pays for his gluttony and voracity with his life.

This is what bee-keepers say, and they convince me.

10. Even among Bees there are some which are lazy, though they do not resemble drones in their habits, for they neither damage the combs nor have designs upon the honey, but feed themselves on the flowers, flying abroad and accompanying the others. But though they have no skill in the making and the gathering of honey, at any rate they are not completely inactive, for some fetch water for their king and for their elders, while the elders themselves attend upon the king and have been set apart to form his bodyguard. Meanwhile others of them have this for their task: they carry the dead bees out of the hive. For it is essential that their honeycombs should be clean, and they will not tolerate a dead bee in the hive. Others again keep watch by night, and their duty is to guard the fabric of honeycombs as though it were some tiny city. *Bees and their various duties*

11. A man may tell the age of Bees in the following way. Those born in the current year are glisten- *Bees and their ages*

---

³ τρέφονται μέν.
⁴ αἵπερ οὖν αἱ πρεσβύτεραι καὶ αὗται τῷ β.
⁵ αὐτῶν τῶν ἀτέχνων.   ⁶ *Gill*: ἀλλὰ καί.

καὶ ἐοίκασιν ἐλαίῳ τὴν χρόαν[1]. αἱ δὲ πρεσβύτεραι
τραχεῖαι καὶ ἰδεῖν καὶ προσψαῦσαι[2] γίνονται,
ῥυσαὶ δὲ ὁρῶνται διὰ τὸ γῆρας· ἐμπειρότεραι δέ
εἰσιν αὗται καὶ τεχνικώτεραι, παιδεύσαντος αὐτὰς
τὴν ἐπὶ τῷ μέλιτι σοφίαν τοῦ χρόνου. ἔχουσι δὲ
καὶ μαντικῶς, ὥστε καὶ ὑετῶν καὶ κρύους ἐπιδημίαν
προμαθεῖν· καὶ ὅταν τούτων τὸ ἕτερον ἢ καὶ
ἀμφότερα ἔσεσθαι συμβάλωσιν, οὐκ ἐπὶ μήκιστον
ἐκτείνουσι τὴν πτῆσιν,[3] ἀλλὰ περιποτῶνται τοῖς
σμήνεσι, καὶ οἱονεὶ περιθυροῦσιν. ἐκ δὴ τούτων οἱ
μελιττουργοὶ οἰωνισάμενοι προλέγουσι τοῖς γεωρ-
γοῖς τὴν μέλλουσαν ἐπιδημίαν τοῦ χειμῶνος. δεδοίκασι δὲ ἄρα οὐ τοσοῦτον τὸ κρύος αἱ μέλιτται,
ὅσον τὸν ὄμβρον τὸν πολὺν καὶ τὸν νιφετόν. ἐναν-
τίαι δὲ πολλάκις τοῦ πνεύματος πέτονται, καὶ βρα-
χεῖαν λίθον ἐν τοῖς ποσὶ κομίζουσι καὶ τοσαύτην
ὅσην εὔφορον αὐταῖς πετομέναις εἶναι, καὶ τρόπον
τινὰ τοῦτο ἕρμα ἑαυταῖς ἐπιτεχνῶνται πρὸς τὸν
ἐμπίπτοντα ἄνεμον τά τε ἄλλα καὶ ἵνα μὴ παρατ-
ρέψῃ τῆς ὁδοῦ ἡ αὔρα αὐτάς.

12. Ἔρωτος δὲ ἰσχὺν καὶ ἰχθύων γένη πολλὰ
ἔγνω, τοῦ τοσούτου θεοῦ μηδὲ τοὺς κάτω καὶ ἐν
τῷ βυθῷ[4] τῆς θαλάττης ὑπεριδόντος καὶ ἀτιμά-
σαντος. λατρεύει γοῦν τῷδε τῷ δαίμονι[5] καὶ
κέφαλος, ἀλλ' οὐ πᾶς, ἐκεῖνος δὲ ὅνπερ οὖν ἀπὸ
τοῦ ὀξέος προσώπου καλοῦσιν οἱ γένη τε καὶ
διαφορὰς ἰχθύων κατεγνωκότες. ἁλίσκονται δέ, ὡς
ἀκούω, περὶ τὸν κόλπον τὸν Ἀχαϊκὸν πολλοί. καὶ
τῆς μὲν κατ' αὐτοὺς ἁλώσεως διαφορότης ἐστί·
μάλιστα δὲ αὐτῶν τὸ λυττῶδες τὸ ἐς τὰ ἀφροδίσια

---

[1] χροιάν.   [2] Gron: ἅψασθαι.   [3] πτῆσιν ἐκ τῆς νομῆς.

26

ing and are the colour of olive oil; the older ones are
rough to the eye and to the touch and appear
wrinkled with age. They have however greater
experience and skill, time having instructed them in
the art of making honey. They have too the faculty ^{as weather-prophets}
of divination, so that they know in advance when rain
and frost are coming. And whenever they reckon
that either or both are on their way, they do not
extend their flight very far, but fly round about their
hives as though they would be close to the door.
It is from these signs that bee-keepers augur the
approach of stormy weather and warn the farmers.
And yet Bees are not so afraid of frost as they are
of heavy rain and snow. Often they fly against the
wind, carrying between their feet a small pebble of
such size as is easy to carry when on the wing.
This is a device which they use to ballast themselves
against a contrary wind, and particularly so that the
breeze may not deflect them from their path.

12. Even among fishes there are many kinds ^{The Mullet (oxyrhynchus)}
which know how strong is love, for that god, powerful
as he is, has not ignored and disdained even the
creatures that dwell below in the depths of the ocean.
One at any rate that pays service to this god is the
Mullet, but not every species, only that to which
men who have observed the different species of fish
have given a name derived from its sharp snout.
These, I am told, are caught in great numbers round
about the Gulf of Achaia, and there are various ways
of catching them. But the following method of
capture proves how madly amorous they are.

---

⁴ ἐν τῷ βυθῷ καὶ κάτω. ⁵ δαιμονίῳ.

κατηγορεῖ ἥδε ἡ ἄγρα. θηράσας ἀνὴρ ἁλιεὺς
θῆλυν,[1] καὶ ἐνδήσας[2] καλάμῳ μακρῷ ἢ σπάρτῳ
καὶ τούτῳ μακρῷ, κατὰ τῆς ᾐόνος ἡσυχῇ βαδίζων
παρανηχόμενον τὸν ἰχθὺν καὶ ἀσπαίροντα ἐπισύρει·
κατ' ἴχνια δὲ αὐτοῦ τις ἔπεται φέρων δίκτυον, καὶ
τὸ μέλλον ὅπῃ τε καὶ ὅπως ἀπαντήσεται φυλάττει
φιλοπόνως ὁ δικτυεὺς οὗτος. οὐκοῦν ἡ μὲν ἄγεται,
ὁπόσοι δὲ ἂν ἴδωσι τῶν ἀρρένων, οἷα δήπου νεανίαι
ἀκόλαστοι μείρακος παραθεούσης εὖ μάλα ὡρικῆς
ἐποφθαλμιάσαντες, ἵενται κατὰ μίξιν[3] οἰστρούμε-
νοι. ὁ τοίνυν τὸ δίκτυον ἔχων ῥίπτει τὸν βόλον,
καὶ πολλάκις ἰχθύων εὐερμίᾳ περιτυγχάνει τῇ τῆς
ἐπιθυμίας ὁρμῇ προσερχομένων. δεῖ δὲ τῷ πρώτῳ
θηρατῇ τὴν αἱρεθεῖσαν ὡραίαν τε εἶναι καὶ εὖ
ἤκουσαν σαρκῶν, ἵνα καὶ πλείους ἐπ' αὐτὴν ὁρμήσω-
σιν, τὸ τῆς ὥρας ἐφολκὸν δέλεαρ λαβόντες. εἰ δὲ
ἄσαρκος εἴη, οἱ πολλοὶ ὑπερφρονήσαντες ᾤχοντο
ἀπιόντες· ὅστις δὲ αὐτῶν ἐστι δύσερως, οὐκ
ἀπαλλάττεται, οὐ τῇ ὥρᾳ, μὰ Δία, ἀλλὰ τῷ τῆς
μίξεως πόθῳ δεδουλωμένος.

13. Ἦσαν δὲ ἄρα καὶ σωφρονεῖν ἰχθύες ἀγαθοί.
ὁ γοῦν αἰτναῖος οὕτω λεγόμενος, ἐπὰν τῇ ἑαυτοῦ
συννόμῳ οἱονεὶ γαμετῇ τινι συνδυασθεὶς κληρώση-
ται τὸ λέχος, ἄλλης οὐχ ἅπτεται, καὶ οὐ δεῖται
συμβολαίων ἐς πίστιν, οὐ προικός, οὐδὲ μὴν δέδοικε
κακώσεως δίκην ὁ αἰτναῖος, οὐδὲ αἰδεῖται Σόλωνα.
ὦ νόμοι γενναῖοι καὶ πολύσεμνοι,[4] οἷς ἀκόλαστοι
ἄνθρωποι οὐκ αἰδοῦνται μὴ πείθεσθαι.

---

[1] θῆλυν ἐκ τῶνδε κέφαλον.   [2] *Reiske*: ἐκ-.
[3] κατὰ τὴν νῆξιν.   [4] *Mein*: πόλεις σεμναί.

A fisherman catches a female Mullet and fastens it _{how caught} to a long rod or a cord (this too must be long); as he walks slowly along the sea-shore he draws the fish, swimming and gasping, after him. In his footsteps there follows one with a net, and this net-fisherman watches diligently to see what is going to happen and where. So the female Mullet is towed along, and all the males that catch sight of her, like (one might say) licentious youths ogling a beautiful girl as she hurries by, come swimming up, mad with sexual desire. Thereupon the man with the net casts it and frequently has good luck, thanks to the urgent lust of the fish that approach. It is essential for the first fisherman's purpose that the captured female should be at her prime and well-fleshed, so that a greater number may be ardent after her and may take the bait which her enticing beauty offers. But should she be lean, most of them will scorn her and go away. Still, if any one of them is madly in love, he will not leave her, because he has been enslaved not by her beauty (that I will swear) but by his desire for sexual intercourse.

13. It seems however that fish are also models of The continence. At any rate when the 'Etna-fish',[a] as 'Etna-fish' it is called, pairs with its mate as with a wife and achieves the married state, it does not touch another female; it needs no covenants to maintain its fidelity, no dowry; it even stands in no fear of an action for ill-usage, nor is Solon[b] to it a name of dread. What noble laws, how worthy of veneration! —And man, the libertine, feels no scruple at disobeying them.

[a] Unidentified.   [b] See 2. 42 n.

14. Κοσσύφῳ δὲ τῷ θαλαττίῳ ἤθη τε καὶ διατριβαὶ αἱ πέτραι καὶ αἱ σηραγγώδεις ὑποδρομαί.
γαμοῦσι δὲ οὗτοι ἕκαστος πολλάς, καὶ τῶν ὀπῶν
οἱονεὶ θαλάμων ⟨ταῖς⟩[1] νύμφαις ἀφίστανται. καὶ
τοῦτο μὲν τὸ τοῦ γάμου θρυπτικὸν καὶ τὸ ἐς πολλὰς
ἔχειν τὴν ὁρμὴν νενεμημένην φαίης[2] ἂν εἶναι τρυ-
φώντων ἐς εὐνὴν βαρβάρων καί, ὡς ἂν εἴποις σὺν
παιδιᾷ σπουδάσας, βίον Μηδικόν τε καὶ Περσικόν.
ἔστι δὲ ἰχθύων ζηλοτυπώτατος καὶ τὴν ἄλλως
μέν,[3] οὐχ ἥκιστα δὲ ὅταν αἱ νύμφαι τίκτωσιν αὐτῷ.
εἰ δὲ λαμυρώτερον ταῦτα τῇ καταχρήσει[4] τῶν
ὀνομάτων εἴρηται, δίδωσιν ἡμῖν τὰ ἐκ τῆς φύσεως
πραττόμενα τὴν τῶν τοιούτων ἐξουσίαν. αἱ μὲν
γὰρ ὠδίνων ἤδη πειρώμεναι ἠρεμοῦσί τε καὶ ἔνδον
μένουσιν, ὁ δὲ ἄρρην, οἷα δήπου γαμέτης, περιθυρῶν
τὰς ἐπιβουλὰς φυλάττει τὰς ἔξωθεν φόβῳ τῶν
βρεφῶν. ἔοικε γὰρ καὶ τὰ μήπω γεννώμενα φιλεῖν
καὶ δέει πατρικῷ ἁλισκόμενος ἐντεῦθεν ὀρρωδεῖν
ἤδη, καὶ διημερεύει μὲν ἐπὶ τῇ φρουρᾷ πάντων
ἄγευστος, καὶ ἡ φροντὶς αὐτὸν τρέφει· δείλης δὲ
ὀψίας γενομένης ἀφεῖται τῆς ἀνάγκης τῆσδε, καὶ
μαστεύει τροφήν, καὶ οὐκ ἀτυχεῖ αὐτῆς. καὶ
ἑκάστη δὲ ἄρα εὑρίσκει τῶν ἔνδον, εἴτε ἐπ᾽ ὠδίσιν
εἴη εἴτε ἤδη λεχώ, φυκία πολλὰ τῶν ἐν ταῖς ὀπαῖς
καὶ περὶ τὰς πέτρας, ἅ οἱ δεῖπνόν ἐστιν.

15. Ἐπιβουλεύειν[5] κοσσύφῳ[6] δεινὸς ἁλιεὺς
ἐφαρμόσας ἀγκίστρῳ μόλυβδον βαρὺν καὶ ἐνείρας
τῷ ἀγκίστρῳ καρίδα μεγάλην καθίησι τὸ δέλεαρ.

---

[1] ⟨ταῖς⟩ add. H.   [2] φαίην most MSS.
[3] τηνάλλως A, καὶ ἄλλως μὲν οὖν most MSS.
[4] Kayser : κράσει.

14. The Wrasse has its haunts and resorts among The Wrasse
the rocks and near cavernous burrows. The males
all have many wives and resign the hollow places, as
though they were women's chambers, to their brides.
This refinement in their mating, and the propensity
which they enjoy for having many wives one might
describe as characteristic of barbarians who luxuriate
in the pleasures of the bed, and (if one may jest on
serious subjects) as living like the Medes and Persians. It is of all fishes the most jealous at all times,
but especially when its wives are producing their
young. (If by excessive use of these expressions
I make my discourse too wanton, the facts of
nature permit me to do things of that sort.) So the
females which are actually facing the strain of birth-
pangs remain quiet in their homes, while the male,
after the manner of a husband, stays about the
entrance to prevent any mischief from outside, being
anxious for his offspring. For it seems that he loves
even those that are yet unborn, and it is his fatherly
concern that causes him these early fears; he even
spends the whole day without touching food: his care
sustains him. But as the afternoon grows late, he
relinquishes his forced watch and seeks for food,
which he does not fail to find. But of course each of
the females within, whether in the act of giving birth
or after it, finds a quantity of seaweed in the hollow
places and about the rocks, and this is their meal.

15. A fisherman who is skilled in angling a Wrasse The Wrasse,
fastens a heavy piece of lead to his hook, wraps how caught
round it a large prawn, and drops the bait. And then

---

⁵ *Jac*: ἐπιβουλεύων.   ⁶ κοσσύφου θήρᾳ.

καὶ ὁ μὲν ὑποκινεῖ τὴν ὁρμιὰν ἐγείρων τε καὶ
θήγων ἐς τὴν τροφὴν τὸ θήραμα, ἡ καρὶς δὲ κινου-
μένη εἶτα μέντοι δόξαν τινὰ ἀποστέλλει μελλούσης
ἐς τὰς ὀπὰς τὰς τοῦ κοσσύφου παριέναι. τῷ δὲ
ἄρα τοῦτο ἔχθιστον· καὶ διὰ ταῦτα αἰσθανόμενος,
ὡς ἔχει θυμοῦ,[1] ἵεται ἀφανίζειν τὴν ἐχθίστην[2] (οὐ
γάρ οἱ μέλει τῆς γαστρὸς τηνικαῦτα), καὶ συνθλά-
σας αὐτὴν ἀπαλλάττεται, προτιμότερον τροφῆς καὶ
πρεσβύτερον τὸ μὴ κατακοιμίσαι τὴν φυλακὴν πεπι-
στευκὼς εἶναι. τῶν δὲ ἄλλων ὅταν τι μέλλῃ τῶν
προσπιπτόντων ἐσθίειν, ὑποθλάσας εἶτα εἴασε
κεῖσθαι· καὶ ἰδὼν τεθνηκός,[3] ἐξ αὐτοῦ τρώγει ἤδη.
οἱ δὲ θήλεις κόσσυφοι, ἕως μὲν ἄρρενα ὁρῶσι προ-
ασπίζοντα, ὡς ἂν εἴποις, μένουσιν ἔνδον καὶ τὸ
τῆς οἰκουρίας φυλάττουσι σχῆμα· ὅταν δὲ ἀφανι-
σθῇ, ἀλύουσιν αἵδε, προάγει τε αὐτὰς καὶ ἐξάγει ἡ
ἀθυμία καὶ ἐνταῦθα ἑαλώκασι. τί πρὸς ταῦτα
⟨οἱ⟩[4] ποιηταὶ λέγουσιν οἱ τήν τε Εὐάδνην ἡμῖν
τὴν Ἴφιδος καὶ τὴν Ἄλκηστιν τὴν Πελίου παῖδα
ἐνδόξως θρυλοῦντες[5];

16. Πατὴρ δὲ ἐν ἰχθύσιν ὁ γλαῦκος οἷός ἐστι.
τὰ γεννώμενα ἐκ τῆς συννόμου παραφυλάττει[6]

---

[1] τοῦ θυμοῦ.
[2] ἀφανίζειν τὴν ἐχθίστην] νομίζων ἐχθράν.
[3] τεθνηκὸς ὅτε μὴ σπαίρει.
[4] ⟨οἱ⟩ add. Jac.
[5] Haupt : θρηνοῦντες.
[6] Schn : παραφυλάττεται.

---

[a] Evadne, wife of Capaneus, one of the 'Seven against
Thebes.' He was slain by Zeus, and when his body was on
the funeral pyre, E. leapt into the flames and perished at his
side.

he moves the line a little, rousing and egging on his prey to take the food, while the prawn by its movement conveys the impression that it intends to enter the Wrasse's den. Now this the Wrasse greatly resents, and therefore, as soon as he observes it, he longs, such is his fury, to demolish the object of his abhorrence, for he is not thinking of his appetite at the moment; and when he has crushed it, he moves off, considering it more honourable and more important that the watchman should not be caught napping than that he should be fed. But when he intends to eat any other creature that comes his way, he crushes it lightly and then lets it lie. As soon as he sees that it is dead, then at length he nibbles at it. But the female Wrasses, so long as they see the male acting as their shield, so to say, ' remain within and with the care of their household ' are occupied. If however the male disappears, they become distraught; their despondency leads them to venture forth, and then they are caught.

What have the poets to say to this—our poets who are for ever extolling Evadne,[a] the daughter of Iphis, and Alcestis,[b] the daughter of Pelias?

16. Among fishes the ' Blue-grey '[c] is a model father. He maintains a strenuous watch over his

The 'Blue-grey' fish

[b] Alcestis, wife of Admetus, undertook to die in place of her husband, but was rescued by Heracles from the clutches of Death.

[c] Not certainly identified.

ἰσχυρῶς, ἵνα ἀνεπιβούλευτά τε καὶ ἀσινῆ ᾖ. καὶ
ἕως μὲν φαιδρὰ καὶ ἔξω δέους διανήχεται, ὁ δὲ
τὴν φρουρὰν οὐκ ἀπολιμπάνει, ἀλλὰ πῇ μὲν οὐραγεῖ,
πῇ δὲ οὔ, ταύτην δὲ παρανήχεται τὴν πλευρὰν ἢ
ἐκείνην· ἐὰν δέ τι δείσῃ τῶν [1] νηπίων, ὁ δὲ χανὼν
ἐσεδέξατο τὸ βρέφος· [2] εἶτα τοῦ φόβου παραδραμόντος τὸν καταφυγόντα ἀνεμεῖ οἷον ἐδέξατο, καὶ
ἐκεῖνος πάλιν νήχεται.

17. Κύων δὲ θαλαττία τεκοῦσα ἔχει συννέοντα
τὰ σκυλάκια ἤδη καὶ οὐκ ἐς ἀναβολάς. ἐὰν δὲ
δείσῃ τι τούτων, ἐς τὴν μητέρα ἐσέδυ αὖθις κατὰ
τὸ ἄρθρον· εἶτα τοῦ δέους παραδραμόντος τὸ δὲ
πρόεισιν, ὥσπερ οὖν ἀνατικτόμενον αὖθις.

18. Θαυμάζουσιν ἄνθρωποι τὰς γυναῖκας ὡς
ἄγαν φιλοτέκνους· ὁρῶ δὲ ὅτι καὶ τεθνεώτων υἱῶν
ἢ θυγατέρων ἔζησαν μητέρες, καὶ τῷ χρόνῳ τοῦ
πάθους εἰλήφασι λήθην τῆς λύπης μεμαρασμένης.
δελφὶς δὲ ἄρα θῆλυς φιλοτεκνότατος ἐς τὰ ἔσχατα
ζῴων ἐστί. τίκτει μὲν γὰρ δύο . . . [3] ὅταν δὲ
ἁλιεὺς ἢ τρώσῃ τὸν παῖδα αὐτῆς τῇ τριαίνῃ ἢ τῇ
ἀκίδι βάλῃ . . . [3] ἡ μὲν ἀκὶς τὰ ἄνω τέτρηται, καὶ
ἐνῆπται σχοῖνος μακρὰ αὐτῇ, οἱ δὲ ὄγκοι ἐσδύντες
ἔχονται τοῦ θηρός. καὶ ἕως μὲν [4] ἔτι ῥώμης ὁ
δελφὶς ὁ τραυματίας μετείληχε, χαλᾷ ὁ θηρατὴς
τὴν σχοῖνον, ἵνα μή ποτε ἄρα ὑπὸ τῆς βίας ἀπορ-
ρήξῃ αὐτήν, καὶ γένηταί οἱ δύο κακώ, ἔχων τε
ἀπέλθῃ τὴν ἀκίδα ὁ δελφὶς καὶ ἀθηρίᾳ περιπέσῃ

[1] δείσῃ τῶν τι. H.
[2] καὶ συνεῖδε τὴν αἰτίαν add. L, del. H.

34

mate's offspring, to ensure that they are not attacked
or injured. And all the while that they are swimming the sea happily and without fear he never
relaxes his vigilance, and sometimes brings up the
rear and sometimes does not, but swims by them now
on this side now on that. And if any of his young is
afraid, he opens his mouth and takes the baby in.
Later, when its fear has passed, he disgorges the one
that took refuge exactly as he received it, and it
resumes its swimming.

17. Directly the Dog-fish has produced its young, The Dog-
it has them swimming by its side, and there is no fish
delay. But if any one of them is afraid, it slips back
into its mother's womb. Later, when its fear has
passed, it emerges, as though it were being born
again.

18. Men admire women for their devotion to The Dolphin
their children, yet I observe that mothers whose sons and its
or whose daughters have died, continued to live and young
in time forgot their sufferings, their grief having
abated. But the female Dolphin far surpasses all
creatures in its devotion to its offspring. It produces two. . . . And when a fisherman either
wounds a young Dolphin with his harpoon or strikes
it with his barb . . . The barb is pierced at the
upper end, and a long line is fastened to it, while
the barbs sink in and hold the fish. So long as the
wounded Dolphin still has any strength, the fisherman leaves the line slack, so that the fish may not
break it by its violence, and so that he himself may
not incur a double misfortune through the Dolphin

---

³ *Lacunae.*   ⁴ μὲν ἀλγῶν.

αὐτός· ὅταν δὲ αἴσθηται καμόντα καί πως παρει-
μένον ἐκ τοῦ τραύματος, ἡσυχῇ παρ' αὐτὴν ἄγει
τὴν ναῦν, καὶ ἔχει τὴν ἄγραν. ἡ δὲ μήτηρ οὐκ
ὀρρωδεῖ τὸ πραχθέν, οὐδὲ ἀναστέλλεται δείσασα,
ἀλλ' ἀπορρήτῳ φύσει τῷ πόθῳ τοῦ παιδὸς ἕπεται·
καὶ δείματα ὁπόσα ἐθέλεις εἰ ἐπάγοις, ἡ δὲ οὐκ
ἐκπλήττεται, τὸν παῖδα οὐχ ὑπομένουσα ἀπολιπεῖν
ἐν ταῖς φοναῖς [1] ὄντα, ἀλλὰ καὶ ἐκ χειρὸς αὐτὴν
πατάξαι πάρεστιν· οὕτως ὁμόσε χωρεῖ τοῖς βάλ-
λουσιν, ὥσπερ οὖν ἀμυνουμένη.[2] καὶ ἐκ τούτων
συναλίσκεται τῷ παιδί, σωθῆναι παρὸν καὶ ἀπελθεῖν αὐτήν. εἰ δὲ ἄμφω τὰ ἔκγονα αὐτῇ παρείη,
καὶ νοήσειε τετρῶσθαι τὸν ἕτερον καὶ ἄγεσθαι, ὡς
προεῖπον, διώκει τὸν ὁλόκληρον καὶ ἀπελαύνει τὴν
τε οὐρὰν [3] ἐπισείουσα καὶ δάκνουσα τῷ στόματι,
καὶ φυσᾷ φύσημά τι ἄσημον [4] μέν, ᾗ δύναται,
σύνθημα δὲ τῆς φυγῆς ἐνδιδοῦσα σωτήριον. καὶ ὁ
μὲν ἀπαλλάττεται, μένει δὲ αὐτὴ [5] ἔστ' ἂν αἱρεθῇ,
καὶ συναποθνήσκει τῷ ἑαλωκότι.

19. Ὁ βοῦς ὁ θαλάττιος ἐν πηλῷ τίκτεται, καὶ
ἔστιν ἐξ ὠδίνων βράχιστος, γίνεται δὲ ἐκ βραχίστου [6] μέγιστος. καὶ τὰ μὲν ὑπὸ τὴν νηδὺν
λευκός ἐστι, τὰ νῶτα δὲ καὶ τὸ πρόσωπον καὶ τὰς
πλευρὰς μέλας δεινῶς.[7] στόμα δὲ αὐτῷ ἐμπέ-
φυκε σμικρόν, οἱ δὲ ὀδόντες, μεμυκότος [8] οὐκ ἂν
αὐτοὺς ἴδοις· ἔστι δὲ [9] μήκιστος καὶ πλατύτατος.

---

[1] τοῖς φόνοις.   [2] ἀμυνομένη.
[3] τῇ τε οὐρᾷ.
[4] *Reiske* : φυσήματι ἀσήμῳ.
[5] *Schn* : αὐτῆ.
[6] βραχύτατος . . . τοῦ βραχίστου.

escaping with the barb and himself failing to catch
anything. As soon as he perceives that the fish is
tiring and is somewhat weakened by the wound, he
gently brings his boat near and lands his catch. But
the mother Dolphin is not scared by what has
occurred nor restrained by fear, but by a mysterious
instinct follows in her yearning for her child. And
though one confront her with terrors never so great,
she is still undismayed, and will not endure to desert
her young one which has come to a bloody end;
indeed, it is even possible to strike her with the hand,
so close does she come to the hunters, as though she
would beat them off. And so it comes about that
she is caught along with her offspring, though she
could save herself and escape. But if both her offspring are by her, and if she realises that one has
been wounded and is being hauled in, as I said
above, she pursues the one that is unscathed and
drives it away, lashing her tail and biting her little
one with her mouth; and she makes a blowing sound
as best she can, indistinct, but giving the signal to
flee, which saves it. So the young Dolphin escapes,
while the mother remains until she is caught and dies
along with the captive.

19. The Horned Ray is born in the mud, and The Horned
though at the time of birth it is very small, it grows Ray
from that size to be enormous. Its belly beneath is
white; its back, its head, and its sides are a deep
black; its mouth however is small, and its teeth—
when it opens its mouth, you cannot see them.

---

7 δεινῶς καὶ ἄναλκίς ἐστι. 8 Jac: μεμυκότες.
9 δὲ καί.

σιτεῖται μὲν οὖν καὶ τῶν ἰχθύων πολλούς, μάλιστα
δὲ σαρκῶν ἀνθρωπείων ἐσθίων ὑπερήδεται. σύνοιδε
δὲ αὑτῷ ὅτι ῥώμην ἥκιστός ἐστι, μόνῳ δὲ ἐπιθαρ-
ρεῖ τῷ μεγέθει. καὶ διὰ τοῦτο ὅταν ἴδῃ τινὰ ἢ
νηχόμενον ἢ ὑποδυόμενον ¹ ἐν ταῖς ὑδροθηρίαις, μετε-
ωρίσας ἑαυτὸν καὶ ἐπικυρτώσας ἐπινήχεταί οἱ ²
βαρὺς ἄνω ἐγκείμενός τε καὶ πιέζων καὶ ἐπαρτῶν
δεῖμά τι,³ ὑπερπετάσας τὸ πᾶν σῶμα τῷ δειλαίῳ
ὡς στέγην, ἀναδῦναί τε καὶ ἀναπνεῦσαι κωλύων
αὐτόν. οὐκοῦν ἐπισχεθέντος οἱ τοῦ πνεύματος, ὁ
μέν, οἷα εἰκός, ἀποθνήσκει, ὁ δὲ ἐμπεσὼν ἔχει τῆς
παραμονῆς μισθὸν ὃ μάλιστα λιχνεύει ⁴ δεῖπνον.

20. Τὰ μὲν ἄλλα τῶν ᾠδικῶν ⁵ [ὀρνέων] ⁶ εὐστομεῖ
καὶ τῇ γλώττῃ φθέγγεται δίκην ἀνθρώπου· οἱ δὲ
τέττιγες κατὰ τὴν ἰξύν εἰσι λαλίστατοι. καὶ σιτοῦνται μὲν τῆς δρόσου, τὰ δὲ ἐξ ἕω ἐς πλήθουσαν
ἀγορὰν σιωπῶσιν, ἡλίου δὲ ὑπαρχομένου τῆς
ἀκμῆς, τὸν ἐξ ἑαυτῶν μεθιᾶσι κέλαδον, φιλόπονοί
τινες ὡς ἂν εἴποις χορευταί, ὑπὲρ κεφαλῆς καὶ
τῶν παρανεμόντων καὶ τῶν ὁδῷ χρωμένων καὶ
τῶν ἀμώντων κατάδοντες. καὶ τοῦτο μὲν τὸ
φιλόμουσον ἔδωκε τοῖς ἄρρεσιν ἡ φύσις· τέττιξ δὲ
θήλεια ἄφωνός ἐστι, καὶ ἔοικε σιωπᾶν δίκην νύμφης
αἰδουμένης.

21. Ὑφαντικὴν καὶ ταλασίαν τὴν θεὸν τὴν
Ἐργάνην ἐπινοῆσαί φασιν ἄνθρωποι· τὴν δὲ ἀράχ-
νην ἡ φύσις σοφὴν ἐς ἱστουργίαν ἐδημιούργησε.
καὶ φιλοτεχνεῖ οὐ κατὰ μίμημα,⁷ οὐδὲ ἔξωθεν

___

¹ ὑποδυόμενον Post, cp. 1. 44, πονούμενον MSS, H.
² οἱ καὶ ἐλλοχᾷ.   ³ Jac: δείματι.

38

Further, it is exceedingly long and flat. While on the one hand it feeds upon a great number of fish, yet its chief delight is to eat the flesh of man. It is conscious of its very small strength: only its great size gives it courage. Hence when it sees a man swimming or diving to catch something in the water, it rises and arching its body attacks him, pressing upon him from above with all its weight; and while causing terror to fasten upon him, the Ray extends all its body over the wretched man like a roof and prevents him from reaching the surface and breathing. When therefore his breathing is arrested, the man naturally dies, and the Ray falls upon him and in the feast which it most greedily desires reaps the reward of its persistence.

20. All other songsters sing sweetly and use their tongue to utter, as men do, but Cicadas produce their incessant chatter from their loins. They feed upon dew, and from dawn until about midday remain silent. But when the sun enters upon his hottest period, they emit their characteristic clamour—industrious members of a chorus, you might call them —and from above the heads of shepherds and wayfarers and reapers their song descends. This love of singing Nature has bestowed upon the males, whereas the female Cicada is mute and appears as silent as some shamefast maiden. *The Cicada*

21. Men say that it was the goddess Ergane who invented weaving and spinning, but it was Nature that trained the Spider to weave. The practice of its craft is not due to any imitation, nor does it *The Spider and its web*

---

4 *Reiske* : ἀνιχνεύει.   5 *Bochart* : Ἰνδικῶν.

6 [ὀρνέων] *del. Warmington.*   7 *Reiske* : νῆμα.

λαμβάνει ⟨τὸ⟩¹ νῆμα, ἀλλ' ἐκ τῆς οἰκείας νηδύος
τοὺς μίτους ἐξάγουσα εἶτα μέντοι τοῖς κούφοις τῶν
πτηνῶν θήρατρα ἀποφαίνει, ὡς δίκτυα ἐκπετάν-
νῦσα. καὶ δι' ὧν ἐξυφαίνει παρὰ τῆς γαστρὸς
λαβοῦσα,² διὰ τῶνδε ἐκείνην ἐκτρέφει πάνυ φιλερ-
γοῦσα, ὡς καὶ τῶν γυναικῶν τὰς μάλιστα εὐχείρας
καὶ νῆμα ἀσκητὸν ἐκπονῆσαι δεινὰς μὴ ἀντιπαρα-
βάλλεσθαι· νενίκηκε γὰρ τῇ λεπτότητι καὶ τὴν
τρίχα.

22. Βαβυλωνίους τε καὶ Χαλδαίους σοφοὺς τὰ
οὐράνια ᾄδουσιν οἱ συγγραφεῖς· μύρμηκες δὲ οὔτε
ἐς οὐρανὸν ἀναβλέποντες οὔτε ³ τὰς τοῦ μηνὸς
ἡμέρας ἐπὶ δακτύλων ἀριθμεῖν ἔχοντες ὅμως δῶρον
ἐκ φύσεως εἰλήχασι παράδοξον· τῇ γὰρ ἡμέρᾳ τοῦ
μηνὸς τῇ νέᾳ ἔσω τῆς ἑαυτῶν στέγης οἰκουροῦσι,
τὴν ὀπὴν οὐχ ὑπερβαίνοντες ἀλλὰ ἀτρεμοῦντες.

23. Οἰκία τῷ σαργῷ τῷ ἰχθύι πέτραι ⁴ τε καὶ
σήραγγες, ἔχουσαι μέντοι διασφάγας μικράς,⁵ ὡς
αὐγὴν ἡλίου ⁶ κατιέναι καὶ φωτὸς ὑποπιμπλάναι
τὰς διαστάσεις τάσδε· χαίρουσι γὰρ οἱ σαργοὶ
φωτὶ μὲν παντί, τῆς δὲ ἀκτῖνος τοῦ ἡλίου καὶ
μᾶλλον διψῶσιν. οἰκοῦσι δὲ ἐν ταὐτῷ πολλοί·
δίαιται δὲ αὐτοῖς καὶ ἤθη ⁷ τὰ τῆς θαλάττης
βράχη, καὶ τῇ γῇ γειτνιῶσι μάλα ἀσμένως.
φιλοῦσι δέ πως ⁸ αἶγας ἰσχυρῶς. ἐὰν γοῦν πλησίον
τῆς ἠόνος νεμομένων ἡ σκιὰ μιᾶς ἢ δευτέρας ἐν
τῇ θαλάττῃ φανῇ, οἱ δὲ ἀσμένως προσνέουσι καὶ

¹ ⟨τό⟩ add. H.   ² Reiske : ἕλκουσα.
³ οὐδέ.   ⁴ πέτρα.
⁵ μικρὰς καὶ τὰς διαστάσεις, v.l. μ. καὶ διεστώσας.

obtain spinning matter from any external source, but
produces the threads from its own belly and then
contrives snares for flimsy winged creatures, spreading them like nets; and it derives its nourishment
from the same material that it extracts from its belly
and weaves. It is so extremely industrious that not
even the most dexterous women, skilled at elaborat-
ing wrought yarn, can be compared to it: its web is
thinner than hair.

22. Historians praise the Babylonians and Chal- The Ant
daeans for their knowledge of the heavenly bodies.
But Ants, though they neither look upwards to the
sky nor are able to count the days of the month on
their fingers, nevertheless have been endowed by
Nature with an extraordinary gift. Thus, on the
first day of the month they stay at home indoors,
never quitting their nest but remaining quietly
within.

23. The fish known as the Sargue has its home The Sargue
among rocks and hollows, which however have in
them narrow clefts so that the rays of the sun can
penetrate within and fill these fissures with light.
For Sargues like all the light there is, but have an
even greater craving for the sunbeams. They live
in great numbers in the same place, and their usual
haunts are the shallows of the sea, and they particularly like to be near the land. For some reason they
have a strong affection for goats. At any rate if the
shadow of one or two goats feeding by the sea-shore
fall upon the water, they swim in eagerly and spring

---

⁶ ἡλίου τε.   ⁷ ἕλη.   ⁸ πως τῶν ἀλόγων.

ἀναπηδῶσιν, ὡς ἡδόμενοι, καὶ προσάψασθαι τῶν
αἰγῶν ποθοῦσιν ἐξαλλόμενοι, καίτοι οὐ πάνυ τι
ὄντες ἁλτικοὶ τὴν ἄλλως· νηχόμενοι δὲ καὶ ὑπὸ
τοῖς κύμασιν ὅμως τῆς τῶν αἰγῶν ὀσμῆς ἔχουσιν αἴσθησιν, καὶ ὑφ' ἡδονῆς προελθεῖν [1] ἐπ' αὐτὰς
σπεύδουσιν. ἐπεὶ τοίνυν δυσέρωτές [2] εἰσιν, ἐξ ὧν
ποθοῦσιν ἐκ τούτων ἁλίσκονται. ἁλιεὺς γὰρ ἀνὴρ
αἰγὸς δορᾷ ἑαυτὸν περιαμπέχει, σὺν αὐτοῖς τοῖς κέρασι δαρείσης αὐτῆς· λαμβάνει ⟨δὲ⟩[3] ἄρα τὸν
ἥλιον κατὰ νώτου ἐπιβουλεύων ὁ θηρατὴς τῇ ἄγρᾳ,
εἶτα καταπάττει τῆς θαλάττης, ὑφ' ἣν οἰκοῦσιν οἱ
προειρημένοι, ἄλφιτα αἰγείῳ ζωμῷ διαβραχέντα.
ἑλκόμενοι δὲ οἱ σαργοὶ ὡς ὑπό τινος ἴυγγος τῆς
ὀσμῆς τῆς προειρημένης προσίασι, καὶ σιτοῦνται μὲν
τῶν ἀλφίτων, κηλοῦνται δὲ ὑπὸ τῆς δορᾶς.[4] αἱρεῖ [5]
δὲ αὐτῶν πολλοὺς ἀγκίστρῳ σκληρῷ καὶ ὁρμιᾷ
λίνου λευκοῦ· ἐξῆπται δὲ οὐχὶ καλάμου, ἀλλὰ
ῥάβδου κρανείας· δεῖ γὰρ τὸν ἐμπεσόντα ἀνασπάσαι ῥᾷστα, ἵνα μὴ τοὺς ἄλλους ἐκταράξῃ. θηρῶν-
ται δὲ καὶ ἀπὸ χειρός, ἐάν τις τὰς ἀκάνθας, ἃς
ἐγείρουσιν ἐς τὸ ἑαυτοῖς ἀμύνειν, ἐς τὸ κάτω μέρος
ἀπό γε τῆς κεφαλῆς ἡσυχῇ κατάγων εἶτα κλίνῃ
καὶ πιέσας τῶν πετρῶν ἐκσπάσῃ, ἐς ἃς ἑαυτοὺς
ὑπὲρ τοῦ λαθεῖν ὠθοῦσιν.

24. Ὁ ἔχις περιπλακεὶς τῇ θηλείᾳ μίγνυται· ἡ
δὲ ἀνέχεται τοῦ νυμφίου καὶ λυπεῖ οὐδὲ ἕν. ὅταν
δὲ πρὸς τῷ τέλει τῶν ἀφροδισίων ὦσι, πονηρὰν
ὑπὲρ τῆς ὁμιλίας τὴν φιλοφροσύνην ἐκτίνει ἡ

---

[1] Abresch : προσ-.
[2] ἐς τὰ προειρημένα δυσ-.
[3] ⟨δὲ⟩ add. H.
[4] δ. βλεπομένης ὡς αἰγός.
[5] αἱρεῖται.

up as though for joy, and in their desire to touch the
goats they leap out of the water, though they are
not in a general way given to leaping. And even
when swimming below the waves they are sensible
of the goats' smell, and for delight in it press in to
be near them. Now since they are thus love-sick,
the object of their love is the means of their capture.
Thus, a fisherman wraps himself in a goatskin which how caught
has been flayed with the horns. Stalking his prey,
the hunter gets the sun behind him and then sprinkles
on the water beneath which the aforesaid fish live,
barley-groats soaked in broth of goats' flesh. And
the Sargues, attracted by the aforesaid smell as
though by some charm, approach and eat the barley-
groats and are fascinated by the goatskin. And
the man catches them in numbers with a stout
hook and a line of white flax attached not to a
reed but to a rod of cornel-wood. For it is essential
to haul in the fish that has taken the bait very
quickly so as to avoid disturbing the others. They
are even to be caught by hand, if by gently
stroking the spines, which they raise in self-pro-
tection, from the head downwards one can lay them,
or by pressure draw the fish out of the rocks
into which they thrust themselves to avoid being
seen.

24. The male Viper couples with the female by Vipers and
wrapping himself round her. And she allows her their mating
mate to do this without resenting it at all. When
however they have finished their act of love, the

43

νύμφη τῷ γαμέτῃ· ἐμφῦσα γὰρ αὐτοῦ τῷ τραχήλῳ,
διακόπτει αὐτὸν αὐτῇ κεφαλῇ· καὶ ὁ μὲν τέθνηκεν,
ἡ δὲ ἔγκαρπον ἔχει τὴν μίξιν καὶ κύει. τίκτει δὲ
οὐκ ᾠά, ἀλλὰ βρέφη, καὶ ἔστιν ἐνεργὰ ἤδη
⟨κατὰ⟩[1] τὴν αὐτῶν φύσιν τὴν κακίστην. διε-
σθίει γοῦν τὴν μητρῴαν νηδύν, καὶ πρόεισι πάραυ-
τα[2] τιμωροῦντα τῷ πατρί. τί οὖν οἱ Ὀρέσται
καὶ οἱ Ἀλκμαίωνες πρὸς ταῦτα, ὦ τραγῳδοὶ
φίλοι;

25. Τὴν ὕαιναν τῆτες μὲν ἄρρενα εἰ θεάσαιο, τὴν
αὐτὴν ἐς νέωτα ὄψει θῆλυν· εἰ δὲ θῆλυν νῦν, μετὰ
ταῦτα ἄρρενα· κοινωνοῦσί τε ἀφροδίτης ἑκατέρας,
καὶ γαμοῦσί τε καὶ γαμοῦνται, ἀνὰ ἔτος πᾶν ἀμεί-
βουσαι τὸ γένος. οὐκοῦν τὸν Καινέα καὶ τὸν Τειρεσίαν ἀρχαίους ἀπέδειξε τὸ ζῷον τοῦτο οὐ
κόμποις ἀλλὰ τοῖς ἔργοις αὐτοῖς.

26. Μάχονται μὲν ὑπὲρ τῶν θηλειῶν ὡς ὑπὲρ
ὡραίων γυναικῶν καὶ οἱ τράγοι πρὸς τράγους καὶ
οἱ ταῦροι πρὸς ταύρους καὶ ὑπὲρ οἰῶν οἱ κριοὶ
πρὸς τοὺς ἀντερῶντας· ὀργῶσι δὲ ἐπὶ τὰς θηλείας
καὶ οἱ θαλάττιοι κάνθαροι. γίνονται δὲ ἐν τοῖς
καλουμένοις ἀσπροῖς[3] χωρίοις, καὶ εἰσὶ ζηλότυποι,
καὶ ἴδοις ἂν μάχην ὑπὲρ τῶν θηλειῶν καρτερὰν·
καὶ ἔστιν ὁ ἀγὼν οὐχ ὑπὲρ πολλῶν, ὡς τοῖς

---

[1] ⟨κατὰ⟩ add. H.   [2] κατ' αὐτά, v.l. κατὰ ταὐτά.
[3] λεπροῖς H after Jac.

---

[a] Orestes slew his mother Clytemnestra in revenge for her
having slain his father Agamemnon.—Alcmaeon slew his
mother Eriphyle who had brought about the death of his
father Amphiaraus.

bride in reward for his embraces repays her husband
with a treacherous show of affection, for she fastens
on his neck and bites it off, head and all. So he dies,
while she conceives and becomes pregnant. But she
produces not eggs but live young ones, which immediately act in accordance with their nature at its
worst. At any rate they gnaw through their
mother's belly and forthwith emerge and avenge
their father.

What then, my dramatist friends, have your
Oresteses [a] and your Alcmaeons to say to this?

25. Should you this year set eyes on a male Hyena, The Hyena
next year you will see the same creature as a female;
conversely, if you see a female now, next time you
will see a male. They share the attributes of
both sexes and are both husband and wife, chang-
ing their sex year by year. So then it is not
through extravagant tales but by actual facts
that this animal has made Caeneus [b] and Teiresias
old-fashioned.

26. As men fight for beautiful women, so do The Black
animals fight for their females, goats with goats, bulls Sea-bream
with bulls, and rams with their rivals in love for
sheep. Even the Black Sea-bream wax wanton for
their females. They are born in what men call
rough places, and are jealous, and one may see them
fighting vigorously for their females. And they do
not contend for several, in the way that Sargues do,

[b] Caeneus, originally a girl named Caenis, was changed by
Poseidon into a man; after death he resumed his female
form. Teiresias likewise changed his sex twice, but the Hyena
does this every year.

σαργοῖς,[1] ἀλλ' ὑπὲρ τῆς ἰδίας συννόμου, ὡς ὑπὲρ
γαμετῆς τῷ Μενέλεῳ πρὸς τὸν Πάριν.

27. Ἑστιᾶται μὲν ⟨ἄλλαις⟩[2] καὶ ἄλλαις τροφαῖς
ὁ πολύπους· ἔστι γὰρ καὶ φαγεῖν δεινὸς καὶ ἐπιβουλεῦσαι σφόδρα πανοῦργος· τὸ δὲ αἴτιον, παμβο-
ρώτατος θηρίων θαλαττίων ἐστί. καὶ ⟨ἡ⟩[3] ἀπό-
δειξις, εἴ τις αὐτῷ γένοιτο ἀθηρία, τῶν ἑαυτοῦ
πλοκάμων παρέτραγε, καὶ τὴν γαστέρα κορέσας
τὴν σπάνιν τῆς ἄγρας ἠκέσατο· εἶτα ἀναφύει τὸ
ἐλλεῖπον, ὥσπερ οὖν τῆς φύσεως τοῦτό[4] οἱ ἐν τῷ
λιμῷ παρασκευαζούσης ἕτοιμον τὸ δεῖπνον.

28. Ἵππος ἐρριμμένος σφηκῶν γένεσίς ἐστιν. ὁ
μὲν γὰρ ὑποσήπεται, ἐκ δὲ τοῦ μυελοῦ ἐκπέτονται
οἱ θῆρες οὗτοι, ὠκίστου ζῴου πτηνὰ ἔκγονα, τοῦ
ἵππου οἱ σφῆκες.

29. Αἱμύλον ζῷον καὶ ἐοικὸς ταῖς φαρμακίσιν ἡ
γλαῦξ. καὶ πρώτους μὲν αἱρεῖ τοὺς ὀρνιθοθήρας
ᾑρημένη. περιάγουσι γοῦν αὐτὴν ὡς παιδικὰ ἢ
καὶ νὴ Δία περίαπτα ἐπὶ τῶν ὤμων. καὶ νύκτωρ
μὲν αὐτοῖς ἀγρυπνεῖ καὶ τῇ φωνῇ οἱονεί τινι
ἐπαοιδῇ γοητείας ὑπεσπαρμένης αἱμύλου τε καὶ
θελκτικῆς τοὺς ὄρνιθας ἕλκει καὶ καθίζει πλησίον
ἑαυτῆς· ἤδη δὲ καὶ ἐν ἡμέρᾳ θήρατρα ἕτερα τοῖς
ὄρνισι προσείει μωκωμένη καὶ ἄλλοτε ἄλλην ἰδέαν
προσώπου στρέφουσα, ὑφ' ὧν κηλοῦνται[5] καὶ
παραμένουσιν ἐνεοὶ[6] πάντες ὄρνιθες, ᾑρημένοι δέει
καὶ μάλα γε ἰσχυρῷ ἐξ ὧν ἐκείνη μορφάζει.

---

[1] *Reiske*: σ. ὁ πόλεμος.   [2] ⟨ἄλλαις⟩ add. H.

but each for its own mate, just as Menelaus fought
for his wife with Paris.

27. The Octopus feeds first on one thing and then The
on another, for it is terribly greedy and for ever Octopus
plotting some evil, the reason being that it is the
most omnivorous of all sea-animals. The proof of
this is that, should it fail to catch anything, it eats
its own tentacles, and by filling its stomach so, finds
a remedy for the lack of prey. Later it renews its
missing limb, Nature seeming to provide this as a
ready meal in times of famine.

28. A horse's carcase is the breeding-place of The Wasp,
Wasps. For as the carcase rots, these creatures fly generated
out of the marrow: the swiftest of animals begets
winged offspring: the horse, Wasps.

29. The Owl is a wily creature and resembles a The Owl
witch. And when captured, it begins by capturing
its hunters. And so they carry it about like a pet
or (I declare) like a charm on their shoulders. By
night it keeps watch for them and with its call that
sounds like some incantation it diffuses a subtle,
soothing enchantment, thereby attracting birds to
settle near it. And even in the daytime it dangles
before the birds another kind of lure to make fools of
them, putting on a different expression at different
times; and all the birds are spell-bound and remain
stupefied and seized with terror, and a mighty terror
too, at these transformations.

---

3 ⟨ἡ⟩ add. H.   4 καὶ τοῦτο.
5 αἱροῦνται.   6 Hemst : οἱ νέοι.

30. Ὁ λάβραξ καρίδος ἥττηται, καὶ εἴη ἂν, ἵνα
τι καὶ παίσας εἴπω,[1] ἰχθύων ὀψοφαγίστατος.
οὐκοῦν ἔλειοι ὄντες τὰς ἐλείους ἐλλοχῶσιν. εἰσὶ
γὰρ τῷ γένει τριτταί· καὶ αἱ μὲν αὐτῶν οἵας
προεῖπον, αἱ δὲ ἐκ φυκίων, πετραῖαί γε μὴν αἱ
τρίται. ἀμύνεσθαι δὲ αὐτοὺς ἀδυνατοῦσαι αἱροῦνται συναποθνήσκειν. καὶ τό γε σόφισμα εἰπεῖν
οὐκ ὀκνήσω αὐτῶν. ὅταν γοῦν αἴσθωνται λαμβανό-
μεναι, τὸ ἐξέχον τῆς κεφαλῆς (ἔοικε δὲ τριήρους
ἐμβόλῳ καὶ μάλα γε ὀξεῖ, καὶ ἄλλως ἐντομὰς ἔχει
δίκην πρίονος) τοῦτο τοίνυν αἱ γενναῖαι σοφῶς
ἐπιστρέψασαι πηδῶσί τε καὶ ἀναθόρνυνται κοῦφα
καὶ ἁλτικά. κέχηνε δὲ ὁ λάβραξ μέγα,[2] καὶ ἔστιν
οἱ τὰ τῆς δέρης ἁπαλά. οὐκοῦν ὁ μὲν συλλαβὼν
τὴν καρίδα καμοῦσαν οἴεται δεῖπνον ἕξειν, ἡ δὲ ἐν
ἐξουσίᾳ τε καὶ εὐρυχωρίᾳ σκιρτᾷ τῆς φάρυγγος ὡς
ἂν εἴποις καταχορεύουσα· εἶτα ἐμπήγνυται τῷ
δειλαίῳ θηρατῇ τὰ κέντρα, καὶ ἑλκοῦταί οἱ τὰ
ἔνδον καὶ ἀνοιδήσαντα αἷμα ἐκβάλλει πολὺ καὶ
ἀποπνίγει, καὶ καινότατα δήπου ἀποκτείνασα
ἀνῄρηται.

31. Ὀνύχων ἀκμαῖς καὶ ὀδόντων διατομαῖς θαρροῦσι καὶ ἄρκτοι καὶ λύκοι καὶ πάρδοι καὶ λέοντες·
τὴν δὲ ὕστριχα ἀκούω ταῦτα μὲν οὐκ ἔχειν, οὐ
μὴν ὅπλων ὑπὸ τῆς φύσεως ἀμυντηρίων ἀπολε-
λεῖφθαι ἐρήμην. τοῖς γοῦν ἐπιοῦσιν ἐπὶ λύμῃ
τὰς ἄνωθεν τρίχας οἱονεὶ βέλη ἐκπέμπει, καὶ
εὐστόχως βάλλει πολλάκις, τὰ νῶτα φρίξασα·

---

[1] ἵνα . . . εἴπω] εἰ καὶ πταίσας ἐρῶ.
[2] καὶ μέγα.

30. The Basse is a victim of the Prawn and is in- Basse and
clined to be (if I may be allowed the jest) the greatest Prawn
gourmet among fish. So being lake-dwellers they
lie in wait for the lake Prawns. These are of three
kinds: the first are such as I have already mentioned;
the second subsist on seaweed, while the third kind
live on the rocks. Being incapable of self-defence
against the Basse, they prefer to die along with it.
And I shall not hesitate to use the word 'stratagem'
of them. For instance, directly they realise that they
are being caught, these precious creatures adroitly
turn outwards the projecting portion of their head,
which resembles the beak of a trireme and is exceedingly sharp and has moreover notches in it like a
saw, and spring and leap lightly and nimbly about.
But the Basse opens its mouth wide, and the flesh
of its throat is tender. So the Basse seizes the
exhausted Prawn and fancies that it is going to
make a meal of it. The Prawn however in this
ample space gambols about and dances in triumph,
so to say, over the Basse's throat. Then it plants its
spikes in its unfortunate pursuer, whose inward
parts are thereby lacerated, so that they swell up
and discharge much blood and choke the Basse, until
in most novel fashion the slayer is himself slain.

31. Strength of claws and sharpness of fangs make The
bears, wolves, leopards, and lions bold, whereas the Porcupine
Porcupine, which (I am told) has not these advan-
tages, none the less has not been left by Nature
destitute of weapons wherewith to defend itself.
For instance, against those who would attack it with
intent to harm it discharges the hairs on its body,
like javelins, and raising the bristles on its back,

καὶ ἐκεῖναί γε πηδῶσιν, ὥσπερ οὖν ἔκ τινος
ἀφειμέναι νευρᾶς.

32. Ἦ δεινὸν κακὸν καὶ νόσημα ἄγριον ἔχθρα
καὶ μῖσος συμφυές, εἴπερ οὖν καὶ τοῖς ἀλόγοις
ἐντέτηκε καὶ αὐτοῖς ἐστι δυσέκνιπτα. μύραινα
γοῦν πολύποδα μισεῖ, καὶ πολύπους καράβῳ πολέμιος, καὶ μυραίνη κάραβος ἔχθιστός ἐστι. μύραινα
μὲν γὰρ ταῖς ἀκμαῖς τῶν ὀδόντων τὰς πλεκτάνας
τῷ πολύποδι διακόπτει, εἶτα μέντοι καὶ ἐς τὴν
γαστέρα ἐσδῦσα αὐτῷ τὰ αὐτὰ δρᾷ, καὶ εἰκότως·
ἡ μὲν γὰρ νηκτική, ὁ δὲ ἔοικεν ἕρποντι· εἰ δὲ καὶ
τρέποιτο τὴν χρόαν κατὰ τὰς πέτρας, ἔοικεν αὐτῷ
τὸ σόφισμα συμφέρειν[1] οὐδὲ ἓν τοῦτο· ἔστι γὰρ
συνιδεῖν ἐκείνη δεινὴ τοῦ ζῴου τὸ παλάμημα.
τούς γε μὴν καράβους αὐτοὶ[2] συλλαβόντες ἐς
πνῖγμα, ὅταν νεκροὺς ἐργάσωνται, τὰ κρέα ἐκμυ-
ζῶσιν αὐτῶν. κέρατα δὲ τὰ ἑαυτοῦ ὁ κάραβος
ἀνεγείρας καὶ θυμωθεὶς ἐς αὐτά, προκαλεῖται μύραιναν.[3] οὐκοῦν ἡ μὲν τοῦ ἀντιπάλου τὰ κέντρα,
ὅσα οἱ προβέβληται, ταῦτα οὐκ ἐννοοῦσα κατα-
δάκνει· ὁ δὲ τὰς χηλὰς οἱονεὶ χεῖρας προτείνας,
τῆς δέρης παρ' ἑκάτερα ἐγκρατῶς ἐχόμενος οὐ
μεθίησιν· ἡ δὲ ἀσχάλλει καὶ ἑαυτὴν ἑλίττει καὶ
περιβάλλει τῶν ὀστράκων ταῖς ἀκμαῖς, ὧνπερ οὖν
ἐς αὐτὴν πηγνυμένων μαλκίει[4] τε καὶ ἀπαγορεύει,
καὶ τελευτῶσα παρειμένη κεῖται· ὁ δὲ τὴν
ἀντίπαλον ποιεῖται δεῖπνον.

---

[1] *Triller*: αἱρεῖν.
[2] αὐτοί *corrupt*, H.
[3] μ. καὶ ὡς εἶναι κατὰ γυναῖκα ὠργισμένην.

frequently makes a good shot. And these hairs leap forth as though sped from a bowstring.

32. Enmity and inborn hate are a truly terrible affliction and a cruel disease when once they have sunk deep into the heart even of brute beasts, and nothing can purge them away. For instance, the Moray loathes the Octopus, and the Octopus is the enemy of the Crayfish, and to the Moray the Crayfish is most hostile. The Moray with its sharp teeth cuts through the tentacles of the Octopus, and then boring into its stomach does the same thing—and very properly, for the Moray swims, while the Octopus is like some creeping thing. And even though it changes its colour to that of the rocks, even this artifice seems to avail it nothing, for the Moray is quick to perceive the creature's stratagem.

As to the Crayfish, the Octopuses strangle them with their grip, and when they have succeeded in killing them, they suck out their flesh. But against the Moray the Crayfish raises its horns and with fury in them challenges it. Thereupon the Moray imprudently tries to bite the prickles which its adversary has thrust forward in self-defence. But the Crayfish reaches out its claws like two hands, and clinging firmly to the Moray's throat on either side, never relaxes its hold, while the Moray in its distress writhes and transfixes itself on the points of the Crayfish's shell; and as these are planted in it, it grows numb and gives up the struggle, finally sinking in exhaustion. And the Crayfish makes a meal off its adversary.

Mutual hatred of Moray, Octopus, and Crayfish

Moray and Octopus

Octopus an Crayfish

Moray and Crayfish

---

[4] μαλακιεῖ.

AELIAN

33. Τὴν μύραιναν [1] τὸν ἰχθὺν τρέφει τὰ πελάγη. ὅταν δὲ αὐτὴν τὸ δίκτυον περιλάβῃ,[2] διανήχεται καὶ ζητεῖ ἢ βρόχον ἀραιὸν ἢ ῥῆγμα τοῦ δικτύου πάνυ σοφῶς· καὶ ἐντυχοῦσα τούτων τινὶ καὶ διεκδῦσα ἐλευθέρα νήχεται αὖθις· εἰ δὲ τύχοι μία τῆσδε τῆς εὐερμίας, καὶ αἱ λοιπαὶ ὅσαι τοῦ αὐτοῦ γένους συνεαλώκασι κατὰ τὴν ἐκείνης φυγὴν ἐξίασιν, ὡς ὁδόν τινα λαβοῦσαι παρ᾽ ἡγεμόνος.

34. Τὴν σηπίαν ὅταν μέλλωσιν αἱρεῖν [3] οἱ τούτων ἀγαθοὶ θηραταί, συνεῖσα ἐκείνη παρῆκε τὸ ἐξ ἑαυτῆς ἀπόσφαγμα,[4] καὶ καταχεῖται ἑαυτῆς, καὶ περιλαμβάνει καὶ ἀφανίζει πᾶσαν, καὶ κλέπτεται τὴν ὄψιν ὁ ἁλιεύς· καὶ ἡ μὲν ἐν ὀφθαλμοῖς ἐστιν, ὁ δὲ οὐχ ὁρᾷ. τοιοῦτόν τι καὶ τῷ Αἰνείᾳ νέφος περιβαλὼν ἠπάτησε τὸν Ἀχιλλέα ὁ Ποσειδῶν, ὡς Ὅμηρος λέγει.

35. Βασκάνων ὀφθαλμοὺς καὶ γοήτων φυλάττεται καὶ τῶν ζῴων τὰ ἄλογα φύσει τινὶ ἀπορρήτῳ καὶ θαυμαστῇ. ἀκούω γοῦν [5] βασκανίας ἀμυντήριον τὰς φάττας δάφνης κλωνία ἀποτραγούσας λεπτὰ εἶτα μέντοι ταῖς ἑαυτῶν καλιαῖς ἐντιθέναι τῶν νεοττίων φειδοῖ· ἰκτῖνοι δὲ ῥάμνον, κίρκοι δὲ πικρίδα, αἵ γε μὴν τρυγόνες τὸν τῆς ἴρεως καρπόν, ἄγνον δὲ κόρακες, οἱ δὲ ἔποπες τὸ ἀδίαντον, ὅπερ οὖν καὶ καλλίτριχον καλοῦσί τινες, ἀριστερεῶνα δὲ κορώνη, καὶ κιττὸν ἄρπη, καρκίνον δὲ ἐρωδιός,

---

[1] *Ges* : σφύραιναν.   [2] περιβάλλῃ.

[3] *Reiske* : αἱρεῖν καὶ λαμβάνειν.

[4] ὑπόσφαγμα *H, cp.* Hippon. 2A(D²).   [5] οὖν.

[a] The genus *picris* embraces a wide variety of plants ; it may here signify *ox-tongue* or *chicory* or *endive* or *Urospermum picroides*.

33. The fish known as the Moray lives in the sea, The Moray
and when the net encircles it, it swims hither and
thither, seeking with great cleverness some weak
mesh or some rent in the net. And when it has
found such a place, it slips through and swims free
once again. And if one of them has this good for-
tune, all the others of its kind that have been caught
along with it escape in the same way, as though
taking their direction from a leader.

34. Whenever fishermen who are skilled in these The
matters plan to catch a Cuttlefish, the fish on realising Cuttlefish
this emits the ink from its body, pours it over itself
and envelops itself so as to be entirely invisible.
The fisherman's sight is deceived: though the fish is
within view, he does not see it. It was by veiling
Aeneas in such a cloud that Poseidon tricked Achilles,
according to Homer [*Il.* 20. 321–].

35. Even brute beasts protect themselves against Birds and
the eyes of sorcerers and wizards by some inexplic- their pro-
able and marvellous gift of Nature. For instance, I against
am told that as a charm against sorcery ring-doves sorcery
nibble off the fine shoots of the bay-tree, and then
insert them in their nests as a protection for their
young. Kites take buck-thorn, falcons picris,^a while
turtle-doves take the fruit ^b of the iris, ravens the
agnus-castus tree, but hoopoes maidenhair fern,
which some call 'lovely hair'; the crow takes
vervain, the shearwater ^c ivy, the heron a crab, the

^b From Thphr. *HP* 3. 3. 4 'it appears that the buds of the
poplar were mistaken for fruit,' Hort *ad loc.* So here perhaps
καρπός should be understood as the *bud* of the iris.

^c ' Ἅρπη . . . prob. *shearwater*,' L-S⁹; but the meaning
is quite uncertain, cp. 12. 4.

πέρδιξ δὲ καλάμου φόβην, θαλλὸν δὲ αἱ κίχλαι
μυρρίνης. προβάλλεται δὲ καὶ κόρυδος ἄγρωστιν,
ἀετοὶ¹ ⟨δὲ⟩² τὸν λίθον, ὅσπερ οὖν ἐξ αὐτῶν
ἀετίτης κέκληται. λέγεται δὲ οὗτος ὁ λίθος καὶ
γυναιξὶ κυούσαις ἀγαθὸν εἶναι, ταῖς ἀμβλώσεσι
πολέμιος ὤν.

36. Ὁ ἰχθὺς ἡ νάρκη ὅτου ἂν καὶ προσάψηται
τὸ ἐξ αὐτῆς ὄνομα ἔδωκέ τε καὶ ναρκᾶν ἐποίησεν.
ἡ δὲ ἐχενηὶς ἐπέχει τὰς ναῦς, καὶ ἐξ οὗ ποιεῖ
καλοῦμεν αὐτήν. κυούσης δὲ ἀλκυόνος ἵσταται
μὲν τὰ πελάγη, εἰρήνην δὲ καὶ φιλίαν ἄγουσιν
ἄνεμοι. κύει δὲ ἄρα χειμῶνος μεσοῦντος, καὶ
ὅμως ἡ τοῦ ἀέρος γαλήνη δίδωσιν εὐημερίαν, καὶ
ἀλκυονείας³ τηνικάδε τῆς ὥρας ἄγομεν ἡμέρας.
ἴχνος δὲ λύκου πατεῖ κατὰ τύχην ἵππος, καὶ νάρκη
περιείληφεν αὐτόν. εἰ δὲ ὑπορρίψειας ἀστράγαλον
λύκου τετρώρῳ⁴ θέοντι, τὸ δὲ ὡς πεπηγὸς ἑστήξε-
ται, τῶν ἵππων τὸν ἀστράγαλον πατησάντων.
λέων δὲ φύλλοις πρίνου τὸ ἴχνος ἐπιβάλλει, καὶ
ναρκᾷ· . . .⁵ δὲ καὶ ὁ λύκος, εἰ καὶ μόνον προ-
σπελάσειε πετήλοις σκίλλης. ταῦτά τοι καὶ αἱ
ἀλώπεκες ἐς τὰς εὐνὰς τῶν λύκων ἐμβάλλουσι, καὶ
εἰκότως· διὰ γὰρ τὴν ἐξ αὐτῶν ἐπιβουλὴν νοοῦσιν
ἔχθιστα αὐτοῖς.

37. Οἱ πελαργοὶ λυμαινομένας αὐτῶν τὰ ὠὰ τὰς
νυκτερίδας ἀμύνονται πάνυ σοφῶς· αἱ μὲν γὰρ

---

¹ αἰετοί MSS always.
² ⟨δέ⟩ add. Jac.
³ εὐημ. καὶ ἀλκ.] σωτηρίαν ἀλκυονίας.
⁴ Jac : καὶ τετρώρῳ.

partridge the hairy head of a reed, thrushes a sprig
of myrtle. The lark protects itself with dog's-tooth
grass; eagles take the stone which is called after
them *aëtite* (eagle-stone). This stone is also said to
be good for women in pregnancy, as a preventive of
abortions.

36. The fish known as Torpedo produces the effect _{The Torpedo}
implied in its name on whatever it touches and
makes it 'torpid' or numb. And the Sucking-fish
clings to ships, and from its action we give it its
name, *Ship-holder*.

While the Halcyon is sitting, the sea is still and the _{The Halcyon}
winds are at peace and amity. It lays its eggs about
mid-winter; nevertheless, the sky is calm and brings
fine weather, and it is at this season of the year that
we enjoy 'halcyon days.'

If a horse chance to tread on the footprint of a _{Objects producing numbness}
Wolf, it is at once seized with numbness. If you
throw the vertebra of a Wolf beneath a four-horse
team in motion, it will come to a stand as though
frozen, owing to the horses having trodden upon the
vertebra. If a Lion put his paw upon the leaves of an
ilex, he goes numb. ⟨And the same thing happens
to⟩ a Wolf, should he even come near the leaves of a
squill. And that is why foxes throw these leaves
into the dens of Wolves, and with good reason,
because their hostility is due to the Wolves' designs
upon them.

37. Storks have a very clever device for warding _{Prophylactics used by birds}
off the bats that would damage their eggs: one

---

⁵ *Lacuna*: ναρκᾷ πατῶν δὲ MSS, ⟨ναρκᾷ⟩ *Jac*, ⟨ὁμοίως⟩ *H*.

προσαψάμεναι μόνον ἀνεμιαῖα ἐργάζονται καὶ
ἄγονα αὐτά. οὐκοῦν τὸ ἐπὶ τούτοις φάρμακον
ἐκεῖνό ἐστι. πλατάνου φύλλα ἐπιφέρουσι ταῖς
καλιαῖς· αἱ δὲ νυκτερίδες ὅταν αὐτοῖς γειτνιάσωσι,
ναρκῶσι καὶ γίνονται λυπεῖν ἀδύνατοι. δῶρον δὲ
ἄρα ἡ φύσις καὶ ταῖς χελιδόσιν ἔδωκεν οἷον. αἱ
σίλφαι καὶ τούτων τὰ ᾠὰ ἀδικοῦσιν. οὐκοῦν αἱ
μητέρες σελίνου κόμην προβάλλονται τῶν βρεφῶν,
καὶ ἐκείναις τὸ ἐντεῦθεν ἄβατά ἐστι. πολύποσι δὲ
εἴ τις ἐπιβάλοι[1] πήγανον, ἀκίνητοι μένουσιν, ὡς
λέγει τις λόγος. ὄφεως δὲ εἰ καθίκοιο καλάμῳ,
μετὰ τὴν πρώτην πληγὴν ἀτρεμεῖ καὶ νάρκῃ[2]
πεδηθεὶς ἡσυχάζει· εἰ δὲ ἐπαγάγοις[3] δευτέραν ἢ
τρίτην, ἀνέρρωσας αὐτόν. καὶ μύραινα δὲ πληγεῖσα
νάρθηκι ἐς ἅπαξ ἡσυχάζει· εἰ δὲ πλεονάκις, ἐς
θυμὸν ἐξάπτεται. λέγουσι δὲ ἁλιεῖς καὶ πολύποδας
ἐς τὴν γῆν προϊέναι, ἐλαίας θαλλοῦ ἐπὶ τῆς ᾐόνος
κειμένου. θηρίων δὲ ἀλεξιφάρμακον ἦν ἄρα πάντων πιμελὴ ἐλέφαντος, ἣν εἴ τις ἐπιχρίσαιτο, καὶ εἰ
γυμνὸς ὁμόσε χωροίη τοῖς ἀγριωτάτοις, ἀσινὴς
ἀπαλλάττεται.

38. Ὀρρωδεῖ ὁ ἐλέφας κεράστην κριὸν καὶ χοί-
ρου βοήν. οὕτω τοι, φασί, καὶ Ῥωμαῖοι τοὺς σὺν
Πύρρῳ τῷ Ἠπειρώτῃ ἐτρέψαντο ἐλέφαντας, καὶ
ἡ νίκη σὺν τοῖς Ῥωμαίοις λαμπρῶς ἐγένετο.
γυναικὸς ⟨δὲ⟩[4] ὡραίας τόδε τὸ ζῷον ἡττᾶται καὶ

---

[1] ἐπιβάλλει.   [2] τῇ νάρκῃ.
[3] ἐπάγοις.   [4] ⟨δὲ⟩ add. H.

---

[a] Σίλφη (rendered 'cockroach' in L-S⁹) here probably
signifies the dipterous insect *Stenopteryx hirundinis*. 'Most

touch from the bats turns them to wind-eggs and makes them infertile. Accordingly, this is the remedy they use to prevent this happening. They lay the leaves of a plane-tree upon their nests, and directly the bats come near the storks, they are benumbed and become incapable of doing harm. On swallows too Nature has bestowed a like gift: cockroaches *a* injure their eggs. Therefore the mother-birds protect their chicks with celery leaves, and hence the cockroaches cannot reach them. If one throws some rue upon an octopus it remains immobile—so the story goes. If you touch a snake with a reed, it will after the first stroke remain still, and in the grip of numbness will lie quiet; if however you repeat the stroke a second or a third time, you at once revive its strength. The moray too, if struck once with a fennel wand, lies still the first time; but if struck several times, its anger is kindled. Fisherfolk assert that even octopuses come ashore if a sprig of olive is laid upon the beach. *Effect of certain herbs on fish and reptiles*

It seems that the fat of an elephant is a remedy against the poisons of all savage creatures, and if a man rub some on his body, even though he encounter unarmed the very fiercest, he will escape unscathed. *Elephant's fat*

38 (i). The Elephant has a terror of a horned ram and of the squealing of a pig. It was by these means, they say, that the Romans turned to flight the elephants of Pyrrhus of Epirus, and that the Romans won a glorious victory. This same animal is over- *The Elephant, fond of perfumes*

of the known *Hippoboscidae* live on birds and are apparently specially fond of the Swallow tribe. They are all winged.' D. Sharp, *Insects*, 519 (Camb. Nat. Hist. 6).

παραλύεται τοῦ θυμοῦ ἐκκωφωθὲν [1] ἐς τὸ κάλλος. καὶ ἀντήρα φασὶν ἐν τῇ Αἰγυπτίᾳ Ἀλεξάνδρου πόλει γυναικὸς στεφάνους πλεκούσης Ἀριστοφάνει τῷ Βυζαντίῳ ἐλέφας.[2] ἀγαπᾷ δὲ ὁ αὐτὸς καὶ εὐωδίαν πᾶσαν, καὶ μύρων καὶ ἀνθέων κηλούμενος τῇ ὀσμῇ.

Ὅστις βούλεται κλὼψ ἢ λῃστὴς κύνας ἄγαν ἀγριωτάτους κατασιγάσαι καὶ θεῖναι φυγάδας, ἐκ πυρᾶς ἀνθρώπου δαλὸν λαβὼν ὁμόσε αὐτοῖς χωρεῖ, φασίν· οἱ δὲ ὀρρωδοῦσιν. ἀκήκοα δὲ καὶ ἐκεῖνον τὸν λόγον. λυκοσπάδα οἶν πέξας ⟨τις⟩[3] καὶ ἐριουργήσας καὶ χιτῶνα ἐργασάμενος λυπεῖ τὸν ἠσθημένον· ὀδαξησμὸν γὰρ ἐργάζεται, ὡς λόγος. ἔριν δὲ εἴ τις καὶ στάσιν ἐθέλοι ἐν τῷ συνδείπνῳ ἐργάσασθαι, δηχθέντα ὑπὸ κυνὸς λίθον ἐμβαλὼν τῷ οἴνῳ λυπεῖ τοὺς συμπότας ἐκμαίνων. κανθάροις δὲ κακόσμοις θηρίοις εἴ τις ἐπιρράνειε [4] μύρου, οἱ δὲ τὴν εὐωδίαν οὐ φέρουσιν, ἀλλ' ἀποθνήσκουσιν. οὕτω τοί φασι καὶ τοὺς βυρσοδέψας συντραφέντας ἀέρι κακῷ βδελύττεσθαι μύρον. λέγουσι δὲ Αἰγύπτιοι καὶ τοὺς ὄφεις πάντας ἴβεων πτερὰ δεδιέναι.

39. Θηρῶσι τὰς τρυγόνας οἱ [5] τούτων ἀκριβοῦντες τὰ θήρατρα, καὶ μάλιστα τῆς πείρας οὐ διαμαρτάνουσι τὸν τρόπον τοῦτον. ἑστήκασιν ὀρ-

---

[1] Reiske : ἐκκωφωθείς.　　　[2] ὁ ἐλέφας.
[3] ⟨τις⟩ add. H.　　　[4] ἐπιρρᾶναι.
[5] οἱ καί.

---

[a] Aristophanes of Byzantium, 3rd/2nd cent. B.C., head of the library at Alexandria, famous as grammarian, literary and

come by beauty in a woman and lays aside its temper, quite stunned by the lovely sight. And at Alexandria in Egypt, they say, an Elephant was the rival of Aristophanes of Byzantium [a] for the love of a woman who was engaged in making garlands. The Elephant also loves every kind of fragrance and is fascinated by the scent of perfumes and of flowers.

(ii) If some thief or robber wants to silence dogs that are too fierce and to make them run away, he takes a brand from a funeral pyre (they say) and goes for them. The dogs are terrified. I have heard too this story: if a man shears a sheep that has been mauled by a wolf, and after working the wool makes himself a tunic, this will irritate him when he puts it on. 'He is weaving a gnawing itch for himself,' as the proverb has it.

How to stop dogs barking

Wool as irritant

(iii) If a man wants to bring about a quarrel and contention at a dinner-party, he will by dropping into the wine a stone that a dog has bitten, vex his fellow-guests to the point of frenzy.

Quarrel at a dinner-party

(iv) If a man sprinkle some perfume upon beetles, which are ill-smelling creatures, they cannot endure the sweet scent, but die. In the same way it is said that tanners, who live all their life in foul air, detest perfumes. And the Egyptians maintain that all snakes dread the feathers of the ibis.

Scents pleasant and unpleasant

39. Those who have a thorough understanding of the matter hunt Sting-rays,[b] and it is chiefly in this way that their efforts are successful. They take their

The Sting-ray, how caught

---

textual critic, especially in the field of Greek poetry. Wrote an epitome of natural history based upon Aristotle; it included ' paradoxa.'

[b] Cp. 17. 18; τρυγών must here stand for τ. θαλαττία.

χούμενοι καὶ ᾄδοντες εὖ μάλα μουσικῶς· αἱ δὲ
καὶ τῇ ἀκοῇ θέλγονται καὶ τῇ ὄψει τῆς ὀρχήσεως
κηλοῦνται καὶ προσίασιν ἐγγυτέρω. οἱ δὲ ὑπανα-
χωροῦσιν ἡσυχῇ καὶ βάδην, ἔνθα δήπου καὶ ὁ
δόλος ταῖς δειλαίαις πρόκειται, δίκτυα ἐκπεπτα-
μένα[1]· εἶτα ἐμπίπτουσιν ἐς αὐτὰ καὶ ἁλίσκονται,
ὀρχήσει καὶ ᾠδῇ ᾑρημέναι πρῶτον.

40. Ὄρκυνος ὄνομα κητώδης ἰχθὺς οὐκ ἄσοφος
ἐς τὰ αὑτοῦ λυσιτελέστατα, δῶρον λαχὼν φύσει
τοῦτο, οὐ τέχνῃ. ὅταν γοῦν περιπαρῇ τῷ ἀγκίστρῳ, καταδύει αὐτὸν ἐς βυθὸν καὶ ὠθεῖ καὶ
προσαράττει τῷ δαπέδῳ καὶ κρούει τὸ στόμα,
ἐκβαλεῖν τὸ ἄγκιστρον ἐθέλων· εἰ δὲ ἀδύνατον
τοῦτο εἴη,[2] εὐρύνει[3] τὸ τραῦμα, καὶ ἐκπτύεται τὸ
λυποῦν αὐτὸν καὶ ἐξάλλεται. πολλάκις δὲ οὐκ
ἔτυχε τῆς πείρας, καὶ ὁ θηρατὴς ἄκοντα ἀνασπάσας
ἔχει τὴν ἄγραν.

41. Δειλότατος ἰχθύων ὁ μελάνουρος, καὶ ἔχει
τῆς δειλίας μάρτυρας τοὺς ἁλιεῖς. οὔτε γοῦν
κύρτῳ λαμβάνονται οὗτοι, οὔτε προσίασιν αὐτῷ·
σαγήνῃ δὲ εἴ ποτε αὐτοὺς περιλάβοι,[4] οἱ δὲ
ἀγνοοῦντες ἑαλώκασι. καὶ ὅταν μὲν ᾖ ὑπεύδια καὶ
λεία ἡ θάλαττα, οἱ δὲ ἄρα κάτω που πρὸς ταῖς
πέτραις ἢ τοῖς φυκίοις ἡσυχάζουσι, καὶ προβάλλονται πᾶν ὅ τι δύνανται, τὸ σῶμα ἀφανίζοντες.
ἐὰν δὲ ᾖ χειμέρια, τοὺς ἄλλους ὁρῶντες καταδύν-
τας ἐκ τῆς τῶν κυμάτων προσβολῆς ἐς τὸν βυθόν,

---

[1] ἐκπεπετασμένα.　　　　[2] Schn : ᾖ.
[3] εὐρύνει οὖν.　　　　[4] περιβάλοι.

stand and dance and sing very sweetly. And the Sting-rays are soothed by the sound and are charmed by the dancing and draw nearer, while the men withdraw gently step by step to the spot where of course the snare is set for the wretched creatures, namely nets spread out. Then the Sting-rays fall into them and are caught, betrayed in the first instance by the dancing and singing.

40. The Great Tunny, as it is called, is a monstrous fish and knows well what is best for it. This gift it has acquired by nature and not by art. For instance, when the hook has pierced it, it dives to the bottom and thrusts and dashes itself against the ground, striking its mouth in its effort to eject the hook. If that fails, it widens the wound and disgorges the instrument of pain and dashes away. Frequently however it fails in the attempt, and the fisherman draws up the reluctant creature and secures his catch. The Great Tunny

41. The *Melanurus* is the most timid of fishes, and to its timidity fishermen bear witness, for it is not caught in weels nor does it go near them; but if by chance a dragnet encircles it, then it is caught without knowing it. And whenever the sea is fairly calm and smooth, these fish lie quiet down below upon the rocks or among the seaweed and cover themselves as best they can, trying to conceal their bodies. But if the weather is stormy, observing other fish diving to the depths out of the buffeting waves, they take courage and approach the shore, The 'Melanurus' (black-tail)

AELIAN

οἱ δὲ ἀναθαρροῦσι,[1] καὶ τῇ γῇ προσπελάζουσι, καὶ ταῖς πέτραις προσνέουσι, καὶ ἡγοῦνταί σφισι πρόβλημα ἱκανὸν εἶναι τὸν ὑπερνηχόμενον ἀφρὸν καλύπτοντά τε αὐτοὺς καὶ ἐπηλυγάζοντα. συνιᾶσι δὲ εὖ μάλα ἀπορρήτως ὅτι τοῖς ἁλιεῦσιν ἐν ἡμέρᾳ τοίᾳ ἢ νυκτὶ ἐς τὴν θάλατταν ἐστιν ἄβατα, ἀγριαινούσης τῆς θαλάττης ⟨καὶ⟩[2] τῶν κυμάτων αἰρομένων μετεώρων τε καὶ φοβερῶν. ἔχουσι δὲ καὶ τροφὴν ἐν χειμῶνι, τοῦ κλύδωνος τὰ μὲν ἀποσπῶντος ἐκ τῶν πετρῶν, τὰ δὲ ἐπισύροντος ἐκ τῆς γῆς· σιτοῦνται δὲ μελάνουροι τὰ ῥυπαρώτερα καὶ ὅσα οὐκ ἂν ῥᾳδίως ἰχθὺς ἄλλος ἂν πάσαιτο, εἰ μὴ πάνυ λιμῷ πιέζοιτο. ἐν γαλήνῃ δὲ ἐπὶ τῆς ἄμμου μόνης σαλεύουσι,[3] καὶ ἐκεῖθεν βόσκονται. ὅπως δὲ ἁλίσκονται, ἐρεῖ ἄλλος.

42. Ἀετὸς δὲ ὀρνίθων ὀξυωπέστατος. καὶ Ὅμηρος αὐτῷ σύνοιδε καὶ τοῦτο, καὶ μαρτυρεῖ ἐν τῇ Πατροκλείᾳ, εἰκάζων τὸν Μενέλεων τῷ ὄρνιθι, ὅτε ἀνεζήτει Ἀντίλοχον, ἵνα ἄγγελον ἀποστείλῃ τῷ Ἀχιλλεῖ, πικρὸν μέν, ἀναγκαῖον δέ, ὑπὲρ τοῦ πάθους τοῦ κατὰ τὸν ἑταῖρον αὐτοῦ, ὃν ἐξέπεμψε μέν, οὐχ ὑπεδέξατο δέ, καίτοι ποθῶν ἐκεῖνος τοῦτο. λέγεται δὲ μὴ ἑαυτῷ μόνῳ χρήσιμος, ἀλλὰ καὶ ἀνθρώπων ὀφθαλμοῖς ὁ ἀετὸς ἀγαθὸς[4] εἶναι. εἰ γοῦν μέλιτί τις Ἀττικῷ τὴν χολὴν αὐτοῦ διαλαβὼν[5] ὑπαλείψαιτο[6] ἀμβλυνόμενος, ὄψεται καὶ ὀξυτάτους γοῦν ἰδεῖν ἕξει τοὺς ὀφθαλμούς.

---

[1] ἀναθαρσοῦσι.
[2] ⟨καὶ⟩ add. Reiske.
[3] Jac : ἁλιεύουσι.
[4] Schn : ἀγαθόν.
[5] ἀναλαβών ? H.
[6] ὑπαλείφοιτο.

swim close to the rocks, and fancy that the foam
floating overhead is sufficient protection while it
conceals and overshadows them. And they know in
some quite inexplicable way that for fishermen the
sea is unnavigable on such a day or such a night, as
it rages with the waves mounting to a terrifying
height. It is in stormy weather that they gather
their food, when the swell drags some off the rocks
and sucks some from the shore. The Melanuruses
feed off the foulest matter, such stuff as no other
fish would readily take, unless it were utterly overcome by hunger. But in calm weather they have only
the sand to ride on, and from there they get their
food. But how they are captured another shall tell.

42. Among birds the Eagle has the keenest sight. The Eagle,
And Homer is aware of this and testifies to the fact its keen sight
in the story of Patroclus when he compares Menelaus
to the bird [*Il.* 17. 674–], at the time when he was
searching for Antilochus, that he might despatch
him to Achilles as a messenger, unwelcome indeed
but necessary, to announce the fate that had be-
fallen his comrade, whom Achilles had sent out ⟨to
battle⟩ but never welcomed home again for all his
yearning. And the Eagle is said to serve not himself alone but to be good for men's eyes as well. At
any rate, if a man whose sight is dim mix an Eagle's
gall with Attic honey and rub it ⟨on his eyes⟩, he
will see and will acquire sight of extreme keenness.

43. Ἀηδὼν ὀρνίθων λιγυρωτάτη τε καὶ εὐμου-
σοτάτη,[1] καὶ κατᾴδει τῶν ἐρημαίων χωρίων
εὐστομώτατα ὀρνίθων καὶ τορώτατα. λέγουσι δὲ
καὶ τὰ κρέα αὐτῆς ἐς ἀγρυπνίαν λυσιτελεῖν.
πονηροὶ μὲν οὖν οἱ τοιαύτης τροφῆς δαιτυμόνες
καὶ ἀμαθεῖς δεινῶς· πονηρὸν δὲ τὸ ἐκ τῆς τροφῆς
δῶρον, φυγὴ ὕπνου, τοῦ καὶ θεῶν καὶ ἀνθρώπων
βασιλέως, ὡς Ὅμηρος λέγει.

44. Τῶν γεράνων αἱ κλαγγαὶ καλοῦσιν ὄμβρους,
ὥς φασιν· ὁ δὲ ἐγκέφαλος γυναικῶν ἐς χάριν
ἀφροδίσιον[2] ἔχει τινὰς ἴυγγας, εἴ τῳ[3] ἱκανοὶ
τεκμηριῶσαι οἱ πρῶτοι φυλάξαντες ταῦτα.[4]

45. Γυπῶν πτερὰ εἰ θυμιάσειέ[5] τις, ὡς ἀκούω,
καὶ ἐκ φωλεῶν καὶ ἐξ εἰλυῶν τοὺς ὄφεις προάξει
ῥᾷστα.

Τὸ ζῷον[6] ὁ δρυοκολάπτης ἐξ οὗ δρᾷ[7] καὶ
κέκληται. ἔχει μὲν γὰρ ῥάμφος ἐπίκυρτον, κολά-
πτει δὲ ἄρα τούτῳ τὰς δρῦς, καὶ ἐνταυθοῖ[8] ὡς
ἐς καλιὰν τοὺς νεοττοὺς ἐντίθησιν, οὐ δεηθεὶς
καρφῶν καὶ τῆς ἐξ αὐτῶν πλοκῆς καὶ οἰκοδομίας
οὐδὲ ἕν. οὐκοῦν εἴ τις λίθον ἐνθεὶς ἐπιφράξειε τῷ
ὀρνέῳ τῷ προειρημένῳ τὴν ἔσδυσιν, ὁ δὲ συμβαλὼν
τὴν ἐπιβουλὴν[9] κομίζει πόαν ἐχθρὰν τῷ λίθῳ
καὶ κατ᾽ αὐτοῦ τίθησιν· ὁ δὲ οἷα βαρούμενος καὶ
μὴ φέρων ἐξάλλεται, καὶ ἀνέῳγεν αὖθις τῷ προειρημένῳ ἡ φίλη ὑποδρομή.

[1] εὐνουστάτη.   [2] ἀφροδισίαν.   [3] που.
[4] αὐτά.   [5] θυμιάσαι.
[6] τὸ ζῷον] ζῷον δέ.   [7] Jac : ἄρα.
[8] ἐνταυθοῖ κοιλάνας τὸν τόπον.

43. Among birds the Nightingale has the clearest The
Nightingale and most musical voice, and fills solitary places with its most lovely and thrilling note. Further, they say that its flesh is good for keeping one awake. But people who feast upon such food are evil and dreadfully foolish. And it is an evil attribute of food that it drives sleep away—sleep, the king of gods and men, as Homer says [*Il.* 14. 233].

44. The screaming of Cranes brings on showers, so The Crane they say, while their brain possesses some kind of spell that leads women to grant sexual favours—if those who first observed the fact are sufficient guarantee.

45. If a man burn the feathers of a Vulture (so I Vulture's
feathers am told), he will have no difficulty in inducing snakes to quit their dens and lurking-places.

The bird ' Woodpecker ' derives its name from what The
Woodpecker it does. For it has a curved beak with which it pecks oak-trees, and deposits its young in them as in a nest; and it has no need at all of dry twigs woven together or of any building. Now if one inserts a stone and blocks up the entrance for the aforesaid bird, it guesses that there is a plot afoot, fetches some herb that is obnoxious to the stone, and places it against the stone. The latter in disgust and unable to endure ⟨the smell⟩ springs out, and once again the bird's caverned home lies open to it.

---

[9] ἐπιβουλὴν τὴν κατ' αὐτοῦ.

AELIAN

46. Οἱ συνόδοντες οὐκ εἰσὶ μονίαι, οὐδὲ τὴν ἀπ᾽ ἀλλήλων ἐρημίαν τε καὶ διαίρεσιν ἀνέχονται. φιλοῦσι δὲ συναγελάζεσθαι καθ᾽ ἡλικίαν. καὶ οἱ μὲν νεώτεροι κατὰ ἴλας νήχονται, οἱ δὲ ἐντελέστεροι πάλιν κοινῇ· καὶ τὸ τοῦ λόγου τοῦτο ἧλιξ ἥλικα καὶ ἐκεῖνοι τέρπουσι, παρόντες παροῦσιν ὡς ἑταίροις καὶ φίλοις ἐκ τῶν αὐτῶν ἐπιτηδευμάτων τε καὶ διατριβῶν. τεχνάζονται δὲ πρὸς τοὺς θηρατὰς ὁποῖα. ὅταν ἁλιεὺς ἀνὴρ τὸ ἐς αὐτοὺς δέλεαρ καθῇ, περιελθόντες πάντες καὶ κυκλόσε γενόμενοι ἐς ἀλλήλους ὁρῶσιν, οἱονεὶ σύνθημα ἕκαστος ἑκάστῳ διδόντες μήτε πλησιάσαι μήτε ἅψασθαι τοῦ καθειμένου δελεάσματος. καὶ οἱ μὲν παρατεταγμένοι ἐς τοῦτο ἀτρεμοῦσιν· ἐκ δὲ [1] ἀλλοτρίας ἀγέλης συνόδων ἀφίκετο, καὶ καταπίνει τὸ ἄγκιστρον, ἐρημίας λαβὼν [2] μισθὸν τὴν ἅλωσιν. καὶ ὁ μὲν ἀνασπᾶται, οἱ δὲ ἤδη θαρροῦσιν ὡς οὐχ ἁλωσόμενοι, καὶ καταφρονήσαντες οὕτω θηρῶνται.

47. Φρύγεται διὰ τοῦ θέρους ὁ κόραξ τῷ δίψει κολαζόμενος, καὶ βοᾷ τὴν τιμωρίαν μαρτυρόμενος, ὥς φασι. καὶ τὴν αἰτίαν λέγουσιν ἐκείνην. ὁ Ἀπόλλων αὐτὸν θεράποντα ὄντα ὑδρευσόμενον ἀποπέμπει· ὁ δὲ ἐντυγχάνει ληΐῳ βαθεῖ μέν, ἔτι δὲ χλωρῷ, καὶ μένει ἔστ᾽ ἂν αὖον γένηται, τῶν πυρῶν παραχναῦσαι βουλόμενος, καὶ τοῦ προστάγματος ὠλιγώρησε. καὶ ὑπὲρ τούτων ἐν τῇ μάλιστα αὐχμηροτάτῃ ὥρᾳ διψῶν δίκας ἐκτίνει. τοῦτο ἔοικε μύθῳ μέν, εἰρήσθω δ᾽ οὖν τῇ τοῦ θεοῦ αἰδοῖ.

[1] δὲ τῆς.   [2] λαχών.

66

46. The Four-toothed Sparus is not solitary nor _{The Four-}
does it endure loneliness and separation from its _{toothed}
_{Sparus}
kind. These fish love to congregate together
according to their age: the younger ones swim
about in shoals, the maturer ones also keep together.
And as the saying is true ' A friend must be of one's
own age,'ᵃ so these creatures delight to be where
others of their kind are, like comrades and friends
sharing the same pursuits and resorts. And these
are the means they devise for evading their pursuers.
Whenever an angler drops a bait for them they all
gather round and forming a ring look at one another
as though each were signalling to each not to
approach and not to touch the bait that has been
lowered. And those that have been posted for this
purpose remain still. But a Sparus from some other,
strange shoal arrives and swallows the bait, and gets
the reward of its solitariness by being caught. So while
he is being drawn up, the rest grow bolder as though
they were not going to be taken, and so through their
scorn ⟨of danger⟩ are caught.

47. All through the summer the Raven is afflicted _{The Raven,}
with a parching thirst, and with his croaking (so they _{its thirst}
say) declares his punishment. And the reason they
give is this. Being a servant he was sent out by
Apollo to draw water. He came to a field of corn,
tall but still green, and waited till it should ripen,
as he wanted to nibble the wheat: to his master's
orders he paid no heed. On that account in the
driest season of the year he is punished with thirst.
This looks like a fable, but let me repeat it out of
reverence for the god.

ᵃ The full phrase is ἧλιξ ἥλικα τέρπει, cp. Pl. *Phaedr.* 240 c.

48. Ὁ κόραξ, ὄρνιν αὐτόν φασιν ἱερόν, καὶ
Ἀπόλλωνος ἀκόλουθον εἶναι λέγουσι. ταῦτά τοι
καὶ μαντικοῖς συμβόλοις ἀγαθὸν ὁμολογοῦσι τὸν
αὐτόν, καὶ ὀττεύονταί γε πρὸς τὴν ἐκείνου βοὴν οἱ
συνιέντες ὀρνίθων καὶ ἕδρας καὶ κλαγγὰς καὶ
πτήσεις αὐτῶν ἢ κατὰ λαιὰν χεῖρα ἢ κατὰ
δεξιάν.

Προσακούω δὲ καὶ ὠὰ κόρακος μελαίνειν τρίχας.
καὶ χρὴ τὸν δολοῦντα τὴν ἑαυτοῦ κόμην ἔλαιον ἐν
τῷ στόματι ἔχειν συμμύσαντα· εἰ δὲ μή, καὶ οἱ
ὀδόντες αὐτῷ σὺν τῇ τριχὶ μελαίνονται δυσέκπλυτοί
τε καὶ δυσέκνιπτοι.

49. Ὁ μέροψ τὸ ὄρνεον ἔμπαλίν φασι τοῖς ἄλλοις ἅπασι πέτεται· τὰ μὲν γὰρ ἐς τοὔμπροσθεν
ἵεται καὶ κατ᾽ ὀφθαλμούς, ὁ δὲ ἐς τοὐπίσω. καὶ
ἔπεισί μοι θαυμάζειν τὴν φύσιν τῆς ἐπισήμου καὶ
παραδόξου καὶ ἀήθους φορᾶς, ἣν ἐκεῖνο ᾄττει [1] τὸ
ζῷον.

50. Ἡ μύραινα ὅταν ὁρμῇς ἀφροδισίου ὑποπλη-
σθῇ, πρόεισιν ἐς τὴν γῆν, καὶ ὁμιλίαν ποθεῖ
νυμφίου καὶ μάλα πονηροῦ· πάρεισι γὰρ εἰς ἔχεως
φωλεόν, καὶ ἄμφω συμπλέκονται. ἤδη δέ φασι
καὶ ὁ ἔχις οἰστρήσας καὶ ἐκεῖνος ἐς μίξιν ἀφικνεῖται
πρὸς τὴν θάλατταν, καὶ οἷον εἰ κωμαστὴς σὺν τῷ
αὐλῷ θυροκοπεῖ, οὕτω τοι καὶ ἐκεῖνος συρίσας τὴν
ἐρωμένην παρακαλεῖ, καὶ αὐτὴ πρόεισι,[2] τῆς
φύσεως τὰ ἀλλήλων διῳκισμένα συναγούσης ἐς
ἐπιθυμίαν τὴν ὁμοίαν καὶ κοῖτον τὸν αὐτόν.

[1] ἄγει.   [2] Ges : πρός-.

48. The Raven, they say, is a sacred bird and attends upon Apollo: that is why men agree that it is also of use in divination, and those who understand the positions of birds, their cries, and their flight whether on the left or on the right hand, are able to divine by its croaking.

I am also informed that Raven's eggs turn the hair black. And it is essential for anyone who is dyeing his hair to keep olive oil in his mouth and his lips closed. Otherwise his teeth also turn black along with his hair, and they are hardly to be washed white again.

49. The Bee-eater flies (so they say) in precisely the opposite way to all other birds, for they move forward in the direction in which they look, while the Bee-eater flies backwards. And I am astonished at the remarkable, incredible, and uncommon character of the motion with which this creature wings its way.

50. Whenever the Moray is filled with amorous impulses it comes out of the sea on to land seeking eagerly for a mate, and a very evil mate. For it goes to a Viper's den and the pair embrace. And they do say that the male Viper also in its frenzied desire for copulation goes down to the sea, and just as a reveller with his flute knocks at the door, so the Viper also with his hissing summons his loved one, and she emerges. Thus does Nature bring those that dwell far apart together in a mutual desire and to a common bed.

The Raven, in divination

its eggs

The Bee-eater

Moray and Viper

51. Ῥάχις ἀνθρώπου νεκροῦ φασιν ὑποσηπόμενον τὸν μυελὸν ἤδη τρέπει ἐς ὄφιν· καὶ ἐκπίπτει τὸ θηρίον, καὶ ἕρπει τὸ [1] ἀγριώτατον ἐκ τοῦ ἡμερωτάτου· καὶ τῶν μὲν καλῶν καὶ ἀγαθῶν τὰ λείψανα ἀναπαύεται, καὶ ἔχει ἆθλον ἡσυχίαν, ὥσπερ οὖν καὶ ἡ ψυχὴ τῶν τοιούτων τὰ ᾀδόμενά τε καὶ ὑμνούμενα ἐκ τῶν σοφῶν· πονηρῶν δὲ ἀνθρώπων ῥάχεις τοιαῦτα τίκτουσι καὶ μετὰ τὸν βίον. ἢ τοίνυν τὸ πᾶν μῦθός ἐστιν, ἤ, εἰ ταῦτ' ἀψευδῶς [2] πεπίστευται, πονηροῦ νεκρός, ὡς κρίνειν ἐμέ, ὄφεως γενέσθαι πατὴρ τοῦ τρόπου μισθὸν ἠνέγκατο.

52. Χελιδὼν δὲ ἄρα τῆς ὥρας τῆς ἀρίστης ὑποσημαίνει τὴν ἐπιδημίαν. καὶ ἔστι φιλάνθρωπος, καὶ χαίρει τῷδε τῷ ζῴῳ ὁμωρόφιος οὖσα, καὶ ἄκλητος ἀφικνεῖται, καὶ ὅτε οἱ φίλον καὶ ἔχει καλῶς, ἀπαλλάττεται. καὶ οἵ γε ἄνθρωποι ὑποδέχονται αὐτὴν κατὰ τὸν τῆς Ὁμηρικῆς ξενίας θεσμόν, ὃς κελεύει καὶ φιλεῖν τὸν [3] παρόντα καὶ ἰέναι βουλόμενον ἀποπέμπειν.

53. Ἔχει τι πλεονέκτημα ἡ αἲξ τὴν τοῦ πνεύματος ἐσροήν, ὡς οἱ νομευτικοὶ λόγοι [4] φασίν. ἀναπνεῖ γὰρ καὶ διὰ τῶν ὤτων καὶ διὰ τῶν μυκτήρων, καὶ αἰσθητικώτατον τῶν διχήλων ἐστί. καὶ τὴν μὲν αἰτίαν εἰπεῖν οὐκ οἶδα, ὃ δὲ οἶδα τοῦτο εἶπον. εἰ δὲ ποίημα Προμηθέως καὶ αἲξ, τί βουλόμενος τοῦτο εἰργάσατο, εἰδέναι καταλιμπάνω αὐτόν.

---

[1] ζῷον τό.
[2] ταῦτα οὑτωσὶ MSS, τ. ὀρθῶς Ges.
[3] ξένον H (1876).   [4] λόγοι καὶ ποιμενικοί.

51. The spine of a dead man, they say, transforms the putrefying marrow into a snake. The brute emerges, and from the gentlest of beings crawls forth the fiercest. Now the remains of those that were fine and noble are at rest and their reward is peace, even as the soul also of such men has the rewards which wise men celebrate in their songs. But it is from the spine of evildoers that such evil monsters are begotten even after life. The fact is, the whole story is either a fable, or if it is to be relied upon as true, then the corpse of a wicked man receives (so I think) the reward of his ways in becoming the progenitor of a snake.

52. A Swallow is a sign that the best season of the year is at hand. And it is friendly to man and takes pleasure in sharing the same roof with this being. It comes uninvited, and when it pleases and sees fit, it departs. Men welcome it in accordance with the law of hospitality laid down by Homer [*Od.* 15. 72–4], who bids us cherish a guest while he is with us and speed him on his way when he wishes to leave.

53. The Goat has a certain advantage ⟨over other animals⟩ in the manner of taking breath, as the narratives of shepherds tell us, for it inhales through its ears as well as through its nostrils, and has a sharper perception than any other cloven-hoofed animal. The cause of this I am unable to tell; I have only told what I know. But if the Goat also was a creation of Prometheus, what the intention of this contrivance was, I leave him to determine.

54. Καὶ ἔχεως δῆγμα καὶ ὄφεως ἄλλου φασὶν
ἀντιπάλων μὴ διαμαρτάνειν φαρμάκων. καὶ τὰ
μὲν αὐτῶν ἀκούω πώματα [1] εἶναι, τὰ δὲ χρίματα [2].
καὶ ἐπαοιδαὶ δὲ ἐπράυναν τὸν [3] ἐγχρισθέντα ἰόν.
ἀσπίδος δὲ ἀκούω μόνης [4] δῆγμα ἀνίατον εἶναι καὶ
ἐπικουρίας κρεῖττον. καὶ μισεῖν ἄξιον τὸ ζῷον
τῆς εὐκληρίας τῆς ἐς τὸ κακόν. ἀλλὰ καὶ τούτου
θηρίον μιαρώτερον καὶ ἀφυλακτότερον γυνὴ φαρμα-
κίς, οἵαν ἀκούομεν καὶ τὴν Μήδειαν καὶ τὴν
Κίρκην· τὰ μὲν γὰρ τῶν ἀσπίδων φάρμακα
δήγματος [5] ἔργα ἐστί, τὰ δὲ ἐκείνων ἀναιρεῖ [6] καὶ
ἐκ μόνης τῆς ἁφῆς, φασίν.

55. Κυνῶν θαλαττίων τρία γένη. καὶ οἱ μὲν
αὐτῶν εἰσι μεγέθει μέγιστοι, καὶ κητῶν ἐν τοῖς
ἀλκιμωτάτοις ἀριθμοῖντο ἄν· γένη δὲ δύο τὰ
λοιπά, πηλαῖοι μὲν τὴν φύσιν, προήκουσι δὲ ἐς
πῆχυν τὸ μέγεθος.[7] καὶ τούτων οἱ μὲν κατεστιγμέ-
νοι καλοῖντο ἂν γαλεοί, κεντρίνας δὲ ὀνομάζων
τοὺς λοιποὺς οὐκ ἂν διαμαρτάνοις. οἱ μὲν οὖν
ποικίλοι καὶ τὴν δορὰν εἰσι μαλακώτεροι καὶ τὴν
κεφαλὴν πλατύτεροι· οἱ δὲ ἕτεροι σκληροὶ [8] τὴν
δορὰν ὄντες [9] τὴν κεφαλὴν δὲ ἀνήκουσαν ἐς ὀξὺ
ἔχοντες τὴν [10] χρόαν ἐς τὸ λευκὸν ἀποκρίνονται.
κέντρα δὲ ἄρα αὐτοῖς συμπέφυκε τὸ μὲν [11] κατὰ
τὴν λοφιάν, ὡς ἂν εἴποις, τὸ δὲ κατὰ τὴν οὐράν·
σκληρὰ δὲ ἄρα τὰ κέντρα καὶ ἀπειθῆ ἐστι, καὶ ἰοῦ

---

[1] πόμ- MSS always.   [2] χρίσματα.
[3] τινων.   [4] μόνον.
[5] Schn : καὶ δήγματος.   [6] ἀναιρεῖν.
[7] μέγεθος καὶ τὸν μὲν αὐτοῖν γαλεὸν τὸν δὲ κεντρίτην φιλοῦσιν
ὀνομάζειν.

54. They say that the bite of the Viper and of Poisonous Snakes other snakes is not without countering remedies. Some, I am told, are to be drunk, others are to be applied; spells too can mitigate poison injected by a sting. But the bite of the Asp [a] alone, I am told, cannot be cured and is beyond help. This creature truly deserves to be hated for being blessed with the power to injure. Yet a monster more abominable and harder to avoid even than the Asp is a sorceress, such as (we are told) Medea and Circe were, for the poison from Asps is the result of a bite, whereas sorceresses kill by a mere touch, so they say.

55. There are three kinds of Sea-hound.[b] The The Shark first is of enormous size and may be reckoned among the most daring of sea monsters.[c] The others are of two kinds, they live in the mud and reach to a cubit The Dog-fish in length. Those that are speckled one may call *galeus* (small shark), and the rest, if you call them Spiny Dog-fish you will not go far wrong. Now the speckled ones have a softer skin and a flatter head, while the others, whose skin is hard and whose head tapers to a point, are distinguished from the rest by the whiteness of their skin. Moreover nature has provided them with spines, one on their crest, so to say, the other in the tail. And these spines are hard and resisting and emit a kind of poison. Of the

---

[a] The Egyptian cobra, *Naia haie*.
[b] The terms θαλάττιος κύων and γαλεός signify both *dog-fish* and *shark*. See INDEX II.
[c] *I.e.* the shark.

---

[8] μικροί τε καὶ σκληροί.
[9] μέντοι ὄντες καί.
[10] καὶ τήν.
[11] τὸ μὲν τῆς κεφαλῆς.

73

τι προσβάλλει. ἁλίσκεται δὲ τῶν κυνῶν τῶν
σμικρῶν τῶνδε ἑκάτερον ⟨τὸ φῦλον⟩ [1] ἐκ τῆς
ἰλύος καὶ τοῦ πηλοῦ, καὶ ἡ ἄγρα, εἰπεῖν αὐτὴν οὐ
χεῖρόν ἐστι. δέλεαρ αὐτῶν καθιᾶσιν ἰχθὺν λευκὸν
ἐκτετμημένον τὴν ῥάχιν. ὅταν τοίνυν εἷς ἁλῷ καὶ
τῷ ἀγκίστρῳ περιπέσῃ, πάντες οἱ θεασάμενοι
ἐμπηδῶσιν [2] αὐτῷ καὶ [3] κάτωθεν ἑλκομένῳ ἕπονται [4] καὶ μέχρι τῆς νεὼς οὐκ ἀναστελλόμενοι, ὡς
εἰκάσαι ζηλοτυπίᾳ δρᾶν ταῦτα αὐτούς, οἷα ἐκείνου
τι τῶν ἐς τροφὴν ἑαυτῷ μόνῳ ποθὲν ἀποσυλήσαν-
τος· καὶ ἐς τὴν ναῦν γε αὐτὴν ἐσεπήδησάν τινες
πολλάκις, καὶ ἑκόντες ἑάλωσαν.

56. Τῆς τρυγόνος τῆς θαλαττίας τὸ κέντρον
ἐστὶν ἀπρόσμαχον. ἐκέντησε γὰρ καὶ ἀπέκτεινε
παραχρῆμα, καὶ πεφρίκασιν αὐτῆς τόδε τὸ ὅπλον
καὶ οἱ τῶν ἁλιέων δεινοὶ τὰ θαλάττια· οὔτε γὰρ
ἄλλος ἰάσεται τὸ τραῦμα οὔτε ἡ τρώσασα· μόνῃ
γάρ, ὡς τὸ εἰκός, τῇ Πηλιώτιδι μελίῃ [5] τοῦτο
ἐδέδοτο.

57. Λεπτὸν [6] θηρίον ὁ κεράστης. ἔστι δὲ ὄφις,
καὶ ὑπὲρ τοῦ μετώπου κέρατα ἔχει δύο, καὶ ἔοικε
τοῖς τοῦ κοχλίου τὰ κέρατα, οὐ μήν ἐστιν ὡς
ἐκείνων ἁπαλά. οὐκοῦν τοῖς μὲν ἄλλοις τῶν
Λιβύων εἰσὶ πολέμιοι· ἔστι δὲ αὐτοῖς πρὸς τοὺς
καλουμένους Ψύλλους ἔνσπονδα, οἵπερ οὖν οὔτε
αὐτοὶ δακόντων ἐπαΐουσι,[7] καὶ τοὺς τῷ τοιούτῳ

---

[1] ⟨φῦλον⟩ add. Reiske, ⟨τό⟩ add. H.
[2] συμπηδῶσιν.   [3] καί τοι.
[4] ἕπονταί τε.
[5] Reiske : βολῇ, v.l. μόνῃ.

small Dog-fish both kinds are caught in the ooze and mud, and the manner of catching them I may as well explain. By way of bait men let down a white fish out of which they have cut the backbone. Directly one of the Dog-fish is caught and hooked, all those that have seen him make a rush for him and follow him as he is drawn upwards, never stopping until they reach the boat. One might imagine that they do this out of envy, as though he had filched some piece of food from somewhere and all for himself. And it often happens that some of them actually leap into the boat and are caught of their own free will.

56. The barb of the Sting-ray nothing can with-stand. It wounds and kills instantly, and even those fishermen who have great knowledge of the sea dread its weapon. For no man can heal the wound, nor will the creature that inflicted it; that was a gift vouchsafed, most probably, to the ashen spear from mount Pelion alone.[a] The Sting-ray

57. The Cerastes is a small creature; it is a snake, and above its brow it has two horns, and these horns are like those of the snail, though unlike the snail's they are not soft. Now these snakes are the enemies of all other Libyans, but towards the Psylli, as they are called, they are gently disposed, for the Psylli are insensible to their bites and have no difficulty The Cerastes and the Psylli

[a] The spear of Achilles was made from an ash-tree on mt Pelion (Hom. *Il.* 16. 143). Telephus, wounded by the spear, was afterwards cured by the rust from it.

---

[6] λευκόν. [7] ἐπαΐουσι τῶν δηγμάτων.

κακῷ περιπεσόντας ἰῶνται ῥᾷστα. καὶ ὁ τρόπος,
ἐὰν πρὶν ἢ πρησθῆναι τὸ πᾶν σῶμα ἀφίκηταί τις
τῶν ἐκεῖθεν κλητὸς ἢ κατὰ τύχην, εἶτα τὸ μὲν
στόμα ὕδατι ἐκκλύσηται,[1] ἀπονίψῃ δὲ τὰς χεῖρας
ἑτέρῳ, καὶ πιεῖν τῷ δηχθέντι δῷ ἑκάτερον, ἀνερ-
ρώσθη τε ἐκεῖνος καὶ κακοῦ παντὸς ἐξάντης τὸ
ἐντεῦθέν ἐστι. διαρρεῖ δὲ καὶ λόγος Λιβυκὸς ὁ
λέγων, Ψύλλον ἄνδρα τὴν ἑαυτοῦ γαμετὴν ὑφο-
ρᾶσθαι καὶ μισεῖν ὡς μεμοιχευμένην καὶ μέντοι
καὶ τὸ ἐξ αὐτῆς βρέφος ὑποπτεύειν ὡς νόθον τε
καὶ τῷ σφετέρῳ γένει κίβδηλον. πεῖραν οὖν
καθεῖναι καὶ μάλα ἐλεγκτικήν φασιν αὐτόν. λάρ-
νακα πληρώσας κεραστῶν ἐμβάλλει[2] τὸ βρέφος,
οἱονεὶ πυρὶ τὸν χρυσὸν τεχνίτης τὸ παιδίον ἐξελέγ-
χων ἐκεῖνος τῇ ἀποθέσει. καὶ οἱ μὲν παραχρῆμα
ἐπανίσταντο καὶ ἠγρίαινον καὶ τὴν συμφυῆ κακίαν
ἠπείλουν· ἐπεὶ δὲ τὸ παιδίον αὐτῶν προσέψαυσεν,
οἱ δὲ ἐμαράνθησαν, καὶ ἐντεῦθεν ὁ Λίβυς ἔγνω οὐ
νόθου ἀλλὰ γόνου γνησίου πατὴρ ὤν. λέγονται δὲ
καὶ τῶν ἑτέρων δακετῶν καὶ φαλαγγίων δὲ
ἀντίπαλοι τόδε τὸ γένος εἶναι. καὶ ταῦτά γε εἰ
τερατεύονται Λίβυες, οὐκ ἐμέ, ἀλλ᾽ αὑτοὺς ἀπατῶν-
τες ἴστωσαν.

58. Μελιττῶν δὲ ἐπίβουλοι καὶ ἐχθροὶ εἶεν ἂν
ἐκεῖνοι, οἵ τε αἰγίθαλοι καλούμενοι καὶ τὰ τούτων
νεόττια καὶ οἱ σφῆκες καὶ αἱ χελιδόνες καὶ οἱ
ὄφεις καὶ αἱ φάλαγγες καὶ αἱ †λύγγαι†.[3] καὶ αἱ

---

[1] ἐπικλύσηται.
[2] Ges : καὶ ἐμβάλλει.
[3] λύγγαι ʻ vox nihili,ʼ φάλλαιναι (or φρῦναι, cp. Arist. HA 626 a
30) Gow.

in curing those who have fallen victims to this venomous creature. Their method is this: if one of that tribe arrive, whether summoned or by chance, before the whole body is inflamed, and if he then rinse his mouth with water and wash the bitten man's hands and give him the water from both to drink, then the victim recovers and thereafter is free from all infection. And there is a story current among the Libyans that, if one of the Psylli suspects his wife and hates her on the ground that she has committed adultery; and if moreover he suspects that the child born from her is a bastard and no true member of his tribe, he then puts it to a very severe test: he fills a chest with Cerastae and drops the baby among them, just as a goldsmith places gold in the fire, and puts the infant to the proof by thus exposing him. And immediately the snakes surge up in anger and threaten the child with their native poison. But directly the infant touches them, they wilt, and then the Libyan knows that he is the father of no bastard but of one sprung of his own race. This tribe is said also to be the enemy of other noxious beasts and of malmignattes.

Well, if the Libyans are here romancing, I would have them know that it is not I but themselves that they are deceiving.

58. The following creatures plot and make war against Bees: the creatures known as Titmice and their young, also Wasps and Swallows and Snakes and Spiders and [Moths?]. Bees are afraid of these, and $\qquad$ Bees and their enemies

77

μὲν δεδίασι ταῦτα, οἱ δ' οὖν μελιττουργοὶ ἐλαύ-
νουσιν αὐτὰ ἀπ' αὐτῶν ἢ κόνυζαν ἐπιθυμιάσαντες
ἢ χλωρὰν ἔτι μήκωνα πρὸ τῶν σίμβλων καταστή-
σαντες ἢ καταστρώσαντες. καὶ ταῦτα μὲν τοῖς
ἄλλοις ἐχθρά ἐστι τοῖς προειρημένοις, σφηκῶν δὲ
ἅλωσις ἐκείνη [1] ἂν εἴη. κύρτον ἀπαρτῆσαι χρὴ
πρὸ τῆς σφηκιᾶς καὶ ἐνθεῖναι αὐτῷ λεπτὴν μεμ-
βράδα ἢ μαινίδα ὀλίγην καὶ σὺν τούτοις ὦπα ἢ
χαλκίδα· οἱ δὲ σφῆκες ὑπὸ τῆς ἐμφύτου γαστριμαργίας ἑλκόμενοι, καλοῦντος αὐτοὺς ⟨τοῦ⟩ [2] δελεάσματος, ἐσπίπτουσιν ἀθρόοι, καὶ περιλαβόντος
αὐτοὺς τοῦ κύρτου οὐκ ἔστιν αὐτοῖς τὴν ὀπίσω
οὐκέτι ἐκπτῆναι.[3] καὶ οἱ σαῦροι δὲ ἐπιβουλεύουσι
ταῖς μελίτταις καὶ οἱ κροκόδιλοι οἱ χερσαῖοι·
ὄλεθρος δὲ καὶ τούτοις ἐπιτετέχνηται ἐκεῖνος.
ἄλφιτα γὰρ ἐλλεβόρῳ δεύσαντες ἢ τιθυμάλλου ὀπῷ
ὑποχέαντες [4] ἢ μαλάχης χυλῷ διασπείρουσι πρὸ
τῶν σίμβλων τὰ ἄλφιτα· ὅπερ οὖν ὄλεθρον φέρει
τοῖς προειρημένοις ἀπογευσαμένοις αὐτῶν. ἐμβαλὼν δὲ ἐς τὴν λίμνην φλόμου φύλλα ἢ κάρυα
ἀπώλεσε τοὺς γυρίνους ὁ τῶν μελιττῶν δεσπότης
ῥᾷστα. αἱ δὲ φάλλαιναι [5] ἀπόλλυνται νύκτωρ, ἐνακ-
μάζοντος [6] λύχνου τεθέντος πρὸ τῶν σμηνῶν
καὶ ἀγγείων ἐλαίου πεπληρωμένων τῷ λύχνῳ
ὑποκειμένων· αἱ δὲ πρὸς τὴν αὐγὴν πετόμεναι ἐμπίπτουσιν ἐς τὸ ἔλαιον καὶ ἀπολώλασιν·
ἑτέρως δὲ οὐκ ἂν αἱρεθεῖεν ῥᾷστα. οἱ δὲ αἰγίθαλοι

---

[1] Schn : ἁλώσεις ἐκεῖνα.
[2] ⟨τοῦ⟩ add. Jac.
[3] ἐκπτῆναι, καὶ ὕδωρ δ' ἂν αὐτῶν κατασκεδάσας ῥᾷον διαφθείραις
ἂν αὐτούς, καὶ πῦρ ἐξάψας καταπρήσαις.
[4] ὑποχέοντες.

so bee-keepers try to drive them away by using flea-
bane as a fumigant or by placing or scattering pop-
pies still green before the hives. Most of the
aforesaid creatures dislike these things, but the way
to catch Wasps is as follows. You should hang up a
cage in front of the Wasps' nest and insert a little
smelt or a small sprat and with them a minnow or a
sardine. And the Wasps, drawn by their natural greed
and lured by the bait, fall into the cage in numbers,
and once they are trapped, it is no longer possible for
them to fly out again. Lizards also have designs upon
Bees, so too have Land-crocodiles.[a] But a means
has been devised of destroying them too, thus:
soak some meal in hellebore, or pour upon it the sap
of spurge or the juice of mallow and scatter it
about in front of the hives. This is death to the
aforesaid creatures, once they have tasted of it. If
a bee-keeper drop the leaves of mullein or nuts [b] into
a pool, he will find it the simplest way of destroying
Tadpoles. But Moths [c] are destroyed at night-
time by the placing of a strong light in front of the
hives and vessels full of oil below the light. And
the Moths fly to the brightness and fall into the oil
and are killed. Otherwise they would not be caught
so very easily. But the Titmice, once they have

[a] 'The "crocodile" is the *Psammosaurus griseus*, a land
lizard, which reaches a size of 3 feet' (How-Wells on Hdt.
4. 192).

[b] Perhaps some word has been lost indicating what kind of
nut is intended.

[c] This may be the Wax-moth, which is found in bees' nests,
its larvae eating the comb; or it may be one of the Hawk-
moths (fam. *Sphingidae*) which enter the nests for honey.

---

[5] *Ges* : φάλαγγες MSS, *H.*   [6] ἐναυγάζοντος.

ἀλφίτων οἴνῳ διαβραχέντων ἀπογευσάμενοι καρη-
βαροῦσιν, εἶτα πίπτουσι, καὶ κείμενοι σπαίρουσι,
καὶ εἰσὶν αἱρεθῆναι † γελοῖοι †,[1] ἀναπτῆναι μὲν
σπεύδοντες, ἀρχὴν δὲ ἀναστῆναι μὴ δυνάμενοι.
οἱ δὲ τὴν χελιδόνα αἰδοῖ τῆς μουσικῆς οὐκ ἀποκτείνουσι, καίτοι ῥᾳδίως ἂν αὐτὴν[2] τοῦτο δράσαντες·
ἀπόχρη δὲ αὐτοῖς κωλύειν τὴν χελιδόνα πλησίον
τῶν σίμβλων καλιὰν ὑποπῆξαι.

Ἀπεχθάνονται δὲ ἄρα αἱ μέλιτται κακοσμίᾳ
πάσῃ καὶ μύρῳ ὁμοίως, οὔτε τὸ δυσῶδες ὑπομέ-
νουσαι οὔτε ἀσπαζόμεναι τῆς εὐωδίας τὸ τεθρυμ-
μένον, οἷα δήπου κόραι ἀστεῖαί τε καὶ σώφρονες
τὸ μὲν βδελυττόμεναι τῆς δὲ ὑπερφρονοῦσαι.

59. Κῦρος μέν, ὥς φασιν, ὁ πρεσβύτερος μέγα
ἐφρόνει ἐπὶ τοῖς βασιλείοις τοῖς ἐν Περσεπόλει,[3]
οἷσπερ οὖν αὐτὸς ᾠκοδομήσατο, Δαρεῖος δὲ ἐπὶ
τῇ κατασκευῇ τῇ τῶν οἰκοδομημάτων τῶν Σου-
σείων[4]· καὶ γὰρ[5] ἐκεῖνος ἐν Σούσοις τὰ ᾀδόμενα
ἐκεῖνα εἰργάσατο. Κῦρος δὲ ὁ δεύτερος ἐν Λυδίᾳ
παράδεισον αὐτὸς κατεφύτευσε ταῖς χερσὶ ταῖς
βασιλικαῖς ἐν[6] τοῖς ἁβροῖς ἐκείνοις χιτῶσι καὶ
τοῖς τερπνοῖς ἐκείνοις καὶ μέγα τιμίοις λίθοις, καὶ
ἐπὶ τούτῳ[7] γε ἐκαλλύνετο καὶ πρὸς ἄλλους μὲν
τῶν Ἑλλήνων, ἀτὰρ οὖν καὶ πρὸς Λύσανδρον τὸν
Λακεδαιμόνιον, ὅτε ἦλθε πρὸς τὸν Κῦρον ὁ
Λύσανδρος ἐς τὴν Λυδίαν. καὶ ὑπὲρ μὲν τούτων

---

[1] ἑτοῖμοι Gow, γε οἷοι Jac, ῥᾴδιοι Lorenz.
[2] Oud : αὐτῇ MSS, H would delete.
[3] Περσαιπόλει.
[4] Reiske : Σούσων.   [5] καὶ γὰρ καί.
[6] σύν.   [7] τούτοις.

tasted the wine-steeped meal, become drowsy; then they fall over and lie quivering and can readily(?) be captured as they struggle to fly and are quite incapable of standing. But the Swallow men refrain from killing out of respect for its music, though they might easily do so. They are content to hinder the Swallow from attaching its nest below the hives.

Again, Bees dislike all bad smells and perfume equally: they cannot endure foul odours nor do they welcome a luxurious fragrance, even as modest, refined girls abhor the former while despising the latter.

59. The elder Cyrus,[a] they say, was filled with pride at the palace in Persepolis which he himself had caused to be built; Darius[b] likewise at the magnificence of his buildings at Susa, for he it was who contrived those far-famed dwelling-places. Cyrus the Second[c] with his own royal hands and clothed in his habitual delicate garments and adorned with his beautiful jewels of great price, planted his Gardens in Lydia and prided himself on the fact before all the Greeks and even before Lysander the Spartan, when Lysander came to visit him in Lydia.

Bees, their combs and hives

 [a] Cyrus I, founder of the Achaemenid Persian empire, 549–29 B.C. City and palace of Persepolis were burned by Alexander the Great.
 [b] Darius, son of Hystaspes, King of Persia, 521–485 B.C., reputed founder of Susa, on the river Choaspes. It was a residence of the Persian kings during the springtime.
 [c] Cyrus II, younger son of Darius II, c. 430–401 B.C., helped Lysander, the Spartan admiral, with sums of money, thereby ensuring the final victory of Sparta in the Peloponnesian war. The 'Gardens' were at Sardes.

AELIAN

ᾄδουσιν οἱ συγγραφεῖς, αἱ δὲ τῶν μελιττῶν
οἰκοδομαὶ σοφώτεραι οὖσαι κατὰ πολὺ καὶ τεχ-
νηέστεραι,[1] ἀλλὰ τούτων γε [2] οὐδὲ ὀλίγην ἔθεντο
ὥραν· ἐκεῖνοι μὲν γὰρ πολλοὺς [3] λυπήσαντες
εἰργάσαντο ὅσα εἰργάσαντο· οὐδὲν δὲ ἄρα ἦν
μελιττῶν εὐχαριτώτερον, ἐπεὶ μηδὲ σοφώτερον
ἦν. πρώτους μὲν γὰρ ἐργάζονται τοὺς θαλάμους
τοὺς τῶν βασιλέων, καὶ εὐρυχωρίαν ἔχουσιν οὗτοι,
καὶ εἰσὶν ἀνώτεροι· καὶ ἕρκος δὲ περιβάλλουσι
τούτοις, οἱονεὶ τεῖχος εἶναι καὶ περίβολον, ἀποσε-
μνύνουσαι καὶ ἐκ τούτου τὴν οἴκησιν τὴν βασίλειον.
διαιροῦσι δὲ αὐτὰς ἐς τρία καὶ οὖν καὶ τὰς
οἰκήσεις τὰς ἑαυτῶν ἐς τοσαῦτα. αἱ μὲν γὰρ
πρεσβύταται [4] γειτνιῶσι τῇ τῶν βασιλέων αὐλῇ,[5]
αἱ δὲ νεώταται [6] μετὰ ταύτας [7] οἰκοῦσιν, αἱ δὲ ἐν
ἥβῃ καὶ ἀκμῇ οὖσαι ἐξωτέρω ἐκείνων, ὡς εἶναι
τὰς μὲν πρεσβυτάτας φρουροὺς τῶν βασιλέων, τὰς
δὲ νεάνιδας ἕρκος τῶν νεωτάτων.

60. Λέγει μέν τις λόγος ἀκέντρους εἶναι τοὺς
τούτων βασιλέας· λέγει δὲ καὶ ἕτερος καὶ πάνυ
ἐρρωμένα τὰ κέντρα συμπεφυκέναι αὐτοῖς καὶ
τεθηγμένα ἀνδρειότατα· οὔτε δὲ ἐπ' ἀνδρί ποτε
χρῆσθαι αὐτοῖς οὔτε ἐπὶ ταῖς μελίτταις, ἀλλὰ συμπε-
πλάσθαι φόβον ἄλλως· μὴ γὰρ θέμις εἶναι τὸν
ἄρχοντα καὶ τῶν τοσούτων ἔφορον κακὸν ἐργάσασθαι. καὶ τὰς μελίττας δὲ τὰς λοιπὰς ὁμολογοῦσιν οἱ τούτων ἐπιστήμονες ἐν ὄψει τῶν ἀρχόντων
τῶν σφετέρων ὑποκλίνειν τὰ κέντρα, οἱονεὶ τῆς

---

[1] Rauw: τὰς δὲ . . . οἰκοδομὰς σοφωτέρας οὖσας . . .
τεχνηεστέρας.
[2] ὑπὲρ τούτων.   [3] πολὺ καὶ πολλούς.

Historians celebrate these constructions, but the dwellings of Bees which are far cleverer and exhibit a greater skill, of these they take not the slightest notice. And yet, while those monarchs wrought what they wrought through the affliction of multitudes, there never was any creature more gracious than the Bee, just as there is none cleverer. The first things that they construct are the chambers of their kings, and they are spacious and above all the rest. Round them they put a barrier, as it were a wall or fence, thereby also enhancing the importance of the royal dwelling. And they divide themselves into three grades, and their dwellings accordingly into the same number. Thus, the eldest dwell nearest the royal palace, and the latest born dwell next to them, while those that are young and in the prime of life are outside the latter. In this way the eldest are the king's bodyguard, and the youthful ones are a protection to the latest born.

60. According to one story the King Bees are stingless; according to another they are born with stings of great strength and trenchant sharpness; and yet they never use them against a man nor against bees: the stings are a pretence, an empty scare, for it would be wrong for one who rules and directs such numbers to do an injury. And those who understand their ways bear witness to the fact that the other Bees when in presence of their rulers withdraw their stings, as though shrinking and giving

The King Bee

---

4 πρεσβύταται καὶ αἱ παλαιόταται.

5 αὐλῇ οἱονεὶ δορύφοροι καὶ φρουροὶ οὗτοι.

6 νεώταται καὶ αἱ αὐτοετεῖς.

7 ταῦτα.

ἐξουσίας ἀφισταμένας καὶ παραχωρούσας. ἑκάτερον δ' ἄν τις ἐκπλαγείη τὸ τῶν βασιλέων ἐκείνων·
εἴτε γὰρ μὴ ἔχουσι πόθεν ἀδικήσουσι, μέγα τοῦτο·
εἴτε καὶ παρὸν ἀδικῆσαι μὴ ἀδικοῦσιν, ἀλλὰ τοῦτό
γε μακρῷ κρεῖττόν ἐστιν.

way before authority. And one might well be
astonished at either of the aforesaid characteristics
in these King Bees: if they have no means of
injuring, this is remarkable; if with all the means of
injuring they do no injury, then this is far more to
their credit.

BOOK II

B

1. Ὅταν τὰ ἤθη τὰ τῶν Θρᾳκῶν καὶ τοὺς κρυμοὺς ἀπολείπωσι τοὺς Θρᾳκίους αἱ γέρανοι, ἀθροίζονται μὲν ἐς τὸν Ἕβρον, λίθον δ' ἑκάστη καταπιοῦσα, ὡς ἔχειν καὶ δεῖπνον καὶ πρὸς τὰς ἐμβολὰς τῶν ἀνέμων ἕρμα, πειρῶνται τοῦ μετοικισμοῦ καὶ τῆς ἐπὶ τὸν Νεῖλον ὁρμῆς, ἀλέας τε καὶ χειμερίου [1] συντροφίας πόθῳ τῆς ἐκεῖθι. μελλουσῶν δὲ αὐτῶν αἴρεσθαι καὶ τοῦ πρόσω ἔχεσθαι, ὁ παλαίτατος γέρανος περιελθὼν τὴν πᾶσαν ἀγέλην ἐς τρίς, εἶτα μέντοι πεσὼν ἀφίησι τὴν ψυχήν. ἐνταῦθα [2] οὖν οἱ λοιποὶ θάπτουσι μὲν τὸν νεκρόν, φέρονται δὲ εὐθὺ τῆς Αἰγύπτου, τὰ μήκιστα πελάγη περαιούμενοι τῷ ταρσῷ τῶν πτερῶν, καὶ οὔτε ὁρμίζονταί που οὔτε ἀναπαύονται. σπείροντας δὲ τοὺς Αἰγυπτίους καταλαμβάνουσι, καὶ τράπεζαν ὡς ἂν εἴποις ἄφθονον τὴν ἐν ταῖς ἀρούραις εὑρόντες εἶτα ἄκλητοι ξενίων μεταλαγχάνουσιν.

2. Τίκτεσθαι μὲν ἐν ὄρεσι ζῷα καὶ ἐν ἀέρι καὶ ἐν θαλάττῃ, θαῦμα οὔπω μέγα· ὕλη γὰρ καὶ τροφὴ καὶ φύσις ἡ τούτων αἰτία· ἔκγονα δὲ πυρὸς πτηνὰ εἶναι τοὺς καλουμένους πυριγόνους, καὶ ἐν αὐτῷ βιοῦν καὶ τεθηλέναι, καὶ δεῦρο καὶ ἐκεῖσε περιποτᾶσθαι, τοῦτο ἐκπληκτικόν. καὶ τὸ ἔτι θαῦμα, ὅταν ἔξω τοῦ πυρὸς τοῦ συντρόφου ἐκνεύ-

---

[1] τῆς χειμερίου.   [2] ἐντεῦθεν.

BOOK II

1. When Cranes are about to leave their Thracian haunts and the frosts of Thrace, they collect on the river Hebrus,[a] and when each one has swallowed a stone by way of food and as ballast against the onslaught of winds, they prepare to emigrate and to set out for the Nile, longing for the warmth and for the food that is to be had there during the winter. And just when they are on the point of rising and moving off, the oldest Crane goes round the entire flock thrice and then falls to the ground and breathes his last. So the others bury the dead body on the spot and fly straight to Egypt, traversing the widest seas on outstretched wing, never landing, never pausing to rest. And they fall in with the Egyptians as they are sowing their fields, and in the ploughlands they find, so to speak, a generous table, and though uninvited partake of the Egyptians' hospitality.

2. That living creatures should be born upon the mountains, in the air, and in the sea, is no great marvel, since matter, food, and nature are the cause. But that there should spring from fire winged creatures which men call 'Fire-flies,'[b] and that these should live and flourish in it, flying to and fro about it, is a startling fact. And what is more extraordinary, when these creatures stray outside the

[a] Mod. Maritza.
[b] Lit. 'fire-born'; these are not what are now called 'fire-flies,' and are unknown to modern science.

89

σωσι καὶ ἀέρος ψυχροῦ μεταλάχωσιν,[1] ἐνταῦθα δὴ
τεθνήκασι. καὶ ἥτις ἡ αἰτία τίκτεσθαι μὲν πυρί,
ἀέρι δὲ ἀπόλλυσθαι, λεγέτωσαν ἄλλοι.

3. Οἱ μὲν ὄρνιθες οἱ ἕτεροι ἀναβαίνονται, ὡς λόγος, αἱ δὲ χελιδόνες οὔ, ἀλλὰ τούτων γε ἐναντία
ἡ μίξις ἐστί. καὶ τὸ αἴτιον οἶδεν ἡ φύσις. λέγει
δὲ ὁ πλείων λόγος ὅτι πεφρίκασι τὸν Τηρέα καὶ
δεδοίκασι μή ποτε ἄρα προσερπύσας λάθρᾳ εἶτα
ἐργάσηται τραγῳδίαν καινήν. ἦν δὲ ἄρα καὶ τοῦτο
χελιδόνι δῶρον ἐκ τῆς φύσεως, ὥς γε ἐμὲ κρίνειν,
τὸ τιμιώτατον· πηρωθεῖσα τὴν ὄψιν περόναις ἐὰν
τύχῃ, ὁρᾷ αὖθις. τί οὖν ἔτι τὸν Τειρεσίαν ᾄδομεν,
καίτοι μὴ ἐνταυθὶ [2] ⟨μόνον⟩,[3] ἀλλὰ καὶ ἐν ᾅδου
σοφώτατον,[4] ὡς Ὅμηρος λέγει;

4. Ζῷα ἐφήμερα οὕτω κέκληται, λαβόντα τὸ
ὄνομα ἐκ τοῦ μέτρου τοῦ κατὰ τὸν βίον· τίκτεται
γὰρ [5] ἐν τῷ οἴνῳ, καὶ ἀνοιχθέντος τοῦ σκεύους
τὰ δὲ ἐξέπτη καὶ εἶδε τὸ φῶς καὶ τέθνηκεν.
οὐκοῦν παρελθεῖν μὲν αὐτοῖς ἐς τὸν βίον ἔδωκεν ἡ
φύσις, τῶν δὲ ἐν αὐτῷ κακῶν ἐρρύσατο τὴν ταχίστην, μήτε τι τῶν ἰδίων συμφορῶν ᾐσθημένοις
μήτε μὴν τινος τῶν ἀλλοτρίων μάρτυσι γεγενημέ-
νοις.

---

[1] μεταλάβωσιν.   [2] ἐνταυθοῖ.
[3] ⟨μόνον⟩ add. H.   [4] σοφώτατον ψυχῶν.
[5] μὲν γάρ.

---

[a] Tereus married Procne and later, under false pretences,
her sister Philomela. To punish him Procne slew their son
Itys and then fled with her sister. When pursued by Tereus

range of the heat to which they are accustomed and take in cold air, they at once perish. And why they should be born in the fire and die in the air others must explain.

3. With other birds the hen is mounted by the cock, so they say; not so Swallows: their manner of coupling is the reverse. Nature alone knows the reason for this. But the common explanation is that the hens are afraid of Tereus,[a] and fear lest one day he steal secretly upon them and enact a fresh tragedy. Now in my opinion the most valuable gift that Nature has bestowed upon the Swallow is this, that if it chance to be blinded with a brooch-pin, it regains its sight.

Swallows
and their
mating

Why then do we continue to sing the praises of Teiresias, even though he was the wisest of men not only on earth but also in Hades, as Homer tells us [*Od.* 10. 493]?

4. There are creatures called *Ephemera* (living only for a day)[b] that take their name from their span of life, for they are generated in wine, and when the vessel is opened they fly out, see the light, and die. Thus it is that Nature has permitted them to come to life, but has rescued them as soon as possible from life's evils, so that they are neither aware of their own misfortune nor are spectators of the misfortune of others.

'Ephemera'

all three were changed into birds, T. into a hoopoe (or hawk), Procne a swallow, Philomela into a nightingale.

  [b] Perhaps the 'Vinegar-fly,' belonging to the genus *Drosophila*.

5. Ἤδη μέντοι τις καὶ ἀσπίδος ἐν μακρῷ τῷ χρόνῳ πληγὴν ἰάσατο ἢ τομὴν παραλαβὼν ἢ πῦρ ὑπομείνας εὖ μάλα τλημόνως ἢ ἀναγκαίοις φαρμάκοις τὸ κακόν, ἵνα μὴ πρόσω ἑρπύσῃ,[1] στήσας ὁ δείλαιος· σπιθαμὴ δὲ βασιλίσκου τὸ μῆκός ἐστι, καὶ μέντοι καὶ θεασάμενος ὁ τῶν ὄφεων μήκιστος αὐτὸν οὐκ ἐς ἀναβολὰς ἀλλὰ ἤδη ἐκ τῆς τοῦ φυσήματος προσβολῆς αὐός ἐστιν. εἰ δὲ ἄνθρωπος κατέχοι ῥάβδον, εἶτα ταύτην ἐκεῖνος ἐνδάκοι,[2] τέθνηκεν ὁ κύριος τῆς λύγου.

6. Τὴν τῶν δελφίνων φιλομουσίαν καὶ τὸ τῶν αὐτῶν ἐρωτικόν, τὸ μὲν ᾄδουσι Κορίνθιοι,[3] καὶ ὁμολογοῦσιν αὐτοῖς Λέσβιοι, τὸ δὲ Ἰῆται·[4] τὰ μὲν Ἀρίονος[5] τοῦ Μηθυμναίου ἐκεῖνοι, τά γε μὴν ἐν τῇ Ἴῳ[6] ὑπὲρ τοῦ παιδὸς τοῦ καλοῦ καὶ τῆς νήξεως αὐτοῦ καὶ τοῦ δελφῖνος οἱ ἕτεροι. λέγει δὲ καὶ Βυζάντιος ἀνήρ, Λεωνίδης ὄνομα, ἰδεῖν αὐτὸς παρὰ τὴν Αἰολίδα πλέων ἐν τῇ καλουμένῃ Ποροσελήνῃ πόλει δελφῖνα ἠθάδα καὶ ἐν λιμένι τῷ ἐκείνων οἰκοῦντα καὶ ὥσπερ οὖν ἰδιοξένοις χρώμενον τοῖς ἐκεῖθι. καὶ ἐπί γε τούτῳ ὁ αὐτὸς λέγει πρεσβῦτίν[7] τινα καὶ γέροντα δὲ συνοικοῦντα αὐτῇ ἐκθρέψαι τόνδε τὸν τρόφιμον δελεάτά[8] οἱ προτείνοντας καὶ μάλα[9] γε ἐφολκά. καὶ μέντοι καὶ ὁμότροφός οἱ ἦν ὁ τῶν πρεσβυτῶν υἱός, καὶ ἐτιθηνοῦντο ἄμφω τὸν δελφῖνα καὶ τὸν παῖδα τὸν

---

[1] Jac : προσερπύσῃ.
[2] δάκοι.
[3] Gron : Αἰγύπτιοι.
[4] Valesius : Τηῖται.
[5] Ἀρίωνος.
[6] Valesius : Τηίῳ.
[7] καὶ πρεσβῦτιν.
[8] δέλεάρ τε.
[9] ἄλλα.

5. Men have, it is true, recovered after a long **The Asp,** while from the bite of an Asp,[a] either by summoning **its bite** excision to their aid or with the utmost fortitude enduring cautery, or they have in their plight prevented the poison from spreading by taking the necessary medicines.

The Basilisk measures but a span, yet at the sight **The** of it the longest snake not after an interval but on **Basilisk** the instant, at the mere impact of its breath, shrivels. And if a man has a stick in his hand and the Basilisk bites it, the owner of the rod dies.

6. The Dolphin's love of music and its affectionate **Dolphin and** nature are a constant theme, the former with the **boy at** people of Corinth (with whom the Lesbians concur), **Poroselene** the latter with the inhabitants of Ios. The Lesbians tell the story of Arion of Methymna; what happened in Ios with the beautiful boy and his swimming and the Dolphin is told by the inhabitants of Ios.

A certain Byzantine, Leonidas by name, declares that while sailing past Aeolis he saw with his own eyes at the town called Poroselene [b] a tame Dolphin which lived in the harbour there and behaved towards the inhabitants as though they were personal friends. And further he declares that an aged couple fed this foster-child, offering it the most alluring baits. What is more, the old couple had a son who was brought up along with the Dolphin, and the pair

[a] But see 1. 54.
[b] Poroselene, island and town, the largest of the Hecatonnesi lying between Lesbos and Asia Minor.

σφέτερον, καί πως ἐκ τῆς συντροφίας ἐλαθέτην
ἐς ἔρωτα ἀλλήλων ὑπελθόντε ὅ τε ἄνθρωπος καὶ
τὸ ζῷον, καί, τοῦτο δὴ τὸ ᾀδόμενον, ὑπέρσεμνος [1]
ἀντέρως ἐτιμᾶτο ἐν τοῖς προειρημένοις. ὁ τοίνυν
δελφὶς ὡς μὲν πατρίδα ἐφίλει τὴν Ποροσελήνην,[2]
ὡς δὲ ἴδιον οἶκον ἠγάπα τὸν λιμένα, καὶ δὴ καὶ
τὰ τροφεῖα τοῖς θρεψαμένοις ἀπεδίδου. καὶ τούτων
γε ἐκεῖνος ἦν ὁ τρόπος. τέλειος ὢν τῆς ἀπὸ
χειρὸς τροφῆς ἐδεῖτο ἥκιστα, ἤδη γε μὴν καὶ
περαιτέρω προνέων καὶ περινηχόμενος καὶ σκοπῶν
ἄγρας ἐναλίους τὰ μὲν ἑαυτῷ δεῖπνον εἶχε, τὰ δὲ
τοῖς οἰκείοις ἀπέφερεν· οἱ δὲ ᾔδεσαν τοῦτο καὶ
μέντοι καὶ ἀνέμενον τὸν ἐξ αὐτοῦ φόρον ἀσμένως.
καὶ μία μὲν ἦν ἥδε ἡ πρόσοδος, ἐκείη δὲ ἄλλη.
ὄνομα [3] τῷ δελφῖνι ὡς τῷ παιδὶ οἱ θρεψάμενοι
ἔθεντο· καὶ ὁ παῖς τῇ συντροφίᾳ θαρρῶν, τοῦτο [4]
αὐτὸν ἐπί τινος προβλῆτος στὰς τόπου ἐκάλει, καὶ
ἅμα τῇ κλήσει καὶ ἐκολάκευεν· ὁ δέ, εἴτε πρὸς
εἰρεσίαν ἡμιλλᾶτό τινα, εἴτ᾽ ἐκυβίστα τῶν ἄλλων
ὅσοι περὶ τὸν χῶρον ἐπλανῶντο ἀγελαῖοι κατα-
σκιρτῶν, εἴτ᾽ ἐθήρα [5] ἐπειγούσης τῆς γαστρὸς
αὐτόν, ἐπανῄει καὶ μάλα γε ὤκιστα δίκην ἐλαυνο-
μένης νεὼς πολλῷ τῷ ῥοθίῳ, καὶ πλησίον τῶν
παιδικῶν γενόμενος συμπαίστης τε ἦν καὶ συνε-
σκίρτα, καὶ πῇ μὲν τῷ παιδὶ παρενήχετο, πῇ δὲ ὁ
δελφὶς οἷα προκαλούμενος εἶτα μέντοι ἐς τὴν
ἅμιλλαν τὴν πρὸς [6] αὐτὸν τὰ παιδικὰ ὑπῆγε. καὶ

---

[1] καὶ μάλα ὑ.
[2] προειρημένῃ.
[3] ὄνομα δὲ καί.
[4] Schn : τοῦτον.
[5] εἴτε ἐς θήραν καὶ μάλα γε.   [6] εἰς.

94

cared for the Dolphin and their own son, and somehow by dint of being brought up together the man-
child and the fish gradually came without knowing it
to love one another, and, as the oft-repeated tag has
it, 'a super-reverent counter-love was cultivated'
by the aforesaid. So then the Dolphin came to
love Poroselene as his native country and grew as
fond of the harbour as of his own home, and what is
more, he repaid those who had cared for him what
they had spent on feeding him. And this was how
he did it. When fully grown he had no need of
being fed from the hand, but would now swim
further out, and as he ranged abroad in his search for
some quarry from the sea, would keep some to feed
himself, and the rest he would bring to his 'relations.'
And they were aware of this and were even glad to
wait for the tribute which he brought. This then
was one gain; another was as follows. As to the
boy so to the Dolphin his foster-parents gave a
name, and the boy with the courage born of their
common upbringing would stand upon some spot
jutting into the sea and call the name, and as
he called would use soothing words. Whereat the
Dolphin, whether he was racing with some oared
ship, or plunging and leaping in scorn of all other
fish that roamed in shoals about the spot, or was
hunting under stress of hunger, would rise to the
surface with all speed, like a ship that raises a
great wave as it drives onward, and drawing near to
his loved one would frolic and gambol at his side;
at one moment would swim close by the boy, at
another would seem to challenge him and even
induce his favourite to race with him. And what was
even more astounding, he would at times even decline

τὸ ἔτι θαῦμα, ἀπέστη καὶ τῆς πρώτης ποτὲ καὶ
δὴ καὶ ὑπενήξατο αὐτῷ, οἷα νικώμενος ἡδέως
δήπου. ταῦτα τοίνυν ἐκεκήρυκτο, καὶ τοῖς πλέου-
σιν ὅραμα ἐδόκει σὺν καὶ τοῖς ἄλλοις ὅσα ἡ πόλις
ἀγαθὰ εἶχε, καὶ τοῖς πρεσβύταις καὶ τῷ μειρακίῳ
πρόσοδος ἦν.

7. Ἐν Λιβύῃ ἡμιόνους [1] ἢ τετρωμένους Ἀρχέ-
λαος λέγει ἢ ἀπειπόντας ὑπὸ δίψους ἐρρῖφθαι
νεκροὺς πολλούς. πολλάκις δὲ ὄφεων ἐπιρρεῦσαν
φῦλον πάμπολυ τῶν κρεῶν ἐσθίειν· ἐπὰν δὲ
βασιλίσκου συρίγματος ἀκούσῃ, τὰ μὲν ὑπὸ τοῖς
εἰλυοῖς [2] καὶ τῇ ψάμμῳ ἀφανίζεσθαι τὴν ταχίστην
καὶ ἀποκρύπτεσθαι, τὸν δὲ προσελθόντα κατὰ
πολλὴν τὴν εἰρήνην δειπνεῖν, εἶτα αὖθις ὑποσυρίζειν
καὶ ἀπαλλάττεσθαι, τοὺς δὲ ἡμιόνους καὶ τὸ
δεῖπνον τὸ ἐξ αὐτῶν σημαίνεσθαι τὸ ἐντεῦθεν, τὸ
τοῦ λόγου τοῦτο, ἄστροις.

8. Λόγοι φασὶν Εὐβοέων δεῦρο φοιτῶντες, τοὺς
ἁλιέας τοὺς ἐκεῖσε τοῖς δελφῖσι τοῖς ἐκεῖθι ἰσομοι-
ρίαν τῆς θήρας ἀπονέμειν· καὶ ἀκούω τὴν ἄγραν
τοιαύτην. γαλήνην εἶναι χρή, καὶ εἰ ταῦθ᾽ οὕτως
ἔχει, τῆς πρῴρας τῶν ἀκατίων κοίλας τινὰς
ἐξαρτῶσιν ἐσχαρίδας πυρὸς ἐνακμάζοντος· καὶ
εἰσὶ διαφανεῖς, ὡς καὶ στέγειν [3] τὸ πῦρ καὶ μὴ
κρύπτειν τὸ φῶς. ἰπνοὺς καλοῦσιν αὐτάς. οἱ
τοίνυν ἰχθῦς δεδίασι τὴν αὐγὴν καὶ τὴν λαμπηδόνα
δυσωποῦνται· καὶ οἱ μὲν οὐκ εἰδότες ὅ τι βούλεται

---

[1] ἡμιόνους τινάς.   [2] ἰλύσι.
[3] Reiske : στέγειν καί.

the winner's place and actually swim second, as
though presumably he was glad to be defeated.

These happenings were noised abroad, and those
who sailed thither reckoned them among the excellent sights which the city had to show; and to the
old people and to the boy they were a source of
revenue.

7. Archelaus tells us that in Libya mules that The Basilisk
have been wounded or which have succumbed from snakes
thirst are thrown out for dead in great numbers.
And frequently a multitude of snakes of all kinds
comes streaming up to eat their flesh, but whenever
they hear the hiss of the Basilisk they disappear as
swiftly as possible into their dens or beneath the
sand, and hide; so the Basilisk on reaching the spot
feasts in complete tranquillity. Then again with a
hiss he is off, and thereafter as to the mules and to
the feast which they provide, ' he marks their place,'
as the saying has it, ' only by the stars.' [a]

8. There are stories which reach us from Euboea of Fishermen
fisher-folk in those parts sharing their catch equally Dolphins
with the Dolphins in those parts. And I am told
that they fish in this way. The weather must be
calm, and if it is, they attach to the prow of their
boats some hollow braziers with fire burning in them,
and one can see through them, so that while retain-
ing the fire they do not conceal the light. They call
them lanterns. Now the fish are afraid of the brightness and are dazzled by the glare, and some of them
not knowing what is the purpose of the thing they see,

[a] *I.e.* he never returns; cp. Jebb on Soph. *OT* 795.

AELIAN

τὸ ὁρώμενον, πλησιάζουσι, μαθεῖν βουλόμενοι τοῦ
φοβοῦντος σφᾶς τὴν αἰτίαν· εἶτα ἐκπλαγέντες ἢ
πρός τινι πέτρᾳ ἡσυχάζουσιν ἀθρόοι παλλόμενοι
τῷ δέει ἢ ἐς τὴν ἠόνα ἐκπίπτουσιν ὠθούμενοι, καὶ
ἐοίκασι τοῖς ἐμβεβροντημένοις. οὕτω γε μὴν
διακειμένους ῥᾷστόν ἐστιν ἤδη καὶ τριαίνῃ πατάξαι.
ἐπειδὰν οὖν θεάσωνται οἱ δελφῖνες τοὺς ἁλιέας τὸ
πῦρ ἐξάψαντας, ἑαυτοὺς εὐτρεπίζουσι. καὶ οἱ
μὲν ἠρέμα ὑπερέττουσιν, οἱ δὲ δελφῖνες τοὺς
ἐξωτέρω τῶν ἰχθύων φοβοῦντες ὠθοῦσι καὶ τοῦ
διαδιδράσκειν ἀναστέλλουσιν. οὐκοῦν ἐκεῖνοι πιε-
ζόμενοι πανταχόθεν καὶ τρόπον τινὰ κεκυκλωμένοι
ἔκ τε τῆς τούτων εἰρεσίας καὶ τῆς νήξεως τῆς
ἐκείνων συνιᾶσιν ἄφυκτα εἶναί σφισι, καὶ παραμένουσι καὶ ἁλίσκονται πάμπολύ τι χρῆμα. καὶ
οἱ δελφῖνες προσίασιν [1] ὡς ἀπαιτοῦντες τοῦ κοινοῦ
πόνου τὴν ἐπικαρπίαν τὴν ὀφειλομένην σφίσιν ἐκ
τῆς νομῆς, καὶ οἵ γε ἁλιεῖς πιστῶς καὶ εὐγνωμόνως
ἀφίστανται τοῖς συνθήροις τοῦ δικαίου μέρους, εἰ
βούλονται καὶ πάλιν σφίσι συμμάχους ἀκλήτους
παρεῖναι καὶ ἀπροφασίστους. πιστεύουσι γὰρ οἱ
ἐκεῖ θαλαττουργοὶ ὅτι παραβάντες ἕξουσιν ἐχθροὺς
οὓς εἶχον πρότερον φίλους.

9. Ἔλαφος ὄφιν νικᾷ, κατά τινα φύσεως δωρεὰν
θαυμαστήν· καὶ οὐκ ἂν αὐτὸν διαλάθοι ἐν τῷ φω-
λεῷ ὢν ὁ ἔχθιστος, ἀλλὰ προσερείσας τῇ κατα-
δρομῇ τοῦ δακετοῦ [2] τοὺς ἑαυτοῦ μυκτῆρας βιαιότατα ἐσπνεῖ, καὶ ἕλκει ὡς ἴυγγι τῷ πνεύματι, καὶ
ἄκοντα προάγει, καὶ προκύπτοντα αὐτὸν ἐσθίειν
ἄρχεται· καὶ μάλιστά γε διὰ χειμῶνος δρᾷ τοῦτο.

[1] Schn: προΐασιν.

98

draw near from a wish to discover what it is that frightens them. Then terror-stricken they either lie still in a mass close to some rock, quivering with fear, or are cast ashore as they are jostled along, and seem thunderstruck. Of course in that condition it is perfectly easy to harpoon them. So when the Dolphins observe that the fishermen have lit their fire, they get ready to act, and while the men row softly the Dolphins scare the fish on the outskirts and push them and prevent any escape. Accordingly the fish pressed on all sides and in some degree surrounded, realise that there is no escaping from the men that row and the Dolphins that swim; so they remain where they are and are caught in great numbers. And the Dolphins approach as though demanding the profits of their common labour due to them from this store of food. And the fishermen loyally and gratefully resign to their comrades in the chase their just portion—assuming that they wish them to come again, unsummoned and prompt, to their aid, for those toilers of the sea are convinced that if they omit to do this, they will make enemies of those who were once friends.

9. A Deer defeats a snake by an extraordinary gift Deer and Snakes that Nature has bestowed. And the fiercest snake lying in its den cannot escape, but the Deer applies its nostrils to the spot where the venomous creature lurks, breathes into it with the utmost force, attracts it by the spell, as it were, of its breath, draws it forth against its will, and when it peeps out, begins to eat it. Especially in the winter does it do this.

ἤδη μέντοι τις [1] καὶ κέρας ἐλάφου ξέσας, εἶτα τὸ
ξέσμα ἐς πῦρ ἐνέβαλε, καὶ ὁ καπνὸς ἀνιὼν διώκει
τοὺς ὄφεις πανταχόθεν, μηδὲ τὴν ὀσμὴν ὑπομέ-
νοντας.

10. Ἔστι μὲν τὴν ἄλλως [2] ὁ ἵππος γαῦρον· καὶ
γὰρ καὶ τὸ μέγεθος καὶ τὸ τάχος αὐτὸν καὶ τοῦ
αὐχένος τὸ ὑψηλὸν καὶ ἡ τῶν σκελῶν ὑγρότης καὶ
ἡ τῶν ὁπλῶν κροῦσις [3] ἐς φρύαγμα καὶ τῦφον
ἀνάγει· μάλιστα δὲ κομῶσα ἵππος ἁβρότατόν τέ
ἐστι καὶ θρυπτικώτατον. ἀτιμάζει γοῦν ἀναβῆναι
τοὺς ὄνους αὐτήν, ἵππῳ δὲ γαμουμένη ἥδεται, καὶ
ἑαυτὴν ἀξιοῖ τῶν μεγίστων. ὅπερ οὖν συνειδότες
οἱ βουλόμενοι ἡμιόνους σφίσι γενέσθαι, ἀποθρί-
σαντες τῆς ἵππου τὴν χαίτην εἰκῆ καὶ ὡς ἔτυχεν,
εἶτα μέντοι τοὺς ὄνους ἐπάγουσιν· ἡ δὲ ὑπομένει
τὸν ἄδοξον ἤδη γαμέτην, πρῶτον αἰδουμένη. καὶ
Σοφοκλῆς δὲ ἔοικε μεμνῆσθαι τοῦ πάθους.

11. Περὶ μὲν τῆς τῶν ἐλεφάντων σοφίας εἶπον
ἀλλαχόθι, καὶ μέντοι καὶ περὶ τῆς θήρας αὐτῶν
καὶ ταύτης [4] εἶπον ὀλίγα ἐκ πολλῶν ὧν ἔφασαν
ἄλλοι. τὸ δὲ νῦν ἔχον ἔοικα [5] ἐρεῖν περί τε
εὐμουσίας αὐτῶν καὶ εὐπειθείας καὶ τῆς ἐς τὰ
μαθήματα εὐκολίας, χαλεπὰ ὅμως ὄντα καὶ
ἀνθρώπῳ τυχεῖν, [6] μή τι γοῦν τοσούτῳ θηρίῳ καὶ
οὕτω τέως ἀγριωτάτῳ συγγενέσθαι. χορείαν γὰρ
καὶ ὀρχηστικὴν καὶ βαίνειν πρὸς ῥυθμὸν καὶ

---

[1] τις after ἐλάφου in MSS.
[2] τὴν ἄλλως] καὶ ἐκ τῶν ἄλλων.
[3] κροῦσις πάντα.
[4] ταῦτα. [5] Schn: ἔθηκα.

Indeed it has even happened that a man has ground a Deer's horn to powder and then has thrown the powder into fire, and that the mounting smoke has driven the snakes from all the neighbourhood: even the smell is to them unendurable.

10. The Horse is generally speaking a proud crea- Mare and ture, the reason being that his size, his speed, his Ass tall neck, the suppleness of his limbs, and the clang of his hooves make him insolent and vain. But it is chiefly a Mare with a long mane that is so full of airs and graces. For instance, she scorns to be covered by an ass, but is glad to mate with a horse, regarding herself as only fit for the greatest ⟨of her kind⟩. Accordingly those who wish to have mules born, knowing this characteristic, clip the Mare's mane in a haphazard fashion anyhow, and then put asses to her. Though ashamed at first, she admits her present ignoble mate. Sophocles also appears to mention this humiliation [*fr.* 659P].[a]

11. Touching the sagacity of Elephants I have The spoken elsewhere; and further, I have spoken too Elephant, of the manner of hunting them, mentioning but a its docility few of the numerous facts recorded by others. For the present I intend to speak of their sense for music and their readiness to obey and their aptitude for learning things which are difficult even for mankind, to say nothing of so huge an animal and one hitherto so fierce to encounter. The movements of a chorus, the steps of a dance, how to march in time, how to

[a] See 11. 18.

[6] τυχεῖν αὐτῶν.

αὐλοῦ ἀσμένως ¹ ἀκούειν καὶ συνιέναι ἤχων διαφοράς, ἢ βραδύνειν ἐνδιδόντων ἢ ταχύνειν παρορ-
μώντων, μαθὼν οἶδεν ἐλέφας, καὶ ἀκριβοῖ καὶ
οὐ σφάλλεται. οὕτως ἄρα ἡ φύσις μεγέθει μὲν
αὐτὸν μέγιστον εἰργάσατο, μάθησις ² δὲ πραότατον
ἀπέφηνε καὶ εὐάγωγον. εἰ μὲν οὖν ἔμελλον τὴν ἐν
Ἰνδοῖς αὐτῶν εὐπείθειαν καὶ εὐμάθειαν ἢ τὴν ἐν
Αἰθιοπίᾳ ἢ τὴν ἐν Λιβύῃ γράφειν, ἴσως ἄν τῳ καὶ
μῦθον ἐδόκουν τινὰ συμπλάσας κομπάζειν, εἶτα
ἐπὶ φήμῃ τοῦ θηρίου τῆς φύσεως καταψεύδεσθαι·
ὅπερ ἐχρῆν δρᾶν φιλοσοφοῦντα ἄνδρα ἥκιστα καὶ
ἀληθείας ἐραστὴν διάπυρον. ἃ δὲ αὐτὸς εἶδον καὶ
ἅτινα πρότερον ἐν τῇ Ῥώμῃ πραχθέντα ἀνέγραψαν
ἄλλοι προειλόμην εἰπεῖν, ἐπιδραμὼν ὀλίγα ἐκ πολλῶν, οὐχ ἥκιστα καὶ ἐντεῦθεν ἀποδεικνὺς τὴν τοῦ
ζῴου ³ ἰδιότητα. ἡμερωθεὶς ⁴ ἐλέφας πραότατόν
ἐστι, καὶ ἄγεται ῥᾷστα ἐς ὅ τί τις ⁵ θέλει. καὶ τά
γε πρεσβύτατα τιμῶν τὸν χρόνον ἐρῶ πρῶτον.
θέας ἐπετέλει Ῥωμαίοις ὁ Γερμανικὸς ὁ Καῖσαρ·
εἴη δ᾽ ἂν ἀδελφιδοῦς Τιβερίου οὗτος. οὐκοῦν
ἐγένοντο ⁶ καὶ ἄρρενες ἐν τῇ Ῥώμῃ τέλειοι πλείους
καὶ θήλειαι, εἶτα ἐξ αὐτῶν ἐτέχθησαν αὐθιγενεῖς.
καὶ ὅτε τὰ κῶλα ὑπήρξαντο πήγνυσθαι, σοφὸς
ἀνὴρ ὁμιλεῖν τοιούτοις θηρίοις ἐπώλευσεν αὐτούς,
δαιμονίᾳ τινὶ καὶ ἐκπληκτικῇ διδασκαλίᾳ μεταχει-
ρισάμενος. προσῆγε δὲ αὐτοὺς ἄρα ἡσυχῇ τήν γε
πρώτην καὶ πράως τοῖς διδάγμασι δελέατα ἄττα

---

¹ αὐλοῦ ἀσμένως] αὐλουμένους.
² Jac : μαθήσει.
³ τῶν ζῴων.
⁴ Schn : ἡμερωθέν.
⁵ ὅ τις. ⁶ ἐγένοντο μέν.

enjoy the sound of flutes, how to distinguish different
notes, when to slacken pace as permitted or when to
quicken at command—all these things the Elephant
has learnt and knows how to do, and does accurately
without making mistakes. Thus, while nature has
created him to be the largest of animals, learning
has rendered him the most gentle and docile. Now
had I set out to write about the readiness to obey
and to learn among elephants in India or in Ethiopia
or in Libya, anyone might suppose that I was concocting some pretentious tale, that in fact I was on
the strength of hearsay about the beast giving a
completely false account of its nature. That is the
last thing that a man in pursuit of knowledge and an
ardent lover of the truth has any right to do. Instead I have preferred to state what I have myself
seen and what others have recorded as having
formerly occurred in Rome, treating summarily a
few facts out of many, which nevertheless sufficiently
demonstrate the peculiar nature of the beast.

The Elephant when once tamed is the gentlest of Performing
Elephants
in Rome
creatures and is easily induced to do whatever one
wants. Now keeping due eye on the time, I shall
state the most important events first. Germanicus
Caesar was about to give some shows for the Romans.
(He would be the nephew [a] of Tiberius.) There were
in Rome several full-grown male and female elephants,
and there were calves born of them in the country;
and when their limbs began to grow firm, a man who
was clever at dealing with such beasts trained them
and instructed them with uncanny and astounding
dexterity. To begin with he introduced them in a
quiet, gentle fashion to his instructions, supplying

[a] Or rather, the adopted son.

ἐπάγων καὶ τροφὰς ἡδίστας καὶ πεποικιλμένας ἐς
τὸ ἐπαγωγόν [1] τε καὶ ἐφολκόν, ὡς εἴ τι μὲν ἦν [2]
ἀγριότητος, τοῦτο ἐκβαλεῖν, ἀπαυτομολῆσαι [3] δὲ
πρὸς τὸ ἥμερον καὶ ἀμωσγέπως ἀνθρώπειον. καὶ
ἦν γε τὰ μαθήματα αὐλῶν [4] ἀκούοντας μὴ ἐκμαί-
νεσθαι, καὶ τυμπάνων ἀράβου κροτοῦντος μὴ
ταράττεσθαι, καὶ κηλεῖσθαι σύριγγι, φέρειν δὲ καὶ
ἤχους ἐκμελεῖς [5] καὶ ποδῶν ἐμβαινόντων ψόφον
καὶ ᾠδὴν συμμιγῆ· ἐξεπονήθησαν δὲ καὶ ἀνθρώπων
πλῆθος μὴ δεδιέναι. ἦν δὲ καὶ ἐκεῖνα διδάγματα
ἀνδρικά, πρὸς τὴν τῆς πληγῆς καταφορὰν μὴ θυ-
μοῦσθαι, μηδὲ μὴν ἀναγκαζομένους λυγίζειν τι
τῶν μελῶν καὶ κάμπτειν ὀρχηστικῶς τε καὶ χορι-
κῶς εἶτα ἐς θυμὸν ἐξάπτεσθαι, καὶ ταῦτα ῥώμης
τε καὶ ἀλκῆς εὖ ἥκοντας. φύσει μὲν οὖν τοῦτο
πλεονέκτημα ἤδη καὶ μάλα γεννικόν, μὴ ἔχειν
ἀτάκτως μηδὲ ἀπειθῶς πρὸς παιδεύματα ἀνθρω-
πικά· ἐπεὶ δὲ ἀπέφηνεν αὐτοὺς ὁ ὀρχηστοδιδάσκα-
λος καὶ μάλα γε σοφούς, καὶ ἠκρίβουν τὰ ἐκ τῆς
παιδεύσεως, οὐκ ἐψεύσαντο τῆς διδασκαλίας τὸν
πόνον, φασίν, ἔνθα ἐπιδείξασθαι τὰ παιδεύματα
αὐτοὺς ἡ χρεία σὺν τῷ καιρῷ παρεκάλει. δώδεκα
μὲν γὰρ τὸν ἀριθμὸν ὅδε ὁ χορὸς ἦσαν· παρῆλθόν
γε μὴν ἐντεῦθεν τοῦ θεάτρου καὶ ἐκεῖθεν νεμηθέντες,
καὶ εἰσῇεσαν ἁβρὰ μὲν βαίνοντες, θρυπτικῶς δὲ
τὸ σῶμα πᾶν διαχέοντες, καὶ ἠμπείχοντο χορευτι-
κὰς στολὰς καὶ ἀνθινάς. καὶ τοῦ γε χορολέκτου
τῇ φωνῇ μόνον ὑποσημήναντος οἱ δὲ ἐπὶ στοῖχον
ᾔεσαν, φασίν, εἰ τοῦτο ἐκέλευσεν ὁ διδάξας· εἶτα

---

[1] ἀγωγόν.   [2] εἰ μέν τι ἐνῆν Cobet.
[3] ἐπαυτομολῆσαι.   [4] καὶ αὐλῶν.

them with delicacies and the most appetising food, varied so as to allure and entice them into abandoning all trace of ferocity and into becoming renegades, that is tame and to some degree human. So what they learnt was not to go wild at the sound of flutes, not to be alarmed at the beating of drums, to be charmed by the pipe and to endure discordant notes, the beat of marching feet, and the singing of crowds. Moreover they were thoroughly trained not to be afraid of men in masses. And further their disciplining was manly in the following respects: they were not to get angry at the infliction of a blow, nor, when obliged to move some limb and to sway in time to dance or song, to burst into a rage, even though they had attained to such strength and courage. Now to refrain by instinct from misbehaving and from flouting the instruction given by a man is a virtue and a mark of nobility. When therefore the dancing-master had brought them to a high degree of proficiency, and they performed accurately what he had taught them, they did not disappoint the labour spent on their training (so they say) in the place where in due time the occasion demanded that they should display what they had been taught. Now this troupe was twelve in number, and they advanced in two groups from the right and the left sides of the theatre. They entered with a mincing gait, swaying their whole body in a delicate manner, and they were clothed in the flowered garments of dancers. And at no more than a word from the conductor they formed into line (so we are told)—supposing that to have been their teacher's order. Then again they

---

[5] *Jac*: ἐμμελεῖς.

αὖ πάλιν περιήρχοντο ἐς κύκλον, ὑποσημήναντος
ἰέναι ταύτῃ· καὶ εἰ ἐξελίττειν ἔδει, ἔπραττον αὐτό,
καὶ ἄνθη μέντοι ῥιπτοῦντες ἐκόσμουν τὸ δάπεδον
οἵδε, μέτρῳ καὶ φειδοῖ δρῶντες, καί τι καὶ [1]
ἐπεκτύπουν τοῖς ποσί, χόρειόν τε [2] καὶ συμμελὲς
ὁμορροθοῦντες οἱ αὐτοί. Δάμωνα μὲν οὖν καὶ
Σπίνθαρον καὶ Ἀριστόξενον καὶ Φιλόξενον καὶ ἄλλους ἐπαΐειν μουσικῆς κάλλιστα καὶ ἐν ὀλίγοις
ἐξετάζεσθαι τήνδε τὴν σοφίαν θαυμαστὸν μέν,
ἄπιστον δὲ καὶ παράλογον οὐδαμῶς· τὸ δὲ αἴτιον,
ἄνθρωπος ζῷόν ἐστι λογικὸν καὶ νοῦ καὶ λογισμοῦ
χωρητικόν· ζῷον δὲ ἄναρθρον συνιέναι καὶ ῥυθμοῦ
καὶ μέλους καὶ φυλάττειν σχῆμα καὶ ἐμμέλειαν
μὴ παρατρέπειν καὶ ἀποπληροῦν τῶν διδαχθέντων
τὴν ἀπαίτησιν, φύσεως δῶρα ταῦτα ἅμα καὶ
ἰδιότης καθ' ἕκαστον ἐκπληκτική. τὰ δὲ ἐπὶ
τούτοις καὶ ἐκμῆναι [3] τὸν θεατὴν ἱκανά· χαμαι-
ζήλων κλινῶν στιβάδες [4] ἐν τῇ ψάμμῳ τοῦ θεάτρου
τεθεῖσαι, εἶτα ἐδέξαντο τυλεῖα [5] καὶ ἐπὶ τούτοις
στρωμνὴν ποικίλην, οἰκίας [6] μέγα εὐδαίμονος καὶ
παλαιοπλούτου σαφῆ [7] μαρτύρια· καὶ κυλίκια ἦν
πολυτελῆ παρακείμενα καὶ κρατῆρες χρυσοῖ καὶ
ἀργυροῖ, καὶ ἐν αὐτοῖς ὕδωρ πάμπολυ, τράπεζαί
τε παρέκειντο θύου τε καὶ ἐλέφαντος εὖ μάλα
σοβαραί, καὶ ἦν ἐπ' αὐτῶν κρέα καὶ ἄρτοι, παμβο-
ρωτάτων ἐμπλῆσαι ζῴων γαστέρας ἱκανὰ ταῦτα.
ἐπεὶ δὲ τὰ τῆς παρασκευῆς ἔκπλεά τε καὶ ἀμφιλαφῆ
ἦν, παρῆλθον οἱ δαιτυμόνες, ἐξ μὲν ἄρρενες, ἰσά-

---

[1] καί τι καί] αὐτίκα δ'.
[2] τι.
[3] ἐκμᾶναι.
[4] *Wytt*: ὡς στιβάδες.
[5] τύλια.

wheeled into a circle when he so ordered them, and
if they had to deploy, that also they did. And then
they sprinkled flowers to deck the floor, but with
moderation and economy, and now and again they
stamped, keeping time in a rhythmical dance.

That Damon therefore, that Spintharus, Aristo-
xenus, Philoxenus, and others should be experts in
music and should be numbered among the few for
their knowledge of it is certainly matter for wonder
but by no means incredible or absurd. The reason
is that man is a rational animal capable of understanding and logical thought. But that an inarticulate animal should comprehend rhythm and
melody, should follow the movements of a tragic
dance without a false step, fulfilling all that its
lessons required of it—these are gifts bestowed by
Nature, and each one is a singularity that fills one
with amazement.

But what followed was enough to send the specta-
tor wild with delight. On the sand of the theatre
were placed mattresses of low couches, and on these
in turn cushions, and over them embroidered cover-
lets, clear evidence of a house of great prosperity and
ancestral wealth. And close at hand were set costly
goblets and bowls of gold and of silver, and in them
a large quantity of water; and beside them were
placed tables of citrus wood and of ivory, of great
magnificence, and they were laden with meat and
bread enough to satisfy the stomachs of the most
voracious animals. So as soon as the preparations
were completed in all their abundance, the ban-
queters came on, six males and an equal number of

Elephants
at a banquet

<hr>

[6] καὶ οἰκίας.　　　　　[7] σαφῶς.

ριθμοὶ δὲ αἱ θήλειαι αὐτοῖς· καὶ οἱ μὲν [1] ἀρρενωπὸν
στολὴν εἶχον, αἱ δὲ θῆλυν, καὶ κατεκλίνησαν [2] σὺν
κόσμῳ συνδυασθέντες ἄρρεν τε καὶ θῆλυ. καὶ
ὑποσημήναντος τὰς προβοσκίδας ὡς χεῖρας κεκο-
λασμένως προύτεινον, καὶ ἐσιτοῦντο εὖ μάλα
σωφρόνως· καὶ οὔτε τις αὐτῶν ἔδοξεν ἀδηφάγος
οὔτε μὴν προτένθης τις ἢ τῆς μοίρας τῆς μείζονος
ἁρπακτικός, ὡς ὁ Πέρσης ὁ παρὰ τῷ Ξενοφῶντι
τῷ χρυσῷ. ἐπεὶ δὲ πίνειν ἔδει, ἑκάστῳ κρατὴρ
παρετέθη, καὶ ἀρυτόμενοι ταῖς προβοσκίσι τὸ
ποτὸν ἔπινον κεκοσμημένως, εἶτα ἀπέρραινον σὺν
παιδιᾷ καὶ οὐχ ὕβρει. πολλὰ δὲ καὶ ἄλλα ἀνέγραψαν [3] τοιαῦτα τῆς ἰδιότητος τῶνδε τῶν ζῴων
σοφὰ καὶ ἐκπληκτικά. ἐγὼ δὲ εἶδον καὶ γράμματα
γράφοντα ἐπὶ πίνακος Ῥωμαῖα ἀστραβῶς τῇ
προβοσκίδι καὶ ἀτρέπτως· πλὴν ἐπέκειτο ⟨ἡ⟩ [4]
χεὶρ τοῦ διδάξαντος ἐς τὴν τῶν γραμμάτων
παιδαγωγοῦσα περιγραφήν, ἔστε ἀπογράψαι τὸ
ζῷον· τὸ δὲ ἀτενὲς ἑώρα κάτω. πεπαιδευμένους
εἶναι τοὺς ὀφθαλμοὺς τῷ ζῴῳ καὶ γραμματικοὺς
εἶπες ἄν.

12. Ἔχει μέντοι καὶ ὁ λαγὼς [5] συμφυεῖς ἰδιό-
τητας. ἐκπεπταμένοις μὲν γὰρ τοῖς βλεφάροις
καθεύδει, κατηγορεῖ δὲ αὐτοῦ τὰ ἔτη τρώγλας
τινὰς ὑποφαίνων. φέρει δὲ καὶ ἐν τῇ νηδύι τὰ
μὲν ἡμιτελῆ, τὰ δὲ ὠδίνει, τὰ δὲ ἤδη οἱ τέτεκται.

---

[1] οἱ μὲν ἐλέφαντες.   [2] κατεκλίθησαν.
[3] Schn : ἀνέγραψα.
[4] ⟨ἡ⟩ add. Schn.
[5] λαγωός.

females; the former were clad in masculine garb, the latter in feminine; and they took their places in orderly fashion in pairs, a male and a female. And at a signal they reached forward their trunks modestly, as though they were hands, and ate with great decorum. And not one of them gave the impression of being a glutton nor yet of trying to forestall others or of being inclined to snatch too large a portion, as the Persian did who occurs in Xenophon the golden.[a] And when they wanted to drink, a bowl was placed by each one, from which they sucked up the water with their trunks and drank it in an orderly manner, and then proceeded to squirt ⟨the attendants⟩[b] in fun, not by way of insult.

Many similar stories have been recorded showing the astounding ingenuity of these animals. And I myself have seen one actually with its trunk writing Roman letters on a tablet in a straight line without any deviation. The only thing was that the instructor's hand was laid upon it, directing it to the shape of the letters until the animal had finished writing; and it looked intently down. You would have said that the animal's eyes had been taught and knew the letters.

12. The Hare has certain innate characteristics. The Hare For one thing it sleeps with its eyelids open; for another it proclaims its age when it half shows certain apertures. Also it carries some of its young half-formed in its womb, some it is in process of bearing, others it has already borne.

---

[a] Xen. *An.* 7. 3. 23; Arystas was however an Arcadian, not a Persian. 'Golden,' cf. Diog. La. 10. 8 Πλάτωνα χρινοῦν, Lucr. 3. 12 [*Epicuri*] *aurea dicta*.

[b] Or 'each other'?

AELIAN

13. Τὰ κήτη τὰ μεγάλα πάντα[1] ἄνευ κυνῶν
δεῖται τοῦ ἡγεμόνος, καὶ τοῖς ὀφθαλμοῖς ἐκείνου
ἄγεται. ἔστι δὲ ἰχθὺς μικρὸς[2] καὶ λεπτός,[3] τὴν
κεφαλὴν προμήκης· στενὸν ⟨δὲ⟩[4] αὐτῷ τὸ
οὐραῖον συμπέφυκεν, ὡς οἱ τούτων λέγουσι σοφοί.
εἴτε δὲ αὐτὸν ἐκεῖνον παρέδωκε τῷ κήτει ἡ φύσις
ἑκάστῳ, εἴτε φιλίᾳ αὐτῷ[5] ἑκὼν πρόσεισιν,[6] οὐκ
οἶδα· φύσεως δὲ ἀνάγκην εἶναι τὸ πραττόμενον
μᾶλλον πεπίστευκα. νήχεται γὰρ ὅδε ὁ ἰχθὺς
οὐδεπώποτε ἑαυτῷ,[7] πρόεισι δὲ τῆς τοῦ κήτους
κεφαλῆς, καὶ ἡγεμών ἐστιν αὐτοῦ, καὶ ὡς εἰπεῖν
οἴαξ. προορᾷ γοῦν ἐκείνῳ τὰ πάντα καὶ προαισθά-
νεται τῷ αὐτῷ, καὶ προδιδάσκει ἕκαστα τῆς
οὐρᾶς τῷ ἄκρῳ, καὶ παρέψαυσε τούτῳ, καὶ ἔδωκε
σύνθημα, καὶ τῶν μὲν φοβερῶν ἀνέστειλεν, ἐπί γε
μὴν τὰ θρέψοντα προάγει, καὶ τὴν ἐκ τῶν θηρατῶν
ἐπιβουλὴν διδάσκει σημείῳ τινὶ ἀτεκμάρτῳ, καὶ
τῶν τόπων ὧν οὐ χρὴ τοσοῦτον θηρίον ἐπιβῆναι
προμηνύει, ἵνα μή ποτε ἄρδην ἐς ἕρμα περισχεθὲν
ἀπόληται. ἡ τοίνυν τοῦ βίου ὑπόθεσις τῷ μεγίστῳ
τὸ βράχιστόν ἐστιν. ἔοικε δὲ καταπιανθὲν τὸ
ζῷον μήτε ὁρᾶν ἔτι μήτε ἀκούειν, εἶναι δὲ πρόβλημα
καὶ τῆς ὄψεως καὶ τῆς ἀκοῆς τῶν σαρκῶν τὸν
ὄγκον. οὐχ ὁρᾶται δὲ τοῦ κήτους ἔρημος, ἀλλὰ
ἀνάγκη, τοῦ πάντων αὐτῷ τῶν προειρημένων
αἰτίου προαπολωλότος, καὶ ἐκεῖνο ἀπολέσθαι.

---

[1] ὀλίγου πάντα.   [2] *Ges* : μακρός.
[3] λευκός.
[4] ⟨δέ⟩ *add. H.*
[5] *Reiske* : αὐτῶν.
[6] *Jac* : πρόεισιν.
[7] *Abresch* : αὐτῷ.

13. All the large fishes, with the exception of the Fishes and their leaders
Shark, require a leader, and are guided by its eyes.
The leader is a small, slim fish with an elongated
head, but its tail is narrow, according to the authorities on the subject. But whether Nature has con-
ferred upon each large fish the aforesaid guide, or
whether it associates with the large fish of its own
free will out of friendliness, I am unable to say, but
I prefer to believe that this is done under the com-
pulsion of Nature, for this fish never swims by itself,
but moves in front of the large fish's head and is its
leader and, as it were, tiller. For instance, it fore-
sees and takes previous notice of everything on behalf of the large fish; it forewarns it of everything
by the tip of its tail, and by its contact signals to the
fish, keeping it away from what is to be feared but
leading it on to what will feed it. And by some
invisible sign it warns the fish that its pursuers have
designs upon it, and gives timely indication of those
spots which a creature of its size ought not to approach, if it is not to be surrounded and perish utterly
on some reef.

So then the first essential for the life of the largest
of creatures is the smallest. And it seems that when
the large fish becomes very fat it can no longer see
nor hear, the vast bulk of its flesh being an obstacle
to sight and to hearing. But the 'leader' is never
seen apart from the large fish; if however, with its
responsibility for the services described above, it
dies first, then the large fish is bound to die
also.

AELIAN

14. Χαμαιλέων τὸ ζῷον ἐς μίαν [1] χρόαν οὐ πέφυκεν οὔτε ὁρᾶσθαι οὔτε γνωρίζεσθαι, κλέπτει δὲ ἑαυτὸν πλανῶν τε ἅμα καὶ παρατρέπων τὴν τῶν ὁρώντων ὄψιν. εἰ γὰρ περιτύχοις μέλανι τὸ εἶδος, ὁ δὲ ἐξέτρεψε τὸ μόρφωμα ἐς χλωρότητα, ὥσπερ οὖν μεταμφιεσάμενος [2]· εἶτα μέντοι ἀλλοῖος ἐφάνη γλαυκότητα [3] ὑποδύς, καθάπερ προσωπεῖον ἕτερον ἢ στολὴν ὑποκριτὴς ἄλλην. ἐπεὶ τοίνυν ταῦθ᾿ οὕτως ἔχει, φαίη τις ἂν καὶ τὴν φύσιν μὴ καθέψουσαν μηδὲ ἐπιχρίουσαν φαρμάκοις, ὥσπερ οὖν ἢ Μήδειάν τινα ἢ Κίρκην, καὶ [4] ἐκείνην φαρμακίδα εἶναι.

15. Πομπίλον [5] πελάγιον καὶ βυθῷ φιληδοῦντα εἰδέναι χρὴ μάλιστα ἰχθύων ὧν ἴσμεν ἀκοῇ· μισεῖ δὲ ἢ αὐτὸς τὴν γῆν ἢ τὸν ἰχθὺν ἐκείνη. τεμνούσας δὲ ἄρα μέσον τὸν πόρον τὰς ναῦς οἶδε οἱ πομπίλοι ὥσπερ οὖν ἐρωμένας προσνέοντες δορυφοροῦσι, καὶ δεῦρο καὶ ἐκεῖσε περιέρχονται χορεύοντες [6] ἅμα καὶ πηδῶντες. οἱ μὲν οὖν περίνεω ὁπόσον [7] ἀφεστᾶσι [8] τῆς γῆς, οὐδὲ ἓν [9] αὐτοῖς εἰδέναι πάρεστι δήπου· οἱ δὲ ναῦται, κλέπτεσθαι [10] μέντοι καὶ αὐτοῖς τὸ ἀληθὲς εἴωθεν. οἵ γε μὴν [11] πομπίλοι μακρόθεν ᾔσθοντο δίκην εὐρίνου κυνὸς τὸ θήραμα ἑλούσης τάχιστα, καὶ οὐκέτι τοσοῦτος αὐτοὺς ἔρως νεὼς καταλαμβάνει, ὡς παραμένειν, ἀλλὰ

---

[1] εἰς ἰδίαν μίαν.
[2] μεταμφιασάμενος.
[3] *Pauw* : λευκότητα.
[4] καὶ μέντοι καί.
[5] *Ges here and below* : πόμφυ-.
[6] καὶ δεῦρο . . . χορεύοντες] καὶ ἐκεῖθι περιέρχονται χορεύοντες καὶ δεῦρο.

112

14. The Chameleon is not disposed to remain of The
one and the same colour for men to see and recognise, Chameleon
but it conceals itself by misleading and deceiving the
eye of the beholder. Thus, if you come across one
that appears black, it changes its semblance to green,
as though it had changed its clothes; then again it
assumes a bluish-grey tint and appears different, like
an actor who puts on another mask or another gar-
ment. This being so, one might say that even
Nature, though she does not boil anyone down nor
apply drugs, like a Medea or a Circe, is also a
sorceress.

15. You must know that the Pilot-fish frequents the The
open sea and loves to dwell in the depths more than Pilot-fish
all others of which we have heard tell. But either it
detests the land or the land detests the fish. Well,
when vessels are cleaving the mid-ocean these Pilot-
fish swim up as though they were in love with them and
attend them like a bodyguard, circling this way and
that as they gambol and leap. Now the passengers
are of course totally unable to tell how far they
are from land, and even the sailors themselves are
frequently mistaken as to the true fact. The
Pilot-fish however can tell from a long way off,
like a keen-scented hound which immediately gets
wind of the prey, and then they are no longer so
captivated by the vessel as to stay at her side, but

---

⁷ *Jac* : πάντες ὁπόταν. ⁸ ἀφεστάναι.
⁹ οὐδέν MSS *always*. ¹⁰ *Jac* : καὶ πταίεσθαι.
¹¹ ἀλλ' οἵ γε μήν.

οἷα ὑπὸ συνθήματι ἀθρόοι γενόμενοι [1] ὤχοντο [2] ἀπιόντες. καὶ ἴσασιν ἐντεῦθεν οἱ τῆς νεὼς ἄρχοντες ὅτι ἄρα χρὴ περιβλέπειν γῆν οὐ πυρσοῖς τεκμαιρομένους ἀλλὰ τοῖς προειρημένοις πεπαιδευμένους.

16. Ἐρύθημα [3] εἴ ποτε ἐπανατέλλει καὶ ὠχρίασις [4] ἐπὶ ψιλῆς τῆς δορᾶς καὶ τριχῶν γυμνῆς, ⟨θαυμαστὸν οὐδέν⟩[5]· τάρανδος δὲ τὸ ζῷον, ἀλλὰ οὗτός γε θριξὶν αὐταῖς τρέπει ἑαυτόν, καὶ πολύχροιαν ἐργάζεται μυρίαν, ὡς ἐκπλήττειν τὴν ὄψιν. ἔστι δὲ Σκύθης, καὶ τὰ †νῶτα†[6] παραπλήσιος ταύρῳ καὶ τὸ μέγεθος. τούτου τοι καὶ τὴν δορὰν ἀγαθὴν ἀντίπαλον αἰχμῇ ταῖς αὐτῶν ἀσπίσι περιτείναντες νοοῦσιν[7] οἱ Σκύθαι.

17. Πελάγιος ἰχθὺς τὴν λῆξιν,[8] τὴν ὄψιν μέλας, τὸ μῆκος κατὰ[9] μεμετρημένην ἔγχελυν, λαβὼν[10] ἐξ ὧν δρᾷ τὸ ὄνομα, θεούσῃ νηὶ καὶ μάλα γε ἐξ οὐρίας[11] προσφθαρεὶς καὶ τῆς πρύμνης τὸ ἄκρον ἐνδακών,[12] ὥσπερ οὖν ἵππον στομίῳ ἀπειθῆ καὶ τραχὺν χαλινῷ σκληρῷ βιαιότατα ἀνακρούσας, ἀναστέλλει τῆς ὁρμῆς καὶ πεδήσας ἔχει· καὶ μάτην μὲν τὰ ἱστία μέσα πέπρησται, ἐς οὐδὲν δὲ φυσῶσιν οἱ ἄνεμοι, ἄχος δ' ἔχει τοὺς πλέοντας. συνιᾶσι δὲ οἱ ναῦται, καὶ τῆς νεὼς γνωρίζουσι τὸ

---

[1] *Reiske*: γένοιντο.
[2] πάντες.
[3] ἐρύθηματα.
[4] ὠχρίασις καὶ γενέσθαι πελιδνὸν καὶ ἀνθρώπῳ ταῦτα καὶ θηρίοις ἀλλά.
[5] ⟨θαυμαστὸν οὐδέν⟩ add. Ges.
[6] νῶτα corrupt.
[7] νοοῦσι καί.

mass as at a signal and are off and away. Thereupon those in control of the vessel know that they must look around for land, not because they judge by beacons but because they have been instructed by the aforesaid fish.

16. If at any time a flush or a pallor appears on a man's bare and hairless skin it causes no astonishment. But the animal known as *Tarandus* (elk?) transforms itself hair and all, and can adopt such an infinite variety of colours as to bewilder the eye. It is a native of Scythia and in its [hide?] [a] and its size resembles a bull; and the Scythians cover their shields with its hide and consider it a good counter to a spear. The 'Tarandus'

17. There is a fish whose province is the open sea, black in appearance, as long as an eel of moderate size, and deriving its name from what it does: with evil purpose it meets a vessel running at full speed before the wind, and fastening its teeth into the front of the prow, like a man vigorously curbing with bit and tightened rein an intractable and savage horse, it checks the vessel's onrush and holds it fast. In vain do the sails belly in the middle, to no purpose do the winds blow, and depression comes upon the passengers. But the sailors understand and realise what ails the ship; and it is from this action that the fish The Sucking-fish

[a] Perhaps 'coats,' *i.e.* summer and winter coats of hair.

---

[8] *Reiske:* νῆξιν.   [9] κατὰ τήν.   [10] λαχών.
[11] οὐρίας καὶ τῶν ἱστίων κεκολπωμένων.
[12] δακών.

πάθος. καὶ ἐντεῦθεν ἐκτήσατο τὸ ὄνομα· ἐχενηίδα
γὰρ καλοῦσιν οἱ πεπειραμένοι.

18. Ἀναβαίνει μὲν ἡ τέχνη καὶ ἐς τριγονίαν
διδασκαλίας παρ' Ὁμήρῳ [1] ἡ περὶ τῶν τετρωμένων
τε καὶ φαρμάκου δεομένων. παιδεύεται μὲν γὰρ
ὁ Μενοιτίου Πάτροκλος ὑπ' Ἀχιλλέως ἰατρικήν,
Ἀχιλλεὺς δὲ ὁ Πηλέως ὑπὸ Χείρωνος τοῦ Κρόνου.
καὶ ἐν ἥρωσί τε καὶ θεῶν παισὶν ἦν τὰ μαθήματα [2]
φύσιν εἰδέναι ῥιζῶν καὶ πόας διαφόρου χρῆσιν καὶ
φαρμάκων κρᾶσιν καὶ ἐπαοιδὰς ἔς τε φλεγμονὴν
ἀντιπάλους, καὶ ἀναστεῖλαι αἷμα, καὶ ὅσα ἄλλα
ἐκεῖνοί γε [3] ᾔδεσαν· καὶ μέντοι καὶ οἱ τοῦ χρόνου
κάτω [4] ἀνίχνευσάν τινα.[5] ἀλλὰ τούτων γε τῶν
σοφισμάτων ἡ φύσις οὐδὲν ἐδεῖτο· καὶ κατηγορεῖ
ὁ ἐλέφας. ὅταν γοῦν ἐς αὐτὸν ἔλθῃ δόρατα καὶ
βέλη πολλά, ἐλαίας πασάμενος [6] ἄνθος ἢ ἔλαιον
αὐτό, εἶτα πᾶν τὸ ἐμπεσὸν ἀπεσείσατο, καὶ ἔστιν
αὖθις ὁλόκληρος.

19. Καὶ τόδε τὸ θαῦμα τοῦδε τοῦ ζῴου ἴδιον.
τεκεῖν βρέφος οὐκ οἶδεν ἄρκτος, οὐδὲ ὁμολογήσει
τις ἐξ ὠδίνων ἰδὼν τὸ ἔκγονον ζωογόνον εἶναι
αὐτήν, ἀλλὰ ἡ μὲν ἐλοχεύθη, τὸ δὲ εἰκῇ κρέας καὶ
ἄσημόν τε καὶ ἀτύπωτον καὶ ἄμορφον. ἡ δὲ ἤδη
φιλεῖ καὶ γνωρίζει ⟨τὸ⟩[7] τέκνον, καὶ ὑπὸ τοῖς

---

[1] παρὰ τῷ Ὁ.
[2] μαθήματα ὁποῖα.
[3] Jac : τε.
[4] Jac : κάτω καὶ ἐν ἥρωσί τε καὶ θεῶν τῷ.
[5] τι.
[6] πάσσων, v.l. πάσας.
[7] ⟨τό⟩ add. H.

has acquired its name, for those who have had experience call it the *Ship-holder*.[a]

18. In Homer skill in treating the wounded and persons in need of medicine goes back as far as the third generation of pupil and master. Thus Patroclus, son of Menoetius, is taught the healing art by Achilles,[b] and Achilles, son of Peleus, is taught by Cheiron, son of Cronus. And heroes and children of the gods learnt about the nature of roots, the use of different herbs, the concocting of drugs, spells to reduce inflammations, the way to staunch blood, and everything else that they knew. And moreover there are discoveries which men of a later age have made. But that Nature really has no need of these ingenuities is proved by the case of the Elephant; for instance, when it is assailed with spears and a shower of arrows, it eats the flower of the olive [c] or the actual oil, and then shakes off every missile that has pierced it and is sound and whole again.

Medicine in the Heroic Age

The Elephant

19. [And here is another strange feature peculiar to this animal.] [d] The Bear is unable to produce a cub, nor would anyone allow, on seeing its offspring immediately after birth, that it had borne a living thing. Yet the Bear has been in labour, though the lump of nondescript flesh has no distinguishing mark, no form, and no shape. But the mother loves it and recognises it as her child, keeps it warm beneath her

The Bear and its cub

---

[a] This is the Sucking-fish or Remora; see Thompson, *Gk. fishes*, p. 70.

[b] Hom. *Il.* 11. 831.

[c] ' Unde Ael. florem oleae duxerit, nescio ' (Schneider).

[d] If these words belong here, the order of the chapters has been confused : ch. 19 should follow one on Bears.

μηροῖς θάλπει, καὶ λεαίνει τῇ γλώττῃ, καὶ ἐκτυποῖ
ἐς ἄρθρα, καὶ μέντοι καὶ κατὰ μικρὰ ἐκμορφοῖ,
καὶ ἰδὼν ἐρεῖς τοῦτο ἄρκτου σκυλάκιον.

20. Κέρατα ἀκλινῆ καὶ ὀρθὰ ἕστηκε ταύροις
ἅπασι, καὶ διὰ ταῦτα ὡς ἐς ὅπλον ὁ ἄνθρωπος,
οὕτω τοι καὶ ἐς κέρας ὁ ταῦρος τεθύμωται. βόες
δὲ Ἐρυθραῖοι κινοῦσι τὰ [1] κέρατα ὡς ὦτα.

21. Γῆ μὲν Αἰθιοπίς (γείτων δὲ καὶ μάλα
ἀγαθὸς καὶ ἀξιόζηλος τὸ τῶν θεῶν λουτρόν, ὃ
Ὅμηρος ἡμῖν Ὠκεανὸν ᾄδει) οὐκοῦν ἥδε ἡ γῆ
δρακόντων μήτηρ ἐστὶ μεγέθει μεγίστων· καὶ
γάρ τοι καὶ ἐς τριάκοντα ὀργυιὰς προήκουσι, καὶ
τὸ ὄνομα μὲν τὸ ἀπὸ γενεᾶς οὐ καλοῦνται, φονέας
δὲ ἐλεφάντων φασὶν αὐτούς, καὶ ἀμιλλῶνται πρὸς
γῆρας τὸ μήκιστον οἵδε οἱ δράκοντες. καὶ λόγοι
μὲν Αἰθίοπες ἐνταῦθά μοι ἵστανται. λέγουσι δὲ
Φρύγιοι λόγοι καὶ ἐν Φρυγίᾳ γίνεσθαι [2] δράκοντας,
καὶ προήκειν αὐτοὺς ἐς δέκα ὀργυιάς, καὶ μεσοῦντος θέρους ὁσημέραι μετὰ πλήθουσαν ἀγορὰν
ἐξέρπειν τῶν φωλεῶν. καὶ παρὰ τῷ ποταμῷ τῷ
καλουμένῳ Ῥυνδάκῳ τὸ μέν τι τῆς σπείρας
ἀπήρεισαν ἐς τὴν γῆν, τὸ λοιπὸν δὲ σῶμα ἀνα-
στήσαντες πᾶν, ἀτρέμα καὶ ἡσυχῇ τὴν φάρυγγα
ἀνατείναντες καὶ μέντοι καὶ τὸ στόμα ἀνοίξαντες,
εἶτα τὰ πτηνὰ ἕλκουσιν οἱονεὶ ἴυγγι τῷ ἄσθματι.
τὰ δὲ ἐς τὰς ἐκείνων ἐσπίπτει γαστέρας ὑπὸ τῆς
παρ' αὐτῶν ἐκπνοῆς συρόμενα αὐτοῖς πτεροῖς.

[1] Jac : καί.   [2] Schn : γένεσθαι.

[a] On the coast of Ionia opposite Chios.

thighs, smooths it with her tongue, fashions it into limbs, and little by little brings it into shape; and when you see it you would say that this is a Bear's cub.

20. All Bulls have inflexible and rigid horns, and this is why, just as a man puts passion into his weapons, so a bull puts passion into its horns. But the oxen of Erythrae [a] can move their horns as they do their ears.

The Oxen of Erythrae

21. The land of Ethiopia (the place where the gods bathe, celebrated by Homer under the name of Ocean,[b] is an excellent and desirable neighbour), this land, I say, is the mother of the very largest Serpents. For, you must know, they attain to a length of one hundred and eighty feet, and they are not called by the name of any species, but people say that they kill elephants, and these Serpents rival the longest-lived animals. Thus far the accounts from Ethiopia. But according to accounts from Phrygia there are Serpents in Phrygia too, and these grow to a length of sixty feet, and every day in midsummer some time after noon they creep out of their lairs. And on the banks of the river Rhyndacus [c] while supporting part of their coils on the ground, they raise all the rest of their body and, steadily and silently extending their neck, open their mouth and attract birds by their breath, as it were by a spell. And the birds descend, feathers and all, into their stomach, drawn in by the Serpents' breathing. And

The Snakes of Ethiopia and Phrygia

---

[b] Hom. *Il.* 1. 423.

[c] The Rhyndacus rises in mt Olympus in Mysia and flows N into the Propontis.

καὶ ταῦτα μὲν ἴδια [1] ἐκείνοις δρᾶται ἐς ἡλίου
δυσμάς· εἶτα ἑαυτοὺς οἱ δράκοντες ἀποκρύψαντες [2]
ἐλλοχῶσι τὰς ποίμνας καὶ ἐκ τῆς νομῆς ἐπὶ τὰ
αὔλια ἰούσας αἱροῦσι,[3] καὶ πολὺν φόνον [4] ἐργασάμενοι καὶ ⟨τοὺς⟩[5] νομέας συνδιέφθειραν πολλάκις,
καὶ ἔχουσι δεῖπνον ἄφθονόν τε καὶ ἀμφιλαφές.

22. Ταῖς ἀφύαις ὁ πηλὸς γένεσίς ἐστι· δι᾽
ἀλλήλων δὲ οὐ τίκτουσιν οὐδὲ ἐπιγίνονται, πηλὸς
δὲ ἐν τῇ θαλάττῃ καὶ πάνυ ἰλυώδης ὅταν συστῇ
καὶ γένηται μέλας, ἀλεαίνεταί τε [6] φύσει τινὶ
ἀπορρήτῳ τε καὶ ζωογόνῳ καὶ μεταβάλλεται καὶ
ἐς ζῷα τρέπεται πάμπολλα. καὶ αἵ γε ἀφύαι
ταῦτά ἐστι, σκωλήκων δίκην ἐν τῷ βορβόρῳ καὶ
τοῖς μυσαροῖς τικτομένων ἐκείνων. γενόμεναι δὲ
αἱ ἀφύαι νηκτικώτατον χρῆμά εἰσι,[7] καὶ δρῶσιν
ὃ πεφύκασιν, εἶτα ἄγονται τινι αἰτίᾳ θαυμαστῇ ἐς
τὰ σωτήρια, ἔνθα ἕξουσι σκέπην καὶ πρόβλημα,
ᾗ μέλλει βιώσιμα αὐταῖς ἔσεσθαι. εἴη δ᾽ ἂν ἡ
καταφυγὴ ἢ πέτρα ἀνεστῶσα ἐπὶ μέγα καὶ
ὑψηλὸν ἢ οἱ καλούμενοι κρίβανοι·[8] εἶεν δ᾽ ἂν
αὗται [9] κολπώδεις πέτραι βρωθεῖσαί [10] τε ὑπὸ τῶν
κυμάτων τῷ χρόνῳ καὶ κοῖλαι γεγενημέναι.
ταῦτα δὲ ἄρα αὐταῖς ἡ φύσις ἔδειξε κρησφύγετα,
ὥστε ὑπὸ τοῦ σάλου μὴ παίεσθαι μηδὲ ἀφανίζεσθαι·
ἀσθενεῖς γάρ εἰσι καὶ ἥκιστα πρὸς τὰς ἐκείνων
ἐμπτώσεις ἀντίτυποι. τροφῆς δὲ δέονται οὐδὲ
ἕν, ἀπόχρη γε μὴν ἀλλήλας περιλιχμήσασθαι.
ἄγρα δὲ αὐτῶν νήματα ἄγαν λεπτὰ καὶ ἐρραφέντα

---

[1] ἰδίᾳ.
[2] ὑποκρύψαντες.
[3] Wytt : αἴρουσιν.
[4] φθόρον.
[5] ⟨τοὺς⟩ add. H.
[6] τε ὑφ᾽ ἡδονῆς ἑαυτοῦ.

these singular practices they continue until sun-
down; next, the Serpents hide and lie in wait for
the flocks, and as they return to the sheepfolds from
the pasture they fall upon them, and after a terrible
slaughter they have frequently killed the herdsmen
as well, thus obtaining a generous and abundant
feast.

22. Sprats are born of mud; they neither beget The Sprat
nor are begotten of one another, but when the mud
in the sea becomes altogether slimy and thick, and
turns black, it is warmed by some inexplicable and
life-giving principle, undergoes a transformation, and
is changed into innumerable living creatures. The
Sprats are these creatures, resembling worms which
are generated in mire and filth. And as soon as born,
Sprats are excellent swimmers, and they do it naturally. Then by some mysterious agency they are led
to safe places where they will find shelter and protection, so that it will be possible for them to live.
And their place of refuge is likely to be either some
rock that rises to a great height or what are called
' baker's pots '; these would be rocks full of embrasures which the waves have in time eaten away
until they have become hollow. These then are the
retreats to which Nature has pointed them so that
they shall not be battered and demolished by the
swell of the sea; for they have little strength and
are powerless to resist the impact of the waves.
They need no food, indeed it is enough for them to
lick one another. The way to catch them is to use
exceedingly fine thread with thin pieces from the

---

7 ἐστι.

9 ἂν αἱ.

8 κρίβανοι ὑπὸ τῶν ἁλιέων.

10 Reiske: βρίθουσαι.

τούτοις ἀραιὰ στημόνια τῶν ἱματίων.[1] καὶ τέχνημα
μὲν εἴη ἂν[2] τοῦτο καὶ μάλα γε ἀρκοῦν[3] ἐς αἵρεσίν
τε καὶ ἅλωσιν αὐτῶν, ἐς δὲ ἄλλων ἰχθύων θήραν
ἥκιστα.

23. Τὸν σαῦρον εἰ παίσας[4] εἴτε ἑκὼν εἴτε καὶ
κατὰ τύχην ῥάβδῳ μέσον διατέμοις, οὐδέτερον[5]
αὐτῷ τῶν μερῶν ἀποτέθνηκεν, ἀλλὰ χωρὶς καὶ
καθ᾽ ἑαυτὸ πρόεισί τε καὶ ζῇ δύο ποσὶν ἐπισυρόμε-
νον τὸ ἡμίτομον[6] καὶ ἐκεῖνο καὶ τοῦτο. εἶτα
ὅταν συνέλθῃ (σύνεισι γὰρ πρὸς τὸ λεῖπον τὸ
ἕτερον πολλάκις), συνδυασθέντε συνηλθέτην ἐκ
τῆς διαιρέσεως· καὶ ἑνωθεὶς ὁ σαῦρος, τοῦ μὲν
πάθους τὸ ἴχνος αὐτῷ κατηγορεῖ ἡ οὐλή, περιθέων
δὲ καὶ τὴν ἀρχαίαν βιοτὴν ἔχων ἔοικε τῶν προειρημένων μὴ πεπειραμένῳ.

24. Ἰὸς μὲν ὁ τῶν ἑρπετῶν δεινός ἐστι, καὶ ὅ
γε τῆς ἀσπίδος ἔτι μᾶλλον. καὶ τούτου[7] ἀντίπαλα
καὶ ἀμυντήρια ῥᾳδίως οὐκ ἂν εὕροι τις, εἰ καὶ
σοφώτατος εἴη κηλεῖν τε ὀδύνας καὶ ἀφανίζειν.
Ἦν δὲ ἄρα καὶ ἐν ἀνθρώπῳ τις ἰὸς ἀπόρρητος,
καὶ πεφώραται τὸν τρόπον ἐκεῖνον. ἔχιν εἰ λάβοις, καὶ πάνυ εὐλαβῶς τε καὶ ἐγκρατῶς τοῦ
τραχήλου κατάσχοις, καὶ διαστήσας τὸ στόμα
εἶτα αὐτῷ[8] προσπτύσειας, ἐς τὴν νηδὺν κατολισθάνει τὸ πτύαλον, καὶ γίνεταί οἱ τοσοῦτον κακὸν
ὡς σήπειν τὸν ἔχιν. ἔνθεν[9] τοι καὶ ἀνθρώπῳ

---

[1] Gow : ἀραιῶν στημονίων τὰ ἱμάτια MSS, ἐρεῶν Bernhardy,
τιλμάτια Haupt.
[2] ταύταις.
[3] ἀρκοῦν μηχάνημα.

warp of garments laced in. This device should be quite sufficient for catching and securing them, though for the capture of other fish it would be utterly inadequate.

23. Should you strike a Lizard with a stick and either on purpose or by accident cut it in two, neither of the two parts is killed, but each moves separately and by itself, and lives, both the one and the other trailing on two feet. Then when the parts meet—for the forepart frequently unites with the hinder—the two join up and coalesce after their separation. And the Lizard, now one body, although a scar gives evidence of what it has suffered, yet runs about and maintains its former method of life exactly like one of its kind that has had no such experience.

The Lizard, its vitality

24. The poison of serpents is a thing to be dreaded, but that of the Asp is far worse. Nor are remedies and antidotes easy to discover, however ingenious one may be at beguiling and dispelling acute pains. Yet after all there is in man also a certain mysterious poison, and this is how it has been discovered. If you capture a Viper and grasp its neck very firmly and with a strong hand, and then open its mouth and spit into it, the spittle slides down into its belly and has so disastrous an effect upon it as to cause the Viper to rot away. From this you see how foul can

The Asp, its poison

Human spittle

---

4 παίσας κατὰ τοῦ βρέγματος *most* MSS, π. κατὰ τὸ μέσον V.

5 οὐθέτερον.

6 ἡμίτομον τῶν ζῴων.

7 ἐπὶ τούτων.

8 ἐπ' αὐτῷ.   9 ὅθεν.

δῆγμα ἀνθρώπου μιαρόν ἐστι καὶ κινδυνῶδες
οὐδενὸς θηρίου μεῖον.

25. Ἐν ὥρᾳ θερείῳ,[1] ἀμητοῦ κατειληφότος καὶ
τῶν σταχύων τριβομένων ἐν τῷ δίνῳ, κατὰ ἴλας
συνίασιν οἱ μύρμηκες, καθ᾽ ἕνα ἰόντες καὶ κατὰ
δύο δέ, ἀλλὰ καὶ ἐς τὸν τρίτον στοῖχον[2] ἔρχονται,
τοὺς ἑαυτῶν οἴκους καὶ τὰς συνήθεις στέγας ἀπο-
λείποντες· εἶτα ἐκλέγουσι τῶν πυρῶν καὶ τῶν
κριθῶν, καὶ τὴν αὐτὴν χωροῦσιν ἀτραπόν. καὶ
οἱ μὲν ἀπίασιν ἐπὶ τὴν τῶν προειρημένων συλ-
λογήν, οἱ δὲ κομίζουσι τὸν φόρτον, καὶ πάνυ
αἰδεσίμως καὶ πεφεισμένως ἀλλήλοις ὑπαφίσταν-
ται[3] τῆς ὁδοῦ, καὶ μᾶλλον τοῖς ἀχθοφόροις οἱ
κοῦφοι· κατελθόντες δὲ ἐς τὰ οἰκεῖα τὰ σφέτερα[4]
καὶ πληρώσαντες τοὺς ἐν τῷ μυχῷ σφίσι σιρούς,[5]
ἑκάστου σπέρματος διατρήσαντες τὸ μέσον, τὸ
μὲν ἐκπεσὸν δεῖπνον γίνεται τῷ μύρμηκι ἐν τῷ
τέως, τὸ δὲ λοιπὸν ἄγονόν ἐστι. παλαμῶνται δὲ
ἄρα οἱ γενναῖοι οἰκονόμοι καὶ φρουροὶ τοῦτο, ἵνα
μὴ τῶν ὄμβρων περιρρευσάντων, εἶτα ἔκφυσιν
ὁλόκληρα ἐκεῖνα ὄντα λάβῃ τινὰ καὶ ἀναθήλῃ, καὶ
τούτων γενομένων ἀτροφίᾳ καὶ λιμῷ διὰ χειμῶνος
περιπέσωσι, καὶ αὐτοῖς ἐξαμβλώσῃ ἡ σπουδή.
φύσεως μὲν δὴ καὶ μύρμηκες λαβεῖν δῶρα εὐτύχη-
σαν καὶ ταῦτα ὡς ἄλλα.

26. Οὐδέποτε ἀετὸς οὔτε πηγῆς δεῖται οὔτε
γλίχεται κονίστρας, ἀλλὰ καὶ δίψους ἀμείνων
ἐστί, καὶ καμάτου φάρμακον οὐκ ἀναμένει πορι-

---

[1] θερείῳ περὶ τὰς ἅλως.   [2] τὸ . . . στοιχεῖον.

be the bite of one man to another and as dangerous as the bite of any beast.

25. In the summertime when the harvest is in and the corn is being threshed on the threshing-floor, Ants assemble in companies, going in single file or two abreast—indeed they sometimes go three abreast—after quitting their homes and customary shelters. Then they pick out some of the barley and the wheat and all follow the same track. And some go to collect the grain, others carry the load, and they get out of each other's way with the utmost deference and consideration, especially those that are not laden for the benefit of those that are. Then they return to their dwellings and fill the pits in their store-chamber after boring through the middle of each grain. What falls out becomes the Ant's meal at the time; what is left is infertile. This is a device on the part of these excellent and thrifty housekeepers to prevent the intact grain from putting out shoots and sprouting afresh when the rains have surrounded them, and to preserve themselves in that case from falling victims during the winter to want of food and to famine, and their zeal from being blunted. It is to Nature then that Ants too owe these and other fortunate gifts.

26. At no time does the Eagle need water or long for a dusting-place; he is on the contrary superior to thirst and looks for no medicine for weariness from

The Ant

The Eagle

---

³ ἀφίστανται H.

⁴ σφέτερα οἱ γενναῖοι.

⁵ Jac: σιροὺς πυρῶν τε καὶ κριθῶν.

AELIAN

σθὲν ἔξωθεν, ὑπερφρονῶν δὲ καὶ τῶν ὑδάτων καὶ
τῆς ἀναπαύσεως τὸν αἰθέριον τέμνει πόλον,[1] καὶ
ὀξύτατα ὁρᾷ ἐκ πολλοῦ τοῦ αἰθέρος καὶ ὑψηλοῦ.
καὶ τόν γε τῶν πτερῶν αὐτοῦ ῥοῖζον καὶ τὸ τῶν
θηρίων ἀτρεπτότατον ὁ δράκων ἀκούσας μόνον
παραχρῆμα[2] κατέδυ καὶ ἀσμένως ἠφανίσθη.
βάσανος δέ οἱ τῶν νεοττῶν τῶν γνησίων ἐκείνη
ἐστίν. ἀντίους τῇ αὐγῇ τοῦ ἡλίου ἵστησιν αὐτοὺς
ὑγροὺς[3] ἔτι καὶ ἀπτῆνας· καὶ ἐὰν μὲν σκαρδαμύξῃ
τις τὴν ἀκμὴν τῆς ἀκτῖνος δυσωπούμενος, ἐξεώσθη
τῆς καλιᾶς, καὶ ἀπεκρίθη τῆσδε τῆς ἑστίας· ἐὰν
δὲ ἀντιβλέψῃ καὶ μάλα ἀτρέπτως, ἀμείνων ἐστὶν
ὑπονοίας καὶ τοῖς γνησίοις ἐγγέγραπται, ἐπεὶ
αὐτῷ πῦρ τὸ οὐράνιον ἡ τοῦ γένους ἀδέκαστός τε
καὶ ἄπρατος[4] ἀληθῶς ἐστιν ἐγγραφή.

27. Ἡ στρουθὸς ἡ μεγάλη λασίοις μὲν τοῖς
πτεροῖς ἐπτέρωται, ἀρθῆναι δὲ καὶ ἐς βαθὺν ἀέρα
μετεωρισθῆναι φύσιν οὐκ ἔχει. θεῖ δὲ ὤκιστα,
καὶ τὰς παρὰ τὴν πλευρὰν ἑκατέραν πτέρυγας
ἁπλοῖ, καὶ ἐμπῖπτον τὸ πνεῦμα κολποῖ δίκην
ἱστίων αὐτάς.[5]

28. Τὴν ὠτίδα ⟨τὸ⟩[6] ζῷον ὀρνίθων εἶναι φιλιπ-
πότατον ἀκούω. καὶ τὸ[7] μαρτύριον, τῶν μὲν
ἄλλων ζῴων καὶ ἐν λειμῶσι καὶ ἐν αὐλῶσι νεμομένων καταφρονεῖ· ἵππον δὲ ὅταν θεάσηται, ἥδιστα
προσπέτεται καὶ πλησιάζει κατὰ τοὺς τῶν ἀνθρώπων ἱππεραστάς.

---

[1] τὸν ἀέρα τέμνει πολύν.
[2] καὶ παραχρῆμα.
[3] Jac : ἀργούς.
[4] Pauw : ἄγραπτος.

any outside source, but scorning water and repose he cleaves the atmosphere and gazes with piercing eye from the vast expanse of heaven on high. And at the mere sound of those rushing wings even that most intrepid of all creatures, the great serpent, dives at once into its den and is glad to disappear. And this is the way in which the Eagle tests the legitimacy of his young ones. He plants them, while they are still tender and unfledged, facing the rays of the sun, and if one of them blinks, unable to endure the brightness of the rays, it is thrust out of the nest and banished from that hearth. If however it can face the sun quite unmoved, it is above suspicion and is enrolled among the legitimate offspring, since the celestial fire is an impartial and uncorrupt register of its origin.

27. The Ostrich is covered with thick feathers, but its nature does not permit it to rise from the ground and mount aloft into the sky. Yet its speed is very great, and when it spreads its wings on either side, the wind meeting them causes them to belly like sails. The Ostrich

28. Among birds the Bustard is, I am told, the most fond of horses. And the proof of this is that it scorns all other animals that live in field or glen, but that when it catches sight of a horse, it delights to fly up to it and to keep it company, just like men who are devoted to horses. The Bustard

---

⁵ αὐτάς, πτῆσιν δὲ οὐκ οἶδεν.
⁶ ⟨τό⟩ add. H.
⁷ τούτου.

29. Μυῖα ἐμπεσοῦσα ἐς ὕδωρ, εἰ καὶ [1] ζώων ἐστὶ θρασυτάτη, ἀλλὰ γοῦν οὔτ᾽ ἐπιτρέχει,[2] οὔτε νηκτική ἐστι, καὶ διὰ ταῦτα ἀποπνίγεται. εἰ δὲ αὐτῆς ἐξέλοις τὸν νεκρόν, καὶ τέφραν ἐμπάσειας καὶ καταθείης [3] ἐν ἡλίου αὐγῇ, ἀναβιώσῃ τὴν μυῖαν.

30. Ἀλεκτρυόνα εἴτε πριάμενος εἴτε δῶρον λαβὼν ἐς τὴν ἀγέλην τὴν σεαυτοῦ καὶ τοὺς ὄρνιθας τοὺς ἠθάδας ἐθέλοις ἀριθμεῖν,[4] οὐκ ἀπολύσεις οὐδὲ ἀφήσεις εἰκῇ καὶ ὡς ἔτυχεν αὐτόν· εἰ δὲ μή, φυγὰς παραχρῆμα οἰχήσεται ἐς τοὺς οἰκείους καὶ τοὺς συννόμους, εἰ καὶ πάνυ πόρρωθεν εἴη οὗτος. δεῖ δὲ ἄρα αὐτῷ φρουρὰν περιβαλεῖν καὶ δεσμὰ ἀφανῆ ὑπὲρ τὰ Ἡφαίστου τὰ Ὁμήρεια. καὶ ὅ γε λέγω τοιοῦτόν ἐστι. τράπεζαν ἐφ᾽ ἧς ἐσθίεις ἐς μέσον καταθεὶς καὶ τὸν ὄρνιθα λαβὼν καὶ τρὶς αὐτὸν τὴν προειρημένην σκηνὴν περιαγαγών, μέθες τὸ ἐντεῦθεν ἄφετον ἀλᾶσθαι σὺν τοῖς ὄρνισι τοῖς οἰκέταις· ὁ δὲ οὐκ ἀπαλλάττεται, ὥσπερ οὖν πεπεδημένος.

31. Ἡ σαλαμάνδρα τὸ ζῷον οὐκ ἔστι μὲν τῶν πυρὸς ἐκγόνων,[5] ὥσπερ οὖν οἱ καλούμενοι πυρίγονοι, θαρρεῖ δὲ αὐτὸ καὶ χωρεῖ τῇ φλογὶ ὁμόσε, καὶ ὡς ἀντίπαλόν τινα σπεύδει καταγωνίσασθαι. καὶ τὸ μαρτύριον,[6] περὶ [7] τοὺς βαναύσους καλινδεῖται καὶ τοὺς χειρώνακτας τοὺς ἐμπύρους. ἐς ὅσον μὲν οὖν ἐνακμάζει τὸ πῦρ αὐτοῖς, καὶ συνερ-

---

[1] εἰ καί] καὶ γὰρ εἰ.   [2] ἀντέχει.
[3] καταθήσεις.   [4] ἀριθμεῖν καὶ ἔχειν.

29. When a Fly falls into the water, though it is The Fly of all creatures the most daring, yet it can neither run upon the surface nor swim, and hence it drowns. If however you pick out the dead body, sprinkle ashes upon it, and place it in the sunshine, you will bring the Fly to life again.

30. If you want to add a Cockerel, whether bought The Cockerel or presented, to your flock of domestic fowls, you must not release him nor let him loose at random and in a casual way; otherwise he will immediately desert and go back to his own kin and mates, however far away from them he be. So you must set upon him a guard and fetters more invisible than those of Hephaestus in Homer [*Od.* 8. 274–]. What I prescribe is this. Place the table at which you eat, in the open, seize the Cockerel, and when you have taken him three times round the aforesaid platform, then let him go free to wander with the fowls of the house. He will not go away any more than if he were chained up.

31. The Salamander is not indeed one of those The Sala-mander fire-born creatures like the so-called 'Fire-flies,'[a] yet it is as bold as they and encounters the flame and is eager to fight it like an enemy. And the proof of this is as follows. Its haunts are among artisans and craftsmen who work at the forge. Now so long as their fire is at full blast and they have it to help

---

[a] See ch. 2.

---

[5] ἐκγόνων οὐδὲ ἐξ αὐτοῦ τίκτεται.
[6] καὶ τούτου τὰ μαρτύρια.
[7] παρά.

γὸν [1] τῇ τέχνῃ ἔχουσιν αὐτὸ καὶ κοινωνὸν τῆς
σοφίας, ὑπὲρ τοῦδε τοῦ ζῴου οὐδὲ ἓν φροντίζουσιν·
ὅταν δὲ τὸ μὲν ἀποσβεσθῇ καὶ μαρανθῇ, μάτην δὲ
αἱ φῦσαι καταπνέωσιν, ἐνταῦθα ἤδη τὸ ζῷον τὸ
εἰρημένον ἀντιπρᾶττόν σφισιν ἴσασι καλῶς. ἀνιχ-
νεύσαντες οὖν τὸ θηρίον καὶ τιμωρησάμενοι, τὸ
πῦρ ἐντεῦθεν αὐτοῖς ἐξάπτεται, καὶ ἔστιν εὐπειθές,
καὶ οὐ σβέννυται τῇ συνηθείᾳ τρεφόμενον.

32. Κύκνος δέ, ὅνπερ οὖν καὶ θεράποντα Ἀπόλ-
λωνι ἔδοσαν ποιηταὶ καὶ λόγοι μέτρων ἀφειμένοι
πολλοί, τὰ μὲν ἄλλα ὅπως μούσης τε καὶ ᾠδῆς
ἔχει εἰπεῖν οὐκ οἶδα· πεπίστευται δὲ ὑπὸ τῶν
ἄνω τοῦ χρόνου ὅτι τὸ κύκνειον οὕτω καλούμενον
ᾄσας εἶτα ἀποθνήσκει. τιμᾷ δὲ ἄρα αὐτὸν ἡ
φύσις καὶ τῶν καλῶν καὶ ἀγαθῶν ἀνθρώπων
μᾶλλον, καὶ εἰκότως· εἴ γε τούτους μὲν καὶ
ἐπαινοῦσι καὶ θρηνοῦσιν ἄλλοι, ἐκεῖνοι δὲ εἴτε
τοῦτο ἐθέλοις εἴτε ἐκεῖνο, ἑαυτοῖς νέμουσιν.

33. Κροκόδιλος μὲν ὅπως ἔχει μεγέθους καὶ ὁ
τέλειος καὶ ὁ ἐκγλυφεὶς πρῶτον, καὶ μέντοι καὶ
γλώττης ὅπως, καὶ εἰ κινεῖ [2] τὴν γένυν, καὶ
ποτέραν τῇ ἑτέρᾳ προσάγει, πολλοὶ λέγουσι.
κατέγνωσαν δὲ ἄρα τοῦ ζῴου τοῦδέ τινες ὅτι
τίκτει [3] τοσαῦτα ᾠὰ ὅσαις ἂν [4] ἡμέραις ἐπῴαζον [5]
εἶτα ἐκγλύψῃ τὰ νεόττια· ἤδη δὲ ἔγωγε ἤκουσα,
ὁ κροκόδιλος ὅταν ἀποθάνῃ,[6] σκορπίον ἐξ αὐτοῦ
τίκτεσθαι, κέντρον δὲ ἄρα οὐραῖον αὐτὸν ἔχειν
λέγουσιν ἰοῦ πεπληρωμένον.

---

[1] συνεργὸν αὐτοῖς.   [2] εἰ κινεῖ] Reiske : ἐπικινεῖ.

their craft and to share their skill, they pay not the smallest attention to this animal. When however the fire goes out or languishes and the bellows blow in vain, then at once they know full well that the aforesaid creature is working against them. Accordingly they track it down and exact vengeance; and then the fire is lit, is easily coaxed up, and does not go out, provided it is kept fed with the usual material.

32. The Swan is assigned by poets and many prose-writers as servant to Apollo, but in what other relation it stands to music and song I do not know. Yet the ancients believed that when it has sung what is called its 'swan-song,' it dies. In that case Nature honours it more highly than it does noble and upright men, and rightly so, for while others praise and lament them, Swans praise or, if you will, lament themselves. *The Swan and its song*

33. Many writers tell us about the size of the Crocodile both when fully grown and when first hatched, and further, about its tongue, and whether it moves its jaw and which jaw it closes upon the other. There are those too who have observed that this animal lays as many eggs as the days during which it sits upon them before hatching out its young. And I have myself heard that when a Crocodile dies a scorpion is born from it; and they do say that it has a sting in its tail which is full of poison. *The Crocodile*

---

³ τίκτει μέν.   ⁴ ἂν καί.
⁵ ἐπῴάζουσιν ὄρνεις.   ⁶ *Jac*: ὅπως ἂν ἀποθάνοι.

34. Εἰ σαφῆ ταῦτα καὶ μὴ ἀμφίλογα, Ἰνδῶν λόγοι πειθέτωσαν· ἃ δὲ νῦν ἐρῶ, τῆς ἐκεῖθεν φήμης διακομιζούσης, ταῦτά ἐστιν. ὁμώνυμον τῷ φυτῷ κιννάμωμον ὄρνιν ἔγωγε τοῦ παιδὸς τοῦ Νικομάχου λέγοντος ἤκουσα. καὶ τὸν μὲν ὄρνιν κομίζειν [1] τὸ φερώνυμον τοῦτο δὴ φυτὸν [2] ἐς Ἰνδούς, εἰδέναι δὲ ἄρα τοὺς ἀνθρώπους ὅπου τε [3] καὶ ὅπως φύεται οὐδὲ ἕν.

35. Αἰγύπτιοι κλύσματα καὶ κάθαρσιν γαστρὸς οὐκ ἔκ τινος ἐπινοίας ἀνθρωπίνης λέγουσι μαθεῖν, διδάσκαλον δέ σφισι τοῦ ἰάματος τοῦδε τὴν ἶβιν ᾄδουσιν. καὶ ὅπως ἐξεπαίδευσε τοὺς πρώτους ἰδόντας, ἐρεῖ ἄλλος· σελήνης δὲ αὔξησιν καὶ μείωσιν ὅτι οἶδε, καὶ τοῦτο ἤκουσα. καὶ ὅτι τὴν τροφὴν ἑαυτῇ ὑφαιρεῖ καὶ προστίθησι κατὰ τὴν τῆς θεοῦ καὶ λῆξιν καὶ πρόσθεσιν, πυθέσθαι ποθὲν οὐκ εἰμὶ ἔξαρνος.

36. Κέντρον πικρότατον καὶ κίνδυνον φέρον ἁπάντων μᾶλλον ἢ τρυγὼν ἢ ἐκ τῆς θαλάττης ἔχει. καὶ τὸ μαρτύριον, εἰ μὲν ἐς δένδρον τεθηλὸς καὶ εὖ μάλα ἀναθέον ἐμπήξειας αὐτό, οὔτε ἐς ἀναβολὰς οὔτε χρόνῳ ὕστερον ἀλλ' ἤδη αὖον τὸ δένδρον· εἰ δέ τι τῶν ζῴων ἀμύξειας, ἀπέκτεινας.

37. Ἡ μυγαλῆ [4] ἐς ὅσον μὲν τὴν ἄλλως πρόεισι, ζῆν ἔχει, καὶ ἐσπείσατο αὐτῇ ἡ φύσις, ἐάν γε μὴ ἄλλῃ τινὶ τύχῃ καταληφθῇ καὶ ἀπόληται· ἐπὰν

---

[1] Bernhardy : κομίζειν ἐντεῦθεν.
[2] τὸ φυτόν.   [3] Reiske : γε.
[4] μυγαλῆ καὶ γὰρ τοῦτο ποίημα ὕλης.

34. If these facts are certain and beyond dispute, The Cinnamon bird
then let this story from India carry conviction.
What I propose to tell has been brought from thence
by report and is as follows. I have learnt from the
son of Nicomachus [Arist. *HA* 616 a 6] that there is a
bird named *Cinnamon* like the plant, and that the
bird brings this plant, which is named after it, to the
Indians, but that these people have no knowledge
where and how the plant grows.[a]

35. The Egyptians assert that a knowledge of The Ibis and clysters
clysters and intestinal purges is derived from no dis-
covery of man's, but they commonly affirm that it
was the Ibis that taught them this remedy. And
how it instructed those who were the first to see it,
some other shall tell. And I have also heard that it
knows when the moon is waxing and when waning;
and I cannot deny that I have learnt from some source
that it diminishes or increases its food according as
the goddess herself diminishes or increases.

36. The Sting-ray in the sea has a far fiercer and The Sting-ray
more dangerous sting than all other creatures. The
proof is that if you fix it in a flourishing tree that has
grown to a great height, then without any delay,
before any time has elapsed, the tree immediately
withers. And if you allow the sting to scratch any
living creature, you kill it at once.

37. So long as the Shrew-mouse proceeds as chance The Shrew-mouse
directs, it can live, and Nature is on friendly terms
with it, unless it is overtaken by misfortune from

[a] See 17. 21.

δὲ ἐς ἁρματοτροχιὰν ἐμπέσῃ, οἱονεὶ πέδῃ κατεί-
ληπται καὶ μάλα ἀφανεῖ, καὶ τέθνηκε. δηχθέντι
δὲ ὑπὸ μυγαλῆς φάρμακον ἐκεῖνο. ἐκ τῆς τῶν
τροχῶν διαδρομῆς ἡ ψάμμος ἀρθεῖσα ἐπεπάσθη
τῷ δήγματι, καὶ ἔσωσε παραχρῆμα.

38. Καὶ ταῦτα δὲ ὑπὲρ τῆς Αἰγυπτίας ἴβεως
προσακήκοα. ἱερὰ τῆς σελήνης ἡ ὄρνις ἐστί.
τοσούτων γοῦν [1] ἡμερῶν τὰ ᾠὰ ἐκγλύφει, ὅσων
ἡ θεὸς αὔξει τε καὶ λήγει. τῆς δὲ Αἰγύπτου
οὔποτε ἀποδημεῖ. τὸ δὲ αἴτιον, νοτιωτάτη χωρῶν
ἁπασῶν Αἴγυπτός ἐστι, καὶ ἡ σελήνη δὲ νοτιωτάτη
τῶν πλανωμένων ἄστρων πεπίστευται. ἑκοῦσα
μὲν οὖν οὐκ ἂν ἀποδημήσειεν ἡ ἶβις· εἰ δέ τις
ἐπιθέμενος αὐτῇ κατὰ τὸ καρτερὸν ἐξαγάγοι, ἡ δὲ
ἀμύνεται τὸν ἐπιβουλεύσαντα, ἐς οὐδὲν αὐτῷ τὴν
σπουδὴν προάγουσα·[2] ἑαυτὴν γὰρ ἀποκτείνει
λιμῷ, καὶ ἀνόνητον τὴν προθυμίαν ἀποφαίνει τῷ
προειρημένῳ. βαδίζει δὲ ἡσυχῇ καὶ κορικῶς, καὶ
οὐκ ἂν αὐτὴν θᾶττον ἢ βάδην προϊοῦσαν θεάσαιτό
τις. καὶ τούτων αἱ μέλαιναι τοὺς πτερωτοὺς
ὄφεις ἐξ Ἀραβίας ἐς Αἴγυπτον παρελθεῖν οὐκ
ἐπιτρέπουσι, τῆς γῆς τῆς φίλης προπολεμοῦσαι·
αἱ δὲ ἕτεραι τοὺς ἐξ Αἰθιοπίας κατὰ τὴν τοῦ
Νείλου ἐπίκλυσιν ἀφικνουμένους ἀπαντῶσαι δια-
φθείρουσιν. ἢ τί ἂν ἐκώλυσε διὰ τῆς ἐκείνων
ἐπιδημίας τοὺς Αἰγυπτίους ἀπολωλέναι;

39. Ἀκούω δέ τι καὶ γένος ἀετῶν, καὶ ὄνομα
αὐτῷ χρυσάετον ἔθεντο, ἄλλοι δὲ ἀστερίαν τὸν

---

[1] Reiske : οὖν.   [2] προαγαγοῦσα.

some other quarter and is killed. When however it falls into a rut, it is caught, so to say, in quite invisible fetters and dies. The remedy for a man who has been bitten by a Shrew-mouse is as follows. Take some sand from the wheel-track, sprinkle it on the bite, and it cures him immediately.

38. Here is another story relating to the Egyptian The Ibis Ibis which I have heard. The bird is sacred to the moon. At any rate it hatches its eggs in the same number of days that the goddess takes to wax and to wane, and never leaves Egypt. The reason for this is that Egypt is the moistest of all countries and the moon is believed to be the moistest of all planets. Of its own free will the Ibis would never quit Egypt, and should some man lay hands upon it and forcibly export it, it will defend itself against its assailant and bring all his labour to nothing, for it will starve itself to death and render its captor's exertions vain. It walks quietly like a maiden, and one would never see it moving at anything faster than a foot's pace. The Black Ibis does not permit the winged serpents from Arabia to cross into Egypt, but fights to protect the land it loves, while the other kind encounters the serpents that come down the Nile when in flood and destroys them. Otherwise there would have been nothing to prevent the Egyptians from being killed by their coming.

39. There is, I am told, a species of eagle to which The Golden men have given the name of ' Golden Eagle,' though Eagle

αὐτὸν καλοῦσιν· ὁρᾶται δὲ οὐ πολλάκις. λέγει
δὲ Ἀριστοτέλης αὐτὸν θηρᾶν καὶ νεβροὺς καὶ
λαγὼς καὶ γεράνους καὶ χῆνας ἐξ αὐλῆς. μέγιστος
δὲ ἀετῶν εἶναι πεπίστευται, καὶ λέγουσί γε [1] καὶ
ταύροις ἐπιτίθεσθαι αὐτὸν κατὰ τὸ καρτερόν, καὶ
περιηγοῦνται τὸ ἔργον τὸν τρόπον τοῦτον. ὁ μὲν
κεκυφὼς κάτω νέμεται ὁ ταῦρος· ὁ δὲ ἀετὸς ἐπὶ
τῷ τένοντι τοῦ ζῴου καθίσας ἑαυτὸν παίει τῷ
στόματι συνεχέσι τε καὶ καρτεραῖς ταῖς πληγαῖς·
ὁ δὲ ὥσπερ οἰστρηθεὶς ἐξάπτεται, καὶ ᾗ ποδῶν
ἔχει φυγῆς ἄρχεται. καὶ ἕως μέν ἐστιν εὐήλατα,
ὁ ἀετὸς ἥσυχός ἐστι καὶ ἐπιποτᾶται παραφυλάτ-
των· ὅταν δὲ τὸν ταῦρον θεάσηται πλησίον
κρημνοῦ γεγενημένα, κυκλώσας τὰ πτερὰ καὶ
ὑπερτείνας αὐτοῦ τῶν ὀφθαλμῶν, ἐποίησε τὰ ἐν
ποσὶ μὴ προϊδόμενον [2] κατενεχθῆναι βιαιότατα.
εἶτα ἐμπεσὼν καὶ ἀναρρήξας τὴν γαστέρα, ῥᾳδίως
χρῆται τῇ ἄγρᾳ, ἐς ὅσον ἐθέλει. θήρας δὲ ἀλλοτρίας οὐχ ἅπτεται κειμένης, ἀλλὰ χαίρει τοῖς
ἑαυτοῦ πόνοις, κοινωνίαν τε τὴν πρὸς ἄλλον ἥκιστα
ἐνδέχεται. κορεσθεὶς δὲ εἶτα τοῦ λοιποῦ πονηρὸν
ἄσθμα καὶ δυσωδέστατον καταπνεύσας, ἄβρωτα
τοῖς ἄλλοις τὰ λείψανα ἐᾷ. καὶ μέντοι καὶ ἀλλήλων ἀπῳκισμένας οἰκοῦσι καλιὰς ὑπὲρ τοῦ μὴ
διαφέρεσθαι ὑπὲρ θήρας [καὶ λυπουμένους λυπεῖν
πολλάκις].[3]

40. Ἦν ⟨δὲ⟩[4] ἄρα γένος ἀετῶν καὶ πρὸς τοὺς
τρέφοντας φιλόστοργον, ὥσπερ οὖν καὶ ὁ τοῦ
Πύρρου. τοῦτόν τοί φασι καὶ ἐπαποθανεῖν [5] τῷ

---

[1] γε εἰς τοὺς κρῆτας.   [2] G. Hoffmann : προειδ-.
[3] [καὶ . . . πολλάκις] del. H.

others call it *Asterias* (starred). And it is seldom
seen. Aristotle says[a] that it hunts fawns, hares,
cranes, and geese of the farmyard. It is believed to
be the largest of eagles; at any rate men say that it
attacks bulls with violence, and its method of attack
they describe as follows. The bull is feeding with
his head down, and the Eagle alights upon his neck its method
and with its beak delivers a rain of powerful blows. of attacking
bulls
And the bull goes wild as though stung by a gadfly,
and sets off to run as fast as he can go. So long
as the land makes going easy the Eagle bides its
time, flying above him and watching. But directly
it sees the bull near a precipice it makes an arch with
its wings, covers the bull's eyes so that he cannot see
what is before him, and down he goes with a fearful
crash. Whereupon the Eagle pounces, rips open his
stomach, and has no difficulty in enjoying its prey to
its heart's content. But the prey killed by some
other creature it will not touch: rather it delights
in its own labours and will not for one moment admit
any other creature to share them. Later when it
has gorged itself, it breathes over the rest of the
carcase a foul and most ill-smelling air, leaving the
remains unfit for any other animal to eat. What is
more, Eagles build their nests far apart from one
another so as to avoid quarrelling over their prey [and
being a constant source of mutual hurt].

40. It seems that Eagles are full of affection even The Eagle,
towards their keepers; witness the Eagle that to its keeper
belonged to Pyrrhus, which (they say) on the death

its devotion

[a] The passage is not to be found in his extant works.

---

[4] ⟨δέ⟩ add. H.　　　　　[5] *Jac*: ἐναποθανεῖν.

δεσπότῃ τροφῆς ἀποστάντα. ἤδη δὲ καὶ ἀνδρὸς
ἰδιώτου ἀετὸς τρόφιμος καομένου τοῦ δεσπότου
ἐς τὴν πυρὰν ἑαυτὸν ἐνέβαλεν· οἱ δὲ οὐκ ἀνδρός,
ἀλλὰ γυναικὸς τὸ θρέμμα εἶναί φασι. ζηλοτυπώτατον δὲ ἄρα ἦν [1] ζῷον ἀετὸς πρὸς τὰ νεόττια. ἐὰν
γοῦν θεάσηταί τινα προσιόντα, ἀπελθεῖν ἀτιμώρη-
τον οὐκ ἐπιτρέπει· παίει γὰρ τοῖς πτεροῖς αὐτὸν
καὶ τοῖς ὄνυξι λυμαίνεται, καὶ ἐπιτίθησίν οἱ
πεφεισμένως τὴν δίκην· οὐ γὰρ χρῆται τῷ στόματι.

41. Ἔστι δὲ θαλαττίων ζῴων τρίγλη λιχνότατον,
καὶ ἐς τὸ ἀπογεύσασθαι παντὸς τοῦ παρατυχόντος
ἀναμφιλόγως ἀφειδέστατον. καί τινες καλοῦνται
λεπρώδεις αὐτῶν, σπάσασαι τὸ ὄνομα ἐκ τῶν
χωρίων, ἅπερ οὖν πέτρας ἔχει λεπράς [2] τε καὶ
ἀραιάς, καὶ φυκία μέσα τούτων δασέα, καί που
καὶ ὑποκάθηται πηλὸς ἢ ψάμμος. φάγοι δ’ ἂν
τρίγλη καὶ ἀνθρώπου νεκροῦ καὶ ἰχθύος· φιληδοῦσι
δὲ μᾶλλον τοῖς μεμιασμένοις καὶ κακόσμοις.

42. Θηρᾶσαι καὶ μάλα γε ἱκανοὶ καὶ οὐδέν τι
μεῖον τῶν ἀετῶν ἱέρακές εἰσιν, ἡμερώτατοι δὲ
ὀρνίθων πεφύκασι καὶ φιλανθρωπότατοι, τὸ μέγεθος
ἀετῶν οὐκ ὄντες ὀλιγώτεροι. ἀκούω δὲ ὅτι ἐν
τῇ Θρᾴκῃ καὶ ἀνθρώποις εἰσὶ σύνθηροι ἐν ταῖς
ἑλείοις ἄγραις. καὶ ὁ τρόπος, οἱ μὲν ἄνθρωποι τὰ
δίκτυα ἁπλώσαντες ἡσυχάζουσιν, οἱ δὲ ἱέρακες
ὑπερπετόμενοι φοβοῦσι [3] τοὺς ὄρνεις [4] καὶ συνωθοῦ-
σιν ἐς τὰς τῶν δικτύων περιβολάς. τῶν οὖν
ᾑρημένων οἱ Θρᾷκες μέρος ἀποκρίνουσι καὶ ἐκείνοις, καὶ ἔχουσιν φίλους [5] πιστούς· μὴ δράσαντες

---

[1] καὶ ζηλοτυπώτατον δὲ ἦν.　　　　[2] Ges : λεπτάς.

of its master abstained from food and died too. And
there was once an Eagle reared by a private citizen
which threw itself on to the pyre where its master's
body was burning. Some say that it had been reared
not by a man but by a woman. The Eagle is apparently the most jealous guardian of its young. At and to its young
any rate if it sees anyone approaching them, it does
not allow him to depart unpunished, for it beats him
with its wings and lacerates him with its talons; and
the punishment it inflicts is moderate, for it does not
use its beak.

41. The Red Mullet is of all sea animals the most The Red Mullet
gluttonous and indisputably the most unrestrained in
tasting everything it comes across. And some of
them are known as 'roughs,' deriving their name
from places where there are rough rocks full of holes
and thick growths of seaweed in them, and where
there is a bottom of mud or sand. A Red Mullet
would eat the dead body of a man or of a fish, and
its special delight is in filthy, ill-smelling food.

42. Falcons are excellent at fowling and are no The Falcon
whit inferior to eagles; they are by nature the tamest
of birds and the most attached to man; in size they
are as large as eagles. And I am told that in Thrace
they even join with men in the pursuit of marsh-fowl.
And this is how they do it. The men spread their
nets and keep still while the Falcons fly over them
and scare the fowl and drive them into the circle of
nets. For this the Thracians allot a portion of their
catch to the Falcons and find them trusty friends;

---

³ καὶ φοβοῦσι.　　　⁴ ὄρνις.　　　⁵ αὐτούς.

δὲ τοῦτο ἑαυτοὺς τῶν συμμάχων ἐστέρησαν.
μάχεται δὲ ὁ τέλειος ἱέραξ καὶ πρὸς ἀλώπεκα καὶ
πρὸς ἀετόν, καὶ γυπὶ μάχεται πολλάκις. καρδίαν
δὲ οὐκ ἂν φάγοι ποτὲ ἱέραξ, τελεστικὸν δήπου
δρῶν καὶ μυστικὸν ἐκεῖνος τοῦτο. νεκρὸν δὲ
ἄνθρωπον ἰδὼν ἱέραξ, ὡς λόγος, πάντως ἐπιβάλλει
γῆς τῷ ἀτάφῳ (καὶ τοῦτο μὲν αὐτῷ οὐ κελεύει
Σόλων[1]), οὐδὲ[2] σώματος ἅψεται. μένει ⟨δὲ⟩[3]
ἄγευστος καὶ ποτοῦ, ἐὰν ἐς αὔλακα ἐποχετεύῃ εἷς
ἄνθρωπος· πεπίστευκε γὰρ αὐτὸν πονούμενον
ζημιοῦν ὑφαιρούμενος ἐκ τῆς ἐκείνου χρείας ὕδωρ·
εἰ δὲ πλείους ἐπάρδοιεν, ἀφθονίαν τοῦ ῥεύματος
ὁρῶν, ὡς φιλοτησίας τινὸς ἐξ αὐτῶν μεταλαμβάνει,
καὶ πίνει ἡδέως.

43. Ἔστι φῦλον ἱεράκων, καὶ καλεῖται κεγχρηίς,
καὶ ποτοῦ δεῖται οὐδὲ ἕν.[4] ὀρείτης δὲ γένος ἄλλο
αὐτῶν· καὶ ἑκάτερός[5] ἐστι δεινῶς φιλόθηλυς, καὶ
ἕπεται κατὰ τοὺς δυσέρωτας, οὐδὲ ἀπολείπεται.
εἰ δὲ ἡ γυνὴ ἀπέλθοι που παραλαθοῦσα, ὁ δὲ
ὑπεραλγεῖ καὶ βοᾷ, καὶ ἔοικε λυπουμένῳ ἐρωτικῶς
εὖ μάλα. καμόντες δὲ τὴν ὄψιν ἱέρακες, εὐθὺ τῶν
αἱμασιῶν ἴασι, καὶ τὴν ἀγρίαν θριδακίνην ἀνασπῶ-
σι, καὶ τὸν ὀπὸν αὐτῆς πικρὸν ὄντα καὶ δριμὺν
ὑπὲρ τῶν ὀφθαλμῶν αἰωροῦσι τῶν σφετέρων, καὶ
λειβόμενον δέχονται, καὶ τοῦτο αὐτοῖς ὑγίειαν
ἐργάζεται. λέγουσι δὲ καὶ τοὺς ἰατρικοὺς χρῆσθαι

---

[1] Σόλων, ὡς Ἀθηναίοις ἐπαίδευσε δρᾶν.
[2] Jac : εἰ δέ.
[3] ⟨δέ⟩ add. Ges.
[4] δέεται οὐδέν.
[5] Schn : ἔκαστος.

if they do not do so, they at once deprive themselves of helpers. Now the full-grown Falcon will fight both with a fox and with an eagle; with a vulture it frequently fights. But a Falcon will never eat the heart, thereby presumably fulfilling some mystic rite. If a Falcon sees the dead body of a man (so it is said), it always heaps earth upon the unburied corpse, though Solon[a] laid no such injunction upon it, and will never touch the body. And it even refrains from drinking if a solitary man is engaged in leading off water into a channel, feeling sure that it will cause damage to the man who so labours if it purloins the water which he needs. But if several men are engaged in irrigating, it sees that the stream is abundant and takes its share from the loving-cup, so to speak, which they offer, and is glad to drink.

43. There is a species of hawk known as the Kestrel which has no need whatever to drink. Another species is the Orites Hawk. Both species are remarkably addicted to the female bird and pursue it after the manner of lovesick men and never cease from the pursuit. But should the female chance to disappear without the male noticing it, he is overcome with grief and cries aloud and is like one in the depths of woe from love.

The Kestrel, the Orites Hawk.

When Hawks are troubled with their eyesight they go straight to some stone wall and pull up some wild lettuce and then holding it above their eyes allow the bitter, astringent juice to drip in; and this restores their health. And men say that doctors use

The Hawk and eye-troubles

---

[a] Solon, of Athens, *c.* 640–*c.* 560 B.C., reformed the laws and constitution.

τῷδε τῷ φαρμάκῳ ἐς τὴν χρείαν τῶν καμνόντων τὴν αὐγήν, καὶ ἐκ τῶν ὀρνίθων ἡ ἴασις κέκληται· καὶ οὐκ ἀρνοῦνται μαθηταὶ ἀκούοντες ὀρνίθων οἱ ἄνθρωποι, ἀλλὰ ὁμολογοῦσι. λέγεται δὲ καὶ θεοσύλην ἐν Δελφοῖς ἐλέγξαι ποτὲ ἱέραξ, ἐμπίπτων τε αὐτῷ καὶ παίων τὴν κεφαλήν. πιστεύονται δὲ εἶναι ἱέρακες καὶ νόθοι, ἀντικριθέντες [1] πρὸς τὰς τῶν ἀετῶν φυλάς. ἦρος δὲ ἀρχομένου οἱ ἐν Αἰγύπτῳ τῶν ἁπάντων δύο προαιροῦνται, καὶ ἀπο-στέλλουσι κατασκεψομένους νήσους τινὰς ἐρήμους, αἵπερ [2] οὖν τῆς Λιβύης πρόκεινται. εἶτα ὑποστρέ-φουσιν οὗτοι, καὶ ἡγοῦνται τῆς πτήσεως τοῖς ἄλλοις. οἱ δὲ ἥκοντες [3] ἑορτὴν ὑπὲρ τῆς ἐπιδημίας τοῖς [4] ἐν τῇ Λιβύῃ παρέχουσι· σίνονται γὰρ οὐδὲ ἕν. παρελθόντες δὲ ἐς τὰς νήσους, ἃς οἱ πρῶτοι θεασάμενοι τῶν ἄλλων ἐπιτηδειοτέρας σφίσιν ἔκριναν, ἐνταῦθα κατὰ πολλὴν τὴν γαλήνην τε καὶ ἡσυχίαν [5] ἀποτίκτουσι καὶ ἐκγλύφουσι, καὶ θηρῶν-ται στρουθοὺς καὶ πελειάδας, καὶ τοὺς νεοττοὺς ἐν ἀφθόνοις ἐκτρέφουσιν· εἶτα ἤδη παγέντας καὶ ἐκπετησίμους γεγενημένους παραλαβόντες ἐς τὴν Αἴγυπτον ἀπάγουσιν, ὥσπερ οὖν ἐς τὰ οἰκεῖα ⟨τὰ⟩[6] πατρῷα τὰς ἐν τοῖς συντρόφοις χωρίοις διατριβάς.

44. Αἱ ἰουλίδες ἰχθῦς εἰσι πέτραις ἔντροφοι, καὶ ἔχουσιν ἰοῦ τὸ στόμα ἔμπλεων καὶ ὅτου ἂν ἰχθύος ἀπογεύσωνται, ἄβρωτον ἀπέφηναν αὐτόν. ἤδη δὲ καὶ οἱ ἁλιεῖς ἡμιβρώτῳ καρίδι περιτυχόντες, καὶ

---

[1] ἀνακριθέντες.
[2] ὅσαιπερ.
[3] Jac : ἑκόντες.
[4] Jac : ἀποδημίας τῆς.
[5] τὴν ἡσυχίαν.
[6] ⟨τά⟩ add. H.

this drug for the benefit of those whose sight is
affected, and the remedy derives its name from these
birds.[a] And men do not refuse to be called the
disciples of birds; rather they admit as much.

It is said that once upon a time a Hawk at Delphi
proved a man guilty of sacrilege by swooping upon
him and striking his head. It is also believed that
Hawks are bastards, if they be compared with the
various kinds of eagles.

At the beginning of spring the Hawks of Egypt
select two from all their number and despatch them
to reconnoitre certain desert islands off the coast of
Libya. When they return they act as leaders to the
rest in their flight. And their arrival is the occasion
of rejoicing on the part of the Libyans at their
sojourn, for they do no damage whatever. And having reached the islands which the original scouts
decided were the most suitable for them, they there
lay and hatch their eggs in complete security and
peace; and they hunt sparrows and pigeons and rear
their young in an abundance of food. Then when
these have grown strong and are able to fly, they
take the young birds with them back to Egypt as
though they were going to their own homes, that
is to their haunts in regions they have grown to
know.

44. Rainbow Wrasses are nurslings of rocks, and
their mouth is full of poison, and whatever fish they
touch they render uneatable. Indeed if it should
happen that fishermen, coming upon a half-eaten
prawn and fancying that their catch is unsaleable,

*Hawk
reveals
sacrilege*

*Hawks of
Egypt*

*The
Rainbow
Wrasse*

---

[a] A certain species with short, round leaves was known as
Hieracion, for the reason stated; cp. Plin. *HN* 20. 7.

ἀξιώσαντες [1] τὸ θήραμα ἄπρατον ὄν, εἰ ἀπογεύ-
σαιντο αὐτοῦ, κλονοῦνται τὴν γαστέρα καὶ στρέ-
φονται. λυποῦσι δὲ καὶ τοὺς ἐν ταῖς ὑδροθηρίαις
ὑποδυομένους τε καὶ νηχομένους, πολλαὶ καὶ
δηκτικαὶ προσπίπτουσαι, ὡς αὐτόχρημα ἐπὶ τῆς
γῆς αἱ μυῖαι· καὶ δεῖ σοβεῖν αὐτὰς ἢ κολάζεσθαι
ἐσθιόμενον· σοβοῦντι δὲ ἐκ τῆς ἀσχολίας ἀπόλωλε
τὸ ἔργον.

45. Λαγὼς δὲ θαλάττιος βρωθεὶς καὶ θάνατον
ἤνεγκε πολλάκις, πάντως δὲ τὴν γαστέρα ὠδύνησεν.
τίκτεται δὲ ἄρα [2] ἐν πηλῷ, καὶ οὐκ ὀλιγάκις ταῖς
ἀφύαις συναλίσκεται· εἴη δ' ἂν κατὰ τὸν κοχλίαν
τὸν γυμνὸν τὸ εἶδος.

46. Γὺψ νεκρῷ πολέμιος. ἐσθίει γοῦν ἐμπεσὼν
ὡς ἐχθρὸν καὶ φυλάττει τεθνηξόμενον. καὶ μέντοι
καὶ ταῖς ἐκδήμοις στρατιαῖς ἕπονται γῦπες, καὶ
μάλα γε μαντικῶς ὅτι [3] ἐς πόλεμον χωροῦσιν
εἰδότες, καὶ ὅτι μάχη πᾶσα ἐργάζεται νεκρούς, καὶ
τοῦτο ἐγνωκότες. γῦπα δὲ ἄρρενα οὔ φασι γίνεσθαι [4] ποτε, ἀλλὰ θηλείας ἁπάσας· ὅπερ ἐπιστά-
μενα τὰ ζῷα καὶ ἐρημίαν τέκνων δεδιότα ἐς
ἐπιγονὴν [5] τοιαῦτα δρᾷ. ἀντίπρῳροι τῷ νότῳ
πέτονται· εἰ δὲ μὴ εἴη νότος, τῷ εὔρῳ κεχήνασι,
καὶ τὸ πνεῦμα ἐσρέον πληροῖ αὐτάς, καὶ κύουσι
τριῶν ἐτῶν. λέγουσι δὲ νεοττιὰν μὴ ὑποπλέκειν

---

[1] ἑαυτῶν ὑπὸ πενίας ἀξιώσαντες.
[2] δὲ ἄρα] γάρ.
[3] γε μαντικῶς ὅτι] μ. ὅτι γε.
[4] γενέσθαι.
[5] Jac : ἐπιγονὴν τέκνων.

should taste it, they are assailed by convulsions and torments in their stomach. And the Wrasses also molest those who dive and swim in pursuit of fish, falling upon them in great numbers and biting them, exactly like flies on land; so that one must either beat them off or be tormented by being eaten up. But while one is busy beating them off, there is no time to attend to one's work.

45. The Sea-hare when eaten has often been the cause even of death; in any case it causes pains in the stomach. It is born in the mud and is not infrequently caught along with sprats. In appearance it is not unlike a snail without its shell. The Sea-hare

46. The Vulture is the dead body's enemy. At any rate it swoops upon it as though it were an adversary and devours it, and watches a man who is in the throes of death. Vultures even follow in the wake of armies in foreign parts, knowing by prophetic instinct that they are marching to war and that every battle provides corpses, as they have discovered. The Vulture

It is said that no male Vulture is ever born: all Vultures are female. And the birds knowing this and fearing to be left childless, take measures to produce them as follows. They fly against the south wind. If however the wind is not from the south, they open their beaks to the east wind, and the inrush of air impregnates them, and their period of gestation lasts for three years. But the Vulture is said never to make a nest. The Aegypius *a* however, which is on the border-line between the vulture and the eagle, is both male and female, and is black in All Vultures are female The 'Aegypius'

a Perhaps the Lämmergeier.

145

γῦπα. τοὺς δὲ αἰγυπιούς, ἐν μεθορίῳ γυπῶν
ὄντας καὶ ἀετῶν, εἶναι καὶ ἄρρενας καὶ τὴν χρόαν
πεφυκέναι μέλανας. καὶ τούτων μὲν ἀκούω καὶ
νεοττιὰς δείκνυσθαι· γῦπας δὲ μὴ ᾠὰ τίκτειν
πέπυσμαι, νεοττοὺς δὲ ὠδίνειν. καὶ ὡς ἀπὸ
γενεᾶς κατάπτεροί εἰσι, καὶ τοῦτο ἤκουσα.

47. Ἰκτῖνος ἐς ἁρπαγὴν ἀφειδέστατος. οἶδε [1]
τῶν μὲν ἐξ ἀγορᾶς ἐμπωληθέντων κρεαδίων ἐὰν
γένωνται κρείττους, ἥρπασαν προσπεσόντες, τῶν
δὲ ἐκ τῆς τοῦ Διὸς ἱερουργίας οὐκ ἂν προσάψαιντο.
Ἡ δὲ ὄρειος ἅρπη τῶν ὀρνίθων προσπεσοῦσα
τοὺς ὀφθαλμοὺς ἀφαρπάζει.

48. Κόρακες Αἰγύπτιοι, ὅσοι τῷ Νείλῳ παρα-
διαιτῶνται,[2] τῶν πλεόντων τὰ πρῶτα ἐοίκασιν
ἱκέται εἶναι, λαβεῖν τι αἰτοῦντες· καὶ λαβόντες
μὲν ἡσυχάζουσιν, ἀτυχήσαντες δὲ ὧν ᾔτουν
συμπέτονται, καὶ ἑαυτοὺς καθίσαντες ἐπὶ τὸ κέρας
τῆς νεὼς τῶν σχοίνων ἐσθίουσί τε καὶ διατέμνουσι
τὰ ἄμματα. Λίβυες δὲ κόρακες, ὅταν οἱ ἄνθρωποι
φόβῳ δίψους ὑδρευσάμενοι πληρώσωσι τὰ ἀγγεῖα
ὕδατος, καὶ κατὰ τῶν τεγῶν θέντες ἐάσωσι τῷ
ἀέρι τὸ ὕδωρ φυλάττειν ἄσηπτον, ἐνταῦθα ἐς
ὅσον μὲν αὐτοῖς τὰ ῥάμφη κάτεισιν ἐγκύπτοντες,
χρῶνται τῷ ποτῷ· ὅταν δὲ ὑπολήξῃ, ψήφους
κομίζουσι καὶ τῷ στόματι καὶ τοῖς ὄνυξι, καὶ
ἐμβάλλουσιν ἐς τὸν κέραμον· καὶ αἱ μὲν ἐκ τοῦ
βάρους ὠθοῦνται καὶ ὑφιζάνουσι, τό γε μὴν ὕδωρ
θλιβόμενον ἀναπλεῖ. καὶ πίνουσιν εὖ μάλα εὐ-

---

[1] οἶδε εἰ δέοι.

colour, and I am told that their nests are pointed out. But I have been informed that Vultures do not lay eggs, but that in their birth-pangs they produce chicks, and that these are feathered from birth I have also heard.

47. There is no limit to the robberies of the Kite. The Kite If they can manage pieces of meat on sale in the market, they pounce upon them and carry them off; on the other hand they will not touch sacrifices offered to Zeus. But the Mountain Kite *a* pounces upon birds and pecks out their eyes.

48. The Ravens in Egypt which live beside the The Raven in Egypt Nile at first appear to be begging of the people sailing on the river, soliciting to be given something. And if they are given, they stop begging; but if their solicitations fail, they fly in a mass and perch on the sailyards of the ship and proceed to eat the ropes and to cut the cords.

But the Ravens of Libya, when men through fear The Raven in Libya of thirst draw water and fill their vessels and place them on the roof so that the fresh air may keep the water from putrefying, the Ravens, I say, help themselves to drink by bending over and inserting their beaks as far as they will go. And when the water gets too low they gather pebbles in their mouth and claws and drop them into the earthenware vessel. Now the pebbles are borne down by their weight and sink, while the water owing to their pressure rises. So the Ravens by a most ingenious

a See 1. 35 n.

---

2 προσδιαιτῶνται, -διαιροῦνται.

μηχάνως οἱ κόρακες, εἰδότες φύσει τινὶ ἀπορρήτῳ
δύο σώματα μίαν χώραν μὴ δέχεσθαι.

49. Λέγει Ἀριστοτέλης εἰδέναι τοὺς κόρακας
διαφορὰν γῆς εὐδαίμονός τε καὶ λυπρᾶς, καὶ ἐν
μὲν τῇ παμφόρῳ τε καὶ πολυφόρῳ κατά τε ἀγέλας
καὶ πλήθη φέρεσθαι, ἐν δὲ τῇ ἀγόνῳ καὶ στερίφῃ
κατὰ δύο. τούς γε μὴν νεοττοὺς τοὺς ἐκτραφέντας [1] τῆς ἑαυτῶν ἔκαστος καλιᾶς φυγάδας ἀποφαί-
νουσιν· ὑπὲρ ὅτου ⟨αὐτοὶ ἑαυτοῖς⟩[2] τροφὴν
μαστεύουσι, καὶ τοὺς γειναμένους σφᾶς μὴ
τρέφουσιν.[3]

50. Ὑπονύξαντες ἰὸν ἀφιᾶσιν ἰχθύων κωβιὸς
καὶ δράκων καὶ χελιδών, οὐ μὴν ἐς θάνατον· ἡ
τρυγὼν δὲ ἀποκτείνει παραχρῆμα τῷ κέντρῳ.
καὶ λέγει γε Λεωνίδης ὁ Βυζάντιος ἰχθύων φύσεώς
τε καὶ κρίσεως ἄπειρον ἄνθρωπον ἁρπάσαντα ἐκ
δικτύου τρυγόνα (ᾤετο δὲ ἄρα ὁ δυστυχὴς ψῆτταν
εἶναι) φέροντα [4] ἐπικόλπιον ἐμβαλεῖν καὶ βαδίζειν,[5]
ὥς τι ἀγαθὸν εὑρόντα καὶ ἐς ἐμπολὴν κερδαλέον
ἑαυτῷ [6] ἅρπαγμα. ἡ δὲ ἄρα ἤλγησε πιεζομένη,
καὶ παίει τῷ κέντρῳ πείρασα,[7] καὶ ἐξέχεε τοῦ
δυστυχοῦς κλέπτου τὰ σπλάγχνα. καὶ ἔκειτο παρὰ
τῇ τρυγόνι νεκρὸς ὁ φώρ, ἐναργὴς ἔλεγχος ὧν
οὐκ εἰδὼς ἔδρασεν.

51. Ὁ κόραξ, οὐκ ἂν αὐτὸν ἐς τόλμαν ἀθυμότε-
ρον εἴποις τῶν ἀετῶν. ὁμόσε γὰρ καὶ αὐτὸς τοῖς

---

[1] ἐκτραφέντας διώκουσι καί.
[2] ⟨αὐτοὶ ἑαυτοῖς⟩ add. Schn.
[3] ἐκτρέφουσιν.
[4] φέροντα ὡς εἶχεν.
[5] βαδίζειν ἵνα λάθῃ.

contrivance get their drink; they know by some mysterious instinct that one space will not contain two bodies.

49. Aristotle asserts [*HA* 618 b 11] that Ravens know the difference between a prosperous and a barren country, and in one that produces all things in plenty they move about in flocks and great numbers, but in a barren and unfruitful country in pairs. As to their young ones, when fully grown, every Raven banishes them from its nest. For that reason they seek their food ⟨for themselves⟩ and neglect to care for their parents.

50. Among fishes the Goby, the Weever, and the Flying Gurnard emit poison when they prick one; not that they are deadly; whereas the Sting-ray with its barb kills on the spot. And Leonidas of Byzantium tells how a man who knew nothing of fishes and could not distinguish them, stole a Sting- ray from a fishing-net—the poor fellow must have taken it for a flounder—, took it and put it in his bosom and walked off as though he had found something good, some spoil whose sale would be profitable to him. But the Sting-ray hurt by the pressure, struck and pierced him with its sting, causing the wretched thief's bowels to gush out. And there the thief lay dead beside the Sting-ray, clear evidence of what he had done in his ignorance.

51. Of the Raven you might say that it has a spirit no less daring than the eagle, for it even attacks

---

[6] ἑαυτῷ ἔχειν. [7] διείρασα.

ζῴοις χωρεῖ, οὐ μέντοι τοῖς βραχυτάτοις, ἀλλ'
ὄνῳ τε καὶ ταύρῳ· κάθηταί τε γὰρ κατὰ τῶν
τενόντων καὶ κόπτει αὐτούς, πολλῶν δὲ καὶ
⟨τοὺς⟩[1] ὀφθαλμοὺς ἐξέκοψεν ὁ κόραξ. μάχεται
δὲ καὶ ὄρνιθι ἰσχυρῷ, τῷ καλουμένῳ αἰσάλωνι·
καὶ ὅταν θεάσηται ἀλώπεκι μαχόμενον, τιμωρεῖται·
πρὸς γὰρ ἐκείνην ἔχει τινὰ φιλίαν. ἦν δὲ ἄρα
ὀρνίθων πολυκλαγγότατός τε καὶ πολυφωνότατος·
μαθὼν γὰρ καὶ ἀνθρωπίνην προΐησι φωνήν.
φθέγμα δὲ αὐτοῦ παίζοντος μὲν ἄλλο, σπουδάζοντος
δὲ ἕτερον· εἰ δὲ ὑποκρίνοιτο τὰ ἐκ τῶν θεῶν,
ἱερὸν ἐνταῦθα καὶ μαντικὸν φθέγγεται. ἴσασι δὲ
διὰ τοῦ θέρους ἐνοχλούμενοι ῥύσει γαστρός, καὶ
διὰ ταῦτα ἑαυτοὺς ὑγρᾶς τροφῆς ἀγεύστους
φυλάττουσιν.

52. Λέγει δὲ Ἀριστοτέλης τῶν ζῴων τὰ μὲν
ζωοτόκα εἶναι, τὰ δὲ ᾠὰ τίκτειν, τὰ δὲ σκώληκας·
καὶ ζῷα μὲν ἀνθρώπους γεννᾶν καὶ τὰ λοιπὰ ὅσα
τριχῶν ἐστιν ἐπίβολα, καὶ τὰ κητώδη τῶν
ἐνύδρων· τούτων δὲ τὰ μὲν αὐλόν, βράγχια δὲ
οὐκ ἔχειν, οἷον δελφῖνα καὶ φάλλαιναν.

53. Μυσοῖς ἄγουσιν ἄχθη βόες, καὶ κεράτων
ἄμοιροί εἰσι. λέγω δὲ τὴν ἀγέλην ἄκερων ὁρᾶσθαι[2] οὐκέτι διὰ κρύος, ἀλλὰ τῶν βοῶν τῶνδε
ἰδίᾳ φύσει,[3] καὶ τὸ μαρτύριον παρὰ πόδας·
γίνονται γὰρ καὶ ἐν Σκύθαις κεράτων[4] οὐκ ἀγέρα-

---

[1] ⟨τοὺς⟩ add. H.
[2] λέγω . . . ὁρᾶσθαι] λέγονται . . . ὁρᾶν.
[3] Reiske : ἰδίᾳ φύσις.

animals, and not the smallest either, but asses and bulls. It settles on their neck and pecks them, and in many cases it actually gouges out their eyes. And it fights with that vigorous bird the merlin, and whenever it sees it fighting with a fox, it comes to the fox's rescue, for it is on friendly terms with the animal.

The Raven must really be the most clamorous of birds and have the largest variety of tones, for it can be taught to speak like a human being. For playful moods it has one voice, for serious moods another, and if it is delivering answers from the gods, then its voice assumes a devout and prophetic tone. *its various tones*

Ravens know that in summer they suffer from looseness of the bowels; for that reason they are careful to abstain from moist food. *its diet*

52. Aristotle tells us [*HA* 489 b 1] that some animals are viviparous, others oviparous, that others again produce grubs. The viviparous are man and all other creatures that have hair, and among marine animals the cetaceans. And of these some have a blow-hole but no gills, like the dolphin and the whale. *Viviparous animals*

53. In Moesia *a* the Oxen draw loads and are hornless. And I maintain that it is not due to the cold that herds are to be seen without horns, but that it is due to the peculiar nature of the Oxen. And the proof is to hand, for even in Scythia there are oxen *Hornless Oxen of Moesia*

a Moesia (Gk. Μυσία), bounded on the N by the Danube, on the S by the Balkan mts, corresponded (roughly speaking) to the northern half of the modern Yugoslavia and Bulgaria.

---

⁴ *Reiske* : κεράτων ἐν Σ.

στοι βόες. ἐγὼ δὲ ἀκούω λέγοντός τινος ἐν συγ-
γραφῇ καὶ μελίττας Σκυθίδας εἶναι, ἐπαίειν τε τοῦ
κρύους οὐδὲ ἕν, καὶ μέντοι καὶ πιπράσκειν ἐς
Μυσοὺς κομίζοντας Σκύθας οὐκ ὀθνεῖόν σφισιν
ἀλλὰ αὐθιγενὲς μέλι καὶ κηρία ἐπιχώρια. εἰ δὲ
ἐναντία Ἡροδότῳ λέγω, μή μοι ἀχθέσθω· ὁ γὰρ
ταῦτα [1] εἰπὼν ἱστορίαν ἀποδείκνυσθαι ἀλλ' οὐκ
ἀκοὴν ᾄδειν ἔφατο ἡμῖν ἀβασάνιστον.

54. Τῶν θαλαττίων πυνθάνομαι μόνον τὸν σκάρον τὴν τροφὴν ἀναπλέουσαν ἐπεσθίειν, ὥσπερ
οὖν καὶ τὰ βληχητά, ἃ δὴ καὶ μαρυκᾶσθαι λέγουσιν.

55. Ὁ γαλεὸς ὠδίνει διὰ τοῦ στόματος ἐν τῇ
θαλάττῃ, πάλιν τε ἐσδέχεται τὰ βρέφη, καὶ
ἀνεμεῖ ταῖς αὐταῖς ὁδοῖς ζῶντα καὶ ἀπαθῆ.

56. Μυὸς ἧπαρ καὶ μάλα ἐκπληκτικῶς τε καὶ
παραδόξως τῆς μὲν σελήνης αὐξανομένης λοβὸν
ἑαυτῷ τινα ἐπιτίκτει ὁσημέραι μέχρι διχομήνου·
εἶτα αὖ πάλιν ὑπολήγει μειουμένου τοῦ μηνὸς τὸν
ἴσον λόγον,[2] ἔστ'[3] ἂν ἐς[4] σῶμα κατολίσθῃ ἀνεί-
δεον. ἀκούω δὲ ἐν τῇ Θηβαΐδι χαλάζης πεσούσης
ἐπὶ τῆς γῆς ὁρᾶσθαι μύας, ὧν τὸ μὲν πηλός
ἐστιν ἔτι, τὸ δὲ σὰρξ ἤδη. ἐγὼ δὲ αὐτὸς ἐκ τῆς
Ἰταλικῆς Νέας πόλεως ἐλαύνων ἐς Δικαιαρχίαν
ὕσθην βατράχοις, καὶ τὸ μὲν μέρος αὐτῶν τὸ πρὸς
τῇ κεφαλῇ εἷρπε, καὶ δύο πόδες ἦγον αὐτό, τὸ δὲ

---

[1] *Schn*: τοιαῦτα.   [2] *Reiske*: λοβόν.
[3] ὑπαφανίζον ἔστ'.   [4] εἰς ἕν.

---

[a] The original Greek name of Puteoli.

not destitute of the glory of horns. And I have learnt from one who records the fact in his history that there are even Bees in Scythia and that they do not mind the cold at all. And what is more, the Scythians bring and sell to the Moesians honey, which is no alien produce but native, and honeycombs of their own country.

If I contradict Herodotus [5. 10], I hope he will not be angry with me, for the man who reported these things vowed that he was presenting the results of his own enquiry and not merely repeating what he had heard and what we could not verify.

54. I learn that of saltwater fishes the Parrot Wrasse alone regurgitates its food and eats it afterwards, as sheep do, which are said to chew the cud.

55. The Shark brings forth its young through its mouth in the sea and takes them back again and then disgorges them by the same channel alive and unharmed.

56. The liver of the Mouse has the most astounding and unexpected habit of growing a lobe day by day as the moon waxes, up to the middle of the month. Then again in proportion as the month declines, so the lobe gradually dwindles until it loses its shape and disappears into the body.

And I am told that when it hails in the Thebaid, mice are to be seen on the earth, and one part of them is still mud while the other is already flesh. And I myself on a journey from Naples to Dicaearchia [a] encountered a shower of frogs, and the forepart of them was crawling, supported by two feet,

ἐπεσύρετο ἔτι ἄπλαστον, καὶ ἐῴκει ἔκ τινος ὕλης
ὑγρᾶς συνεστῶτι.

57. Τὸ τῶν βοῶν ἄρα πάγχρηστον ἦν γένος [1]
καὶ ἐς γεωργίας κοινωνίαν καὶ ἐς ἀγωγὴν φόρτου
διαφόρου. καὶ γαυλοὺς [2] ἐμπλῆσαι βοῦς ἀγαθός
ἐστι, καὶ βωμοὺς κοσμεῖ, καὶ ἀγάλλει πανηγύρεις,
καὶ πανθοινίαν παρέχει. καὶ ἀποθανὼν δὲ βοῦς
γενναῖόν τι χρῆμα καὶ ἀξιέπαινον. μέλιτται γοῦν
ἐκ τῶν ἐκείνου λειψάνων ἐκφύονται, ζῷον φιλεργό-
τατον καὶ τῶν καρπῶν τὸν ἄριστόν τε καὶ γλύκιστον
ἐν ἀνθρώποις παρασκευάζον, τὸ μέλι.

---

[1] γένος καὶ ἀνθρώποις ζῷον λυσιτελέστατον.
[2] Reiske : γάλακτος

while the other part trailed behind, still formless, seeming to consist of some moist substance.

57. Oxen are after all the most serviceable crea- The Ox and
tures. At sharing the farmer's labours, at carrying its services to man
loads of various kinds, at filling the milk-pail—at all
these things the Ox is excellent. He graces the
altars, gladdens festivals, and provides a solemn
banquet. And even when dead the Ox is a splendid
creature deserving our praise. At any rate bees are
begotten of his carcase—bees, the most industrious of
creatures, which afford the best and sweetest of fruits
that man has, namely honey.

BOOK III

Γ

1. Μαυρουσίῳ δὲ ἀνδρὶ ὁ λέων καὶ ὁδοῦ κοινωνεῖ καὶ πίνει τῆς αὐτῆς πηγῆς ὕδωρ. ἀκούω δὲ ὅτι καὶ ἐς τὰς οἰκίας τῶν Μαυρουσίων οἱ λέοντες φοιτῶσιν, ὅταν αὐτοῖς ἀπαντήσῃ ἀθηρία καὶ λιμὸς αὐτοὺς ἰσχυρὸς περιλάβῃ. καὶ ἐὰν μὲν παρῇ ⟨ὁ⟩[1] ἀνήρ, ἀνείργει τὸν λέοντα καὶ ἀναστέλλει διώκων ἀνὰ κράτος· ἐὰν δὲ ὁ μὲν ἀπῇ, μόνη δὲ ἡ γυνὴ καταλειφθῇ, λόγοις αὐτὸν ἐντρεπτικοῖς ἴσχει τοῦ πρόσω καὶ ῥυθμίζει, σωφρονίζουσα ἑαυτοῦ κρατεῖν καὶ μὴ φλεγμαίνειν ὑπὸ τοῦ λιμοῦ. ἐπαΐει δὲ ἄρα λέων φωνῆς Μαυρουσίας, καὶ ὁ νοῦς τῆς ἐπιπλήξεως τῇ γυναικὶ τῆς πρὸς τὸ θηρίον τοιόσδε ἐστίν, ὡς ἐκεῖνοι λέγουσι· 'σὺ δὲ οὐκ αἰδῇ λέων ὢν ὁ τῶν ζῴων βασιλεὺς ἐπὶ τὴν ἐμὴν καλύβην ἰών, καὶ γυναικὸς δεόμενος ἵνα τραφῇς, καὶ δίκην ἀνθρώπου λελωβημένου τὸ σῶμα ἐς χεῖρας γυναικείας ἀποβλέπεις, ἵνα οἴκτῳ καὶ ἐλέῳ τύχῃς ὧν δέῃ; ὃν[2] δέον ἐς ὀρείους ὁρμῆσαι διατριβὰς ἐπί τε ἐλάφους καὶ βουβαλίδας καὶ τὰ λοιπὰ ὅσα λεόντων δεῖπνον ἔνδοξον. κυνιδίου δὲ ἀθλίου φύσει[3] ἀγαπᾷς παρατραφῆναι.' καὶ ἡ μὲν ἐπᾴδει τοιαῦτα, ὁ δὲ ὥσπερ οὖν πληγεὶς τὴν ψυχὴν καὶ ὑποπλησθεὶς αἰδοῦς ἡσυχῇ καὶ κάτω βλέπων ἀπαλλάττεται, ἡττηθεὶς τῶν δικαίων. εἰ δὲ ἵπποι καὶ κύνες διὰ τὴν συντροφίαν ἀπειλούντων

---

[1] ⟨ὁ⟩ add. Jac.   [2] ὃν del. Cobet.   [3] φύσει προσεοικώς.

BOOK III

1. A Lion will accompany a Moor on his journey and will drink water from the same spring. And I am told that Lions even resort to the houses of Moors when they fail to find any prey and are overtaken by the pangs of hunger. And if the master of the house happens to be there, he keeps the Lion off and drives him away, pursuing him vigorously. If however he is out and his wife is left all alone, then with words that put the Lion to shame she checks his approach, restrains him, and admonishes him to control himself and not to allow his hunger to incense him. The Lion, it seems, understands the Moorish tongue; and the sense of the rebuke which the woman administers to the animal is (so they say) as follows. ' Are not you ashamed, you, a Lion, the king of beasts, to come to my hut and to ask a woman to feed you, and do you, like some cripple, look to a woman's hands hoping that thanks to her pity and compassion you may get what you want?—You who should be on your way to mountain haunts in pursuit of deer and antelopes and all other creatures that lions may eat without discredit. Whereas, like some sorry lap-dog, you are content to be fed by another.' Such are the spells she employs, whereupon the Lion, as though his heart smote him and he were filled with shame, quietly and with downcast eyes moves off, overcome by the justice of her words.

Now if horses and hounds through being reared in

ἀνθρώπων συνιᾶσι καὶ καταπτήσσουσι, καὶ Μαυρουσίους οὐκ ἂν θαυμάσαιμι λεόντων ὄντας συντρόφους καὶ ὁμοτρόφους αὐτοῖς ὑπ᾿ αὐτῶν ἐκείνων ἀκούεσθαι. τοῖς γάρ τοι βρέφεσι τοῖς ἑαυτῶν μαρτυροῦσιν ὅτι τοὺς σκύμνους τῶν λεόντων τῆς ἴσης τε καὶ ὁμοίας διαίτης ἀξιοῦσι καὶ κοίτης μιᾶς καὶ στέγης· καὶ ἐκ τούτων καὶ φωνῆς τῆς προειρημένης ἀκούειν τοὺς θῆρας, οὐδὲν οὔτε ἄπιστον οὔτε παράδοξον.

2. Ἵππου δὲ τῆς Λιβύσσης πέρι Λιβύων λεγόντων ἀκούω τοιαῦτα. ὤκιστοι μέν εἰσιν ἵππων, καμάτου δὲ ἢ [1] τι αἰσθάνονται [2] ⟨ἢ⟩ [3] οὐδὲ ἕν. λεπτοὶ δὲ καὶ οὐκ εὔσαρκοι, ἐπιτήδειοί γε μὴν καὶ φέρειν ὀλιγωρίαν δεσπότου εἰσίν. οὔτε γοῦν αὐτοῖς κομιδὴν προσφέρουσιν οἱ δεσπόται, οὐ καταψῶντες, [4] οὐ καλινδήθραν ἐργασάμενοι, οὐχ ὁπλὰς ἐκκαθαίροντες, οὐ κόμας κτενίζοντες, οὐ χαίτας ὑποπλέκοντες, οὐ λούοντες καμόντας, ἀλλὰ ἅμα τε διήνυσαν τὸν προκείμενον δρόμον, καὶ ἀποβάντες νέμεσθαι ἴασι. καὶ λεπτοὶ μὲν καὶ αὐχμώδεις οἱ Λίβυες, ἐπὶ τοιούτων δὲ καὶ ἵππων ὀχοῦνται. σοβαροὶ δὲ Μῆδοι καὶ ἁβροί, καὶ μέντοι καὶ οἱ ἐκείνων [5] ἵπποι. φαίης ἂν αὐτοὺς τρυφᾶν σὺν τοῖς δεσπόταις καὶ τῷ μεγέθει τοῦ σώματος καὶ τῷ κάλλει, ἤδη δὲ καὶ τῇ χλιδῇ καὶ τῇ θεραπείᾳ τῇ ἔξωθεν. [6] ταῦτά τοι καὶ περὶ τῶν κυνῶν ἔπεισι νοεῖν μοι. κύων Κρῆσσα κούφη καὶ ἁλτικὴ καὶ ὀρειβασίαις σύντροφος· καὶ μέντοι

---

[1] Reiske : δή. [2] Schn : αἰσθονται.
[3] ⟨ἢ⟩ add. Reiske. [4] καταψῶντες καμόντας.
[5] ἐκείνων τοιοῦτοι.

their company understand and quail before the
threats of men, I should not be surprised if Moors
too, who are reared and brought up along with Lions,
are understood by these very animals. For the
Moors profess to treat lion-cubs to the same kind of
food, the same bed, and the same roof as their own
children. Consequently there is nothing incredible
or marvellous in Lions understanding human speech
as described above.

2. Concerning the Libyan Horse this is what I have The Horses
learnt from accounts given by the Libyans. These of Libya
Horses are exceedingly swift and know little or nothing of fatigue; they are slim and not well-fleshed but
are fitted to endure the scanty attention paid to them
by their masters. At any rate the masters devote
no care to them: they neither rub them down nor
roll them nor clean their hooves nor comb their
manes nor plait their forelocks nor wash them when
tired, but as soon as they have completed the journey
they intended they dismount and turn the Horses
loose to graze. Moreover the Libyans themselves
are slim and dirty, like the Horses which they ride.
The Persians on the other hand are proud and deli- of Persia
cate, and what is more, their Horses are like them.
One would say that both horse and master prided
themselves on the size and beauty of their bodies and
even on their finery and outward adornment.

And here is a point which occurs to me to note in Hounds of
connexion with Hounds. The Cretan Hound is different
nimble and can leap and is brought up to range the countries

---

6 ἔξωθεν καὶ τῇ θρύψει ἐοίκασιν αἰσθανομένοις μεγέθους τε τοῦ
σφετέρου καὶ κάλλους καὶ ὅτι χλιδῶσι τῷ κόσμῳ.

161

καὶ αὐτοὶ Κρῆτες τοιούτους αὐτοὺς παραδεικνύασι,[1]
καὶ ᾄδει ἡ φήμη. θυμικώτατος δὲ κυνῶν Μολοσ-
σός, ἐπεὶ θυμωδέστατοι καὶ οἱ ἄνδρες. ἀνὴρ δὲ
Καρμάνιος καὶ κύων ἀμφότεροι ἀγριωτάτω καὶ
μειλιχθῆναι ἀτέγκτω,[2] φασίν.

3. Ἴδια δὲ ἄρα φύσεως ζῴων καὶ ταῦτα ἦν.
ὗν οὔτε ἄγριον οὔτε ἥμερον ἐν Ἰνδοῖς γίνεσθαι[3]
λέγει Κτησίας, πρόβατα δὲ τὰ ἐκείνων οὐρὰς
πήχεως ἔχειν τὸ πλάτος πού φησιν.

4. Οἱ μύρμηκες οἱ Ἰνδικοὶ ⟨οἱ⟩[4] τὸν χρυσὸν
φυλάττοντες οὐκ ἂν διέλθοιεν τὸν καλούμενον
Καμπύλινον ποταμόν. Ἰσσηδόνες δὲ τούτοις συνοι-
κοῦντες[5] τοῖς μύρμηξι . . .[6] καλοῦνταί τε καί
εἰσιν.

5. Φαγοῦσα ὄφεως χελώνη καὶ ἐπιτραγοῦσα
ὀριγάνου ἐξάντης γίνεται τοῦ κακοῦ, ὃ πάντως
αὐτὴν[7] ἀνελεῖν ἔμελλεν.

Περιστερὰν δὲ ὀρνίθων σωφρονεστάτην καὶ κε-
κολασμένην ἐς ἀφροδίτην μάλιστα ἀκούω λεγόν-
των· οὐ γάρ ποτε ἀλλήλων διασπῶνται, οὔτε ἡ
θήλεια, ἐὰν μὴ ἀφαιρεθῇ τύχῃ τινὶ τοῦ συννόμου,
οὔτε ὁ ἄρρην, ἐὰν[8] μὴ χῆρος γένηται.

Πέρδικες δὲ ἀκράτορές εἰσιν ἀφροδίτης· οὐκοῦν
τὰ ᾠὰ τὰ γεννώμενα ἀφανίζουσιν, ἵνα μὴ ἄγωσιν

---

[1] περιδεικνῦσι.
[2] Schn : ἀγριώτατοι . . . ἄτεγκτα.
[3] Schn : γενέσθαι.
[4] ⟨οἱ⟩ add. Jac.
[5] συνοικοῦντές γε.
[6] Lacuna.

mountains. Moreover the Cretans show the same qualities, such is the common report. Among Hounds the Molossian is the most high-spirited, for the men also of Molossia are hot-tempered. In Carmania too both men and Hounds are said to be most savage and implacable.

3. The following also are examples of the peculiarities of animal nature. Ctesias reports that neither the wild nor the domestic Pig exists in India, and he says somewhere that Indian Sheep have tails one cubit in width. *India, devoid of pigs* *its sheep*

4. The Ants of India which guard the gold will not cross the river Campylinus.[a] And the Issedonians[b] who inhabit the same country as the Ants . . . they are called, and so they are. *The Ants of India*

5. If a Tortoise eats part of a snake and thereafter some marjoram, it becomes immune from the poison which was bound to be quite fatal to it. *Marjoram, antidote to snake poison*

I have heard people say that the Pigeon is of all birds the most temperate and restrained in its sexual relations. For Pigeons never separate, neither the female bird unless by some mishap she is parted from her mate, nor the male unless he is widowed. *The Pigeon, its continence*

Partridges on the other hand are unrestrained in their indulgence. For that reason they destroy the eggs that have been laid, in order that the female *The Partridge, its incontinence*

---

[a] Not identified.
[b] The Issedonians appear to have inhabited a region to the NE of the Caspian Sea.

---

[7] αὐτὴν ἐκ τῆς τροφῆς.   [8] ἤν.

αἱ θήλειαι παιδοτροφοῦσαι τῆς πρὸς αὐτοὺς
ὁμιλίας ἀσχολίαν.

6. Λύκοι ποταμὸν διανέοντες, ὑπὲρ τοῦ μὴ πρὸς
βίαν ἐκ τῆς τοῦ ῥεύματος ἐμβολῆς ἀνατρέπεσθαι
ἕρμα ἴδιον αὐτοῖς ἡ φύσις συμπλάσασα ἐδιδάξατο
σωτηρίαν ἐξ ἀπόρων καὶ μάλα εὔπορον. τὰς
οὐρὰς τὰς ἀλλήλων ἐνδακόντες, εἶτα ἀντιπίπτουσι
τῷ ῥεύματι, καὶ ἀλύπως [1] διενήξαντο καὶ ἀσφαλῶς.

7. Ὄνοις θηλείαις βρώμησιν ἡ φύσις οὐκ
ἔνειμε, φασί. κύνας δὲ ἀφώνους ἀποφαίνειν ταῖς
ὑαίναις [2] ἡ αὐτὴ παρέσχεν. εὐωδία δὲ καὶ μύρον
γυψὶν αἴτια θανάτου. κύκνων δὲ κώνειον ὄλεθρος.
κάμηλον δὲ ὡς δέδοικεν ἵππος ἔγνω Κῦρός τε καὶ
Κροῖσος, ὥς φασιν.

8. Τὰ βρέφη τὰ τῶν ἵππων ὅταν αἱ μητέρες
καταλίπωσι πρὸ τῆς ἐκείνων ἐκθρέψεως οἷον
ὀρφανά, ἐκτρέφουσι μετὰ τῶν οἰκείων παιδίων
οἰκτείρουσαι αἱ ἄλλαι αὐτά.

9. Κορῶναι ἀλλήλαις εἰσὶ πιστόταται, καὶ ὅταν
ἐς κοινωνίαν συνέλθωσι, πάνυ σφόδρα ἀγαπῶσι
σφᾶς, καὶ οὐκ ἂν ἴδοι τις μιγνύμενα ταῦτα τὰ ζῷα
ἀνέδην καὶ ὡς ἔτυχεν. λέγουσι δὲ οἱ τὰ ὑπὲρ
τούτων ἀκριβοῦντες ὅτι ἂν [3] ἀποθάνῃ τὸ ἕτερον,
τὸ λοιπὸν χηρεύει. ἀκούω δὲ τοὺς πάλαι καὶ ἐν
τοῖς γάμοις μετὰ τὸν ὑμέναιον 'τὴν κορώνην'

---

[1] ἀλύπως γε MSS, ἀ. τε Reiske.
[2] τὰς ὑαίνας ὅταν αὐταῖς τὴν σκιὰν ἐπιβάλῃ.
[3] κἄν.

birds may not be too busy with nursing their chicks to have time for sexual intercourse.

6. When Wolves swim across a river Nature has devised for them an original safeguard to prevent them from being forcibly carried away by the impact of the stream and has taught them how to escape from difficulties, and that with ease. Fastening their teeth in one another's tails they then breast the stream and swim across without harm or danger.

Wolves cross a river

7. It is said that Nature has not bestowed the power of braying upon she-Asses. Nature too has enabled Hyenas to stop hounds from barking. The fragrance of perfumes causes death to Vultures; hemlock is the bane of Swans; Cyrus and Croesus learned how Horses dread camels, so the story goes.

Animal antipathies

8. When Mares desert their foals and leave them, like orphans, before they are fully weaned, other Mares take compassion on them and bring them up with their own foals.

Mares and foals

9. Crows are exceedingly faithful to each other, and when they enter into partnership they love one another intensely, and you would never see these creatures indulging freely in promiscuous intercourse. And those who are accurately informed about them assert that if one dies, the other remains in widowhood. I have heard too that men of old used actually at weddings to sing 'the Crow'[a] after the bridal

The Crow and conjugal fidelity

---

[a] Cp. *Carm. pop.* 31 (Diehl, *Anth. lyr. Gr.*) and L-S⁹ s.v. ἐκκορέω.

ᾄδειν,[1] σύνθημα ὁμονοίας τοῦτο τοῖς συνιοῦσιν
ἐπὶ [2] παιδοποιίᾳ διδόντας. οἱ δὲ [3] ἕδρας ὀρνίθων
καὶ πτήσεις παραφυλάττοντες οὐκ εὐσύμβολον [4]
ὀπυίουσιν [5] εἶναί φασιν ὑπακοῦσαι κορώνης μίας.[6]
ἐπεὶ δὲ ἡ γλαῦξ ἐστιν αὐτῇ πολέμιον, καὶ νύκτωρ
ἐπιβουλεύει τοῖς ᾠοῖς τῆς κορώνης, ἡ δὲ μεθ᾽
ἡμέραν ἐκείνην ταὐτὸ δρᾷ τοῦτο, εἰδυῖα ἔχειν τὴν
ὄψιν τηνικαῦτα τὴν γλαῦκα ἀσθενῆ.

10. Ἐχῖνον τὸν χερσαῖον οὐκ ἄσοφον οὐδ᾽
ἀμαθῆ ταμιείας τῆς ἐς τὴν χρείαν ἡ φύσις ἐποίησεν.
ἐπεὶ γὰρ δεῖται τροφῆς διετησίου, τὰ δὲ ὡραῖα οὐ
πᾶσα ὥρα δίδωσιν, ἑαυτὸν ἐν ταῖς τρασιαῖς κυλίει,[7]
φασί, καὶ τῶν ἰσχάδων τὰς περιπαρείσας, αἳ
πολλαὶ ἐμπήγνυνται [8] ταῖς ἀκάνθαις, ἡσυχῇ κομίζει
καὶ ἀποθησαυρίσας φυλάττει, καὶ ἔχει λαβεῖν ἐκ
τοῦ φωλεοῦ, ὅτε πορίσαι οὐχ οἷόν τε ἔξωθέν ἐστιν.

11. Ἤδη μέντοι [9] καὶ τῶν ζῴων τὰ ἀγριώτατα
πρὸς τὰ ὀνῆσαι δυνάμενα εἰρηναῖα καὶ ἔνσπονδά
ἐστι, τῆς συμφυοῦς κακίας ἐς τὴν χρείαν παραλυθέντα. ὁ γοῦν κροκόδιλος νήχεταί τε ἅμα καὶ
κέχηνεν. ἐμπίπτουσιν οὖν αἱ βδέλλαι ἐς αὐτὸν
καὶ λυποῦσιν. ὅπερ εἰδὼς ἰατροῦ δεῖται τοῦ
τροχίλου· πλήρης γὰρ αὐτῶν γενόμενος, ἐπὶ τὴν
ὄχθην προελθὼν κατὰ τῆς ἀκτῖνος κέχηνεν. ὁ
τοίνυν τρόχιλος ἐμβαλὼν τὸ ῥάμφος ἐξάγει τὰς
προειρημένας, καρτερεῖ δὲ ὠφελούμενος ὁ κροκόδι-

---

[1] καλεῖν.   [2] ἐπὶ τῇ.   [3] τε.
[4] εὐσύμβολον εἰς μαντείαν.
[5] *Pierson*: ὀττεύουσιν MSS and H, *who regards* ὑπακοῦσαι *as corrupt.*

song by way of pledging those who came together
for the begetting of children to be of one mind.
While those who observe the quarters from which
birds come and their flight, declare that to hear a
single Crow is an evil omen at a wedding. Since the
Owl is an enemy of the Crow and at night has designs Owl and
upon the Crow's eggs, the Crow by day does the same Crow
to her, knowing that at that time the Owl's sight is
feeble.

10. Nature has made the Hedgehog prudent and The
experienced in providing for its own wants. Thus, Hedgehog
since it needs food to last a whole year, and since
every season does not yield produce, it rolls among
fig-crates (they say), and such dried figs as are pierced
—a great number become fixed upon its prickles—it
quietly removes, and after laying up a store, keeps
them and can draw from its nest when it is impossible
to obtain food out of doors.

11. It is a fact that the fiercest of animals will, The
when the need arises, lay aside their natural savagery Crocodile
and be peaceful and gently disposed towards those
that can be of service to them. For instance, the
Crocodile swims with its jaws open; accordingly
leeches fall into them and cause it pain. Knowing
this it needs the Egyptian Plover as doctor. For and the
when it is infested with leeches, it moves to the bank Egyptian
and opens its jaws to face the sun. Whereupon the Plover
Egyptian Plover inserts its beak and draws out the
aforesaid creatures, while the Crocodile endures this

---

⁶ Gow : κορώνη μία MSS, H.   ⁷ Reiske : κυλίειν.
⁸ πήγνυνται.   ⁹ μέν.

λος καὶ ἀτρεμεῖ. καὶ ὁ μὲν ἔχει δεῖπνον τὰς
βδέλλας, ὁ δὲ ὀνίναται, καὶ τὸ μηδὲν ἀδικῆσαι τὸν
τροχίλον λογίζεταί οἱ μισθόν.

12. Κολοιοὺς δὲ εὐεργέτας νομίζουσι καὶ Θετ-
ταλοὶ καὶ Ἰλλυριοὶ καὶ Λήμνιοι, καὶ δημοσίας γε
αὑτοῖς τροφὰς ἐψηφίσαντο,[1] ἐπεὶ τῶν ἀκρίδων, αἳ
λυμαίνονται[2] τοὺς καρποὺς τοῖς προειρημένοις, τὰ
ᾠὰ ἀφανίζουσί τε οἱ κολοιοὶ καὶ διαφθείρουσι τὴν
ἐπιγονὴν αὐτοῖς. μειοῦται δὴ κατὰ πολὺ τὰ τῶν
ἀκρίδων νέφη, καὶ τοῖς προειρημένοις μένει τὰ
ὡραῖα ἀσινῆ.

13. Αἱ γέρανοι γίνονται μὲν ἐν Θρᾴκῃ, ἣ δὲ
χειμεριώτατον χωρίων ἐστὶ καὶ κρυμωδέστατον
ὧν ἀκούω. οὐκοῦν φιλοῦσι τὴν χώραν ἐν ᾗ
γεγόνασι, φιλοῦσι δὲ καὶ ἑαυτάς, καὶ νέμουσι τὸ
μέν τι τοῖς ἤθεσι τοῖς πατρῴοις, τὸ δέ τι τῇ
σφῶν αὐτῶν σωτηρίᾳ. τοῦ μὲν γὰρ θέρους κατὰ
χώραν μένουσι, φθινοπώρου δὲ ἤδη μεσοῦντος ἐς
Αἴγυπτόν τε καὶ Λιβύην ἀπαίρουσι καὶ Αἰθιοπίαν,
ὥσπερ οὖν γῆς περίοδον εἰδυῖαι καὶ φύσεις ἀέρων
καὶ ὡρῶν διαφοράς. καὶ χειμῶνα ἠρινὸν διαγα-
γοῦσαι, πάλιν ὅταν ὑπεύδια ἄρξηται καὶ εἰρηναῖα
τὰ τοῦ ἀέρος, ὑποστρέφουσιν ὀπίσω. ποιοῦνται
δὲ ἡγεμόνας τῆς πτήσεως τὰς ἤδη τῆς ὁδοῦ
πεπειραμένας· εἶεν δ' ἂν ὡς τὸ εἰκὸς αἱ πρεσβύτεραι. καὶ οὐραγεῖν δὲ τὰς τηλικαύτας ἀποκρίνουσι·
μέσαι δὲ αὐτῶν αἱ νέαι τετάχαται. φυλάξασαι
δὲ ἄνεμον οὖρον καὶ φίλον σφίσι καὶ κατόπιν
ῥέοντα, χρώμεναί οἱ πομπῷ καὶ ἐπωθοῦντι ἐς τὸ
πρόσω, εἶτα μέντοι τρίγωνον ὀξυγώνιον τὸ σχῆμα

service and remains motionless. So the bird gets a
feast of leeches, while the Crocodile is benefited and
reckons the fact that it has not injured it as the bird's
fee.

12. The inhabitants of Thessaly, of Illyria, and of The
Lemnos regard Jackdaws as benefactors and have and Locusts
decreed that they be fed at the public expense,[1] seeing that Jackdaws make away with the eggs and
destroy the young of the locusts which ruin the crops
of the aforesaid people. The clouds of locusts are
in fact considerably reduced and the season's produce
of these people remains undamaged.

13. Cranes have their birthplace in Thrace, which Cranes and
is the most wintry and the coldest region that I know migrations
of. Well, they love the country of their birth, but
they love themselves too; so they devote part of
their time to their ancestral haunts and part to
their own preservation. In summer they remain
in their country, but in mid-autumn they leave for
Egypt, Libya, and Ethiopia, appearing to know the
map of the earth, the disposition of the winds, and
the variations of the seasons. And after spending a winter like spring, when again conditions
are becoming tolerably settled and the sky is calm,
they return. To lead their flight they appoint those
that have already had experience of the journey;
these would naturally be the older birds, and they
select others of the same age to bring up the rear,
while the young ones are ranged in their midst.
Having waited for a fair and favouring wind from

---

[1] ἐψηφίσαντο αἵδε αἱ πόλεις.   [2] *Reiske*: ἐλυμαίνοντο.

AELIAN

τῆς πτήσεως ἀποφῆναι, ἵνα ἐμπίπτουσαι τῷ
ἀέρι διακόπτωσιν αὐτὸν ῥᾷστα, τῆς πορείας
ἔχονται. οὕτω μὲν δὴ θερίζουσί τε καὶ χειμάζουσι
γέρανοι· σοφίαν δὲ ἥγηνται ἄνθρωποι θαυμαστὴν
τοῦ Περσῶν βασιλέως ἐς ἐπιστήμην ἀέρων
κράσεως,[1] Σοῦσα καὶ Ἐκβάτανα ᾄδοντες καὶ τὰς
δεῦρο καὶ ἐκεῖσε τοῦ Πέρσου τεθρυλημένας μετα-
βάσεις. ὅταν δὲ προσφερόμενον ἀετὸν αἱ γέρανοι
θεάσωνται, γενόμεναι κυκλόσε[2] καὶ κολπωσάμε-
ναι[3] ἀπειλοῦσιν ὡς ἀντιταξόμεναι· ὁ δὲ[4] κρούεται
τὸ πτερόν. ἀλλήλων δὲ τοῖς πυγαίοις ἐπερείδουσαι
τὰ ῥάμφη, εἶτα μέντοι τρόπον τινὰ τὴν πτῆσιν
συνδέουσι, καὶ τὸν κάματόν σφισιν εὐκάματον
ἀποφαίνουσι, πεφεισμένως ἀναπαυόμεναι ἐς ἀλλήλας αἱ αὐταί. ἐν δὲ γῇ μηκίστῃ . . .[5] πηγῆς
ὅταν τύχωσιν, ἀναπαύονται νύκτωρ[6] καὶ καθεύ-
δουσι, τρεῖς δὲ ἢ τέτταρες προφυλάττουσι τῶν
λοιπῶν καὶ ὑπὲρ τοῦ μὴ κατακοιμίσαι τὴν φυλακὴν
ἑστᾶσι μὲν ἀσκωλιάζουσαι, τῷ γε μὴν μετεώρῳ
ποδὶ λίθον κατέχουσι τοῖς ὄνυξι μάλα ἐγκρατῶς τε
καὶ εὐλαβῶς, ἵνα ἐάν ποτε λάθωσιν ἑαυτὰς ἐς
ὕπνον ὑπολισθάνουσαι, πεσὼν καὶ ὑποκτυπήσας ὁ
λίθος ἀποδαρθάνειν καταναγκάσῃ. γέρανος δὲ
λίθον ὅνπερ οὖν καταπίνει ὑπὲρ τοῦ ἔχειν ἔρμα,[7]
χρυσοῦ βάσανός ἐστιν, ὅταν οἷον ὁρμισαμένη καὶ
καταχθεῖσα[8] εἶτα μέντοι ἀνεμέσῃ αὐτόν.

[1] χρήσεως.
[2] Lobeck : κύκλος.
[3] κολπωσάμενοι μηνοειδὲς τὸ μέσον ἀποφῆναι.
[4] ὁ δὲ ἀναχωρεῖ καί.
[5] Lacuna.
[6] νύκτωρ αἱ λοιπαί.
[7] ἔρμα πετομένη.

170

behind, and using it as an escort to speed them forward, they then form their order of flight into an acute-angled triangle, in order that as they encounter the air they may cleave it with the least difficulty, and so hold on their way. This then is how Cranes spend their summer and winter. (But mankind regards as marvellous the Persian king's comprehension of temperature, and harps on Susa and Ecbatana[a] and the repeated stories of the Persian's journeyings to and fro.) When however the Cranes observe an eagle bearing down upon them, they form a circle and in a bellying mass threaten him with attack; and he retires. Resting their bills upon each other's tail-feathers they form in a sense a continuous chain of flight, and sweeten their labour[b] as they repose gently one upon another. And in some distant land . . . when they light upon some water-spring they rest for the night and sleep, while three or four mount guard for all the others; and in order to avoid falling asleep during their watch they stand on one leg, but with the other held up they clutch a stone firmly and securely in their claws. Their object is that, if they should inadvertently drop off to sleep, the stone should fall and wake them with the sound.[s]

Now the stone which a Crane swallows to give itself ballast is a touchstone for gold when regurgitated by the Crane after it has, so to say, come to anchor and reached land.

[a] Identified with the modern Hamadan; it lay at the foot of mt Orontes, some 200 miles N of Susa, and was a summer residence of the Achaemenid kings.

[b] Eur. *Bacc.* 66 κάματον εὐκάματον.

---

[s] καταχθεῖσα ἔνθα ἥκει.

14. Κυβερνήτης ἰδὼν ἐν πελάγει μέσῳ γεράνους
ὑποστρεφούσας καὶ τὴν ἔμπαλιν πετομένας, συνεῖ-
δεν ἐναντίου προσβολῇ πνεύματος ἐκείνας ἀποστῆναι τοῦ πρόσω· καὶ τῶν ὀρνέων ὡς ἂν εἴποις
μαθητὴς γενόμενος παλίμπλους ἦλθε, καὶ τὴν
ναῦν περιέσωσε. καὶ τοῦτο πρῶτον γενόμενον
μάθημά τε ὁμοῦ καὶ παίδευμα ⟨ὑπὸ⟩[1] τῶνδε
⟨τῶν⟩[2] ὀρνίθων τοῖς ἀνθρώποις παρεδόθη.

15. Περιστεραὶ ἐν μὲν ταῖς πόλεσι τοῖς ἀνθρώποις συναγελάζονται, καί εἰσι πρᾱόταται, καὶ
εἰλοῦνται περὶ τοῖς ποσίν, ἐν δὲ τοῖς ἐρήμοις
χωρίοις ἀποδιδράσκουσι, καὶ τοὺς ἀνθρώπους οὐχ
ὑπομένουσι. θαρροῦσι μὲν γὰρ τοῖς πλήθεσι, καὶ
ὅτι μηδὲν πείσονται δυσχερὲς ἴσασι κάλλιστα.
ὅπου δὲ ὀρνιθοθῆραι καὶ δίκτυα καὶ ἐπιβουλαὶ
κατ' αὐτῶν, ἄτρεστα οἰκοῦσιν οὐκέτι, ἵνα εἴπω τὸ
ἐπ' αὐτῶν ἐκείνων λεχθὲν Εὐριπίδῃ.

16. Ὅταν μέλλωσι πέρδικες πρὸς τῷ τίκτειν
εἶναι, παρασκευάζουσιν ἑαυτοῖς ἔκ τινων καρφῶν
τὴν καλουμένην ἅλω. πλέγμα δέ ἐστι κοῖλον καὶ
ἐγκαθίσαι μάλα ἐπιτήδειον. καὶ κόνιν ἐγχέαντες,
καὶ μαλακήν τινα οἱονεὶ κοίτην ἐργασάμενοι, καὶ
ἐνδύντες, εἶτα ἐπηλυγάσαντες ἑαυτοὺς ἄνωθεν
κάρφεσιν ὑπὲρ τοῦ καὶ τοὺς ὄρνιθας λαθεῖν τοὺς
ἁρπακτικοὺς καὶ τῶν ἀνθρώπων τοὺς θηρευτάς,
κατὰ πολλὴν τὴν εἰρήνην ἀποτίκτουσιν[3]· εἶτα τὰ
ᾠὰ οὐ πιστεύουσι τῇ χώρᾳ τῇ αὐτῇ, ἀλλ' ἑτέρᾳ,

---

[1] ⟨ὑπό⟩ add. H.   [2] ⟨τῶν⟩ add. Reiske.
[3] Reiske : κατακλίνουσιν.

14. If a pilot observes on the high seas a flock of Cranes turning and flying back, he realises that they have refrained from advancing further owing to the assault of a contrary wind. And taught, as you might say, by the birds he sails home again and preserves his vessel. So the pilot's art, being a lesson and a discipline first acquired by these birds, has been handed on to mankind.

Cranes give warning of storms

15. In cities Pigeons congregate with human beings; they are extremely tame and swarm about one's feet; but in lonely places they flee away and cannot endure human beings. For it is crowds that give them courage, and they are well aware that they will be unmolested. Where however there are bird-catchers, nets, and schemes to take them, ' they dwell ' no more ' without fear,' to quote what Euripides says [*Ion* 1198] of those same birds.

The Pigeon

16. When Partridges are about to lay they make themselves what is called a ' threshing-floor ' (*i.e.* nest) out of dry twigs. It is plaited, hollow, and well-suited for sitting in. They pour in dust and construct as it were a soft bed; they enter and after screening themselves over with dry twigs so as to avoid being seen by birds of prey and by human hunters, they lay their eggs in complete tranquillity. Next, they do not entrust their eggs to the same place but to some other, emigrating *a* as it were, because

The Partridge and its nest

a Cp. Arist. *HA* 613 b 15.

οἱονεὶ μετοικιζόμενοι [1]· δεδοίκασι γὰρ [2] μή ποτε
ἄρα φωραθῶσιν. νεοττεύοντες δὲ [3] τοὺς νεοττοὺς
ὄντας ἁπαλοὺς ὑποθάλπουσι καὶ τοῖς ἑαυτῶν
πτεροῖς ἀλεαίνουσιν, οἱονεὶ σπαργάνοις τοῖς πτίλοις
περιαμπέχοντες· οὐ λούουσι δὲ αὐτούς, ἀλλὰ
κονίοντες ἐργάζονται φαιδροτέρους. ἐὰν δὲ πέρδιξ
ἴδῃ τινὰ προσιόντα καὶ ἐπιβουλεύοντα καὶ αὐτῷ
καὶ τοῖς βρέφεσιν, ἐνταῦθα αὐτὸς μὲν ἑαυτὸν πρὸ
τῶν ποδῶν κυλίει τῶν τοῦ θηρατοῦ, καὶ ἐνδίδωσιν
ἐλπίδα τοῦ δύνασθαι συλλαβεῖν εἰλούμενον. καὶ ὁ
μὲν ἐπικύπτει ἐς τὴν ἄγραν, ὁ δὲ ἐξελίττει
ἑαυτόν· καὶ διαδιδράσκει καὶ γίνεται πρὸ ὁδοῦ
⟨τὰ βρέφη⟩.[4] ὅπερ οὖν συννοήσας ὁ πέρδιξ,
θαρρῶν ἤδη τῆς ἀσχολίας τῆς ματαίας ἀπαλλάττει
τὸν ὀρνιθοθήραν ἀναπτάς, καὶ ἐᾷ [5] τὸν ἄνδρα
κεχηνότα. εἶτα ἐν ἀδείᾳ ἡ μήτηρ γενομένη καὶ
ἐν καλῷ στᾶσα τὰ βρέφη καλεῖ. οἱ δὲ αὐτῇ
προσπέτονται γνωρίσαντες τὸ φώνημα. πέρδιξ δὲ
ὠδῖνα ἀπολύειν μέλλων πειρᾶται λαθεῖν τὸν σύννομον, ἵνα μὴ τὰ ᾠὰ συντρίψῃ· λάγνος γὰρ ὢν
οὐκ ἐᾷ τῇ παιδοτροφίᾳ σχολάζειν τὴν μητέρα.
οὕτω δέ ἐστιν ἀκόλαστον τὸ τῶν περδίκων γένος.
ὅταν αὐτοὺς ἀπολιποῦσαι εἶτα ἐπῳάζωσιν αἱ
θήλειαι, οἱ δὲ ἐπίτηδες ἐς ὀργὴν ἀλλήλους ἐξά-
πτουσι, καὶ παίουσί τε καὶ παίονται πικρότατα·
καὶ ὅ γε ἡττηθεὶς ὀχεύεται [ὡς ὄρνις],[6] καὶ δρᾷ
τοῦτο ἀνέδην ⟨ὁ κρατήσας⟩,[7] ἔστ᾽ ἂν ὑφ᾽ ἑτέρου
καὶ αὐτὸς ἡττηθεὶς εἶτα ἐς τὰς ὁμοίας λαβὰς
ἐμπέσῃ.

[1] μετοικιζόμενοι ἐκεῖνά τε ἐπάγονται.
[2] γὰρ ἐν ταὐτῷ διατρίβοντες.
[3] δὲ ἐν χώροις ἑτέροις ἀπαίροντές τε αὖ.

they are afraid that they may perhaps be detected. And when they hatch their young they impart heat to them, being callow, and warm them with their wings, enveloping them in their feathers, as it might be swaddling-clothes. They do not however wash them, but render them more sleek by putting dust on them.

If a Partridge sees someone approaching with evil and its intent against itself and its young, it thereupon rolls young about in front of the hunter's feet and fills him with the hope of seizing it as it moves this way and that. And the man bends down to catch his prey, but it eludes him. Meantime the young ones slip away and get some distance ahead. So when the Partridge is aware of this, it takes courage and releases the bird-catcher from his fruitless occupation by flying off, leaving the man gaping. Then when the mother-bird is secure and advantageously placed, she calls her chicks, and they recognising her voice flutter towards her.

The Partridge when about to lay her eggs en- The male deavours to hide from her mate for fear that he may bird crush them, because he is lustful and tries to prevent the mother from devoting her time to rearing her young. So incontinent a creature is the Partridge. When the females leave the males and brood their eggs, the male birds of set purpose provoke one another to anger and deal and receive the most violent blows; and the vanquished bird gets trodden, the victor performing unsparingly, until he in his turn is vanquished and is caught in like clutches.

---

⁴ ⟨τὰ βρέφη⟩ add. H.
⁵ καὶ ἐᾷ] καὶ τοὺς νεοττοὺς καταλαβὼν καὶ ἐάσας.
⁶ [ὡς ὄρνις] ' verba suspecta,' H.
⁷ ⟨ὁ κρατήσας⟩ add. Jac.

AELIAN

17. Λέγει μὲν οὖν Εὐριπίδης δυσώνυμον τὸν[1] φθόνον· οὗτος δὲ ἄρα ἐνοικεῖ καὶ τῶν ζῴων ἔστιν οἷς. ὁ γοῦν γαλεώτης, ὥς φησι Θεόφραστος, ὅταν ἀποδύσηται τὸ γῆρας, ἐπιστραφεὶς εἶτα μέντοι καταπιὼν ἀφανίζει αὐτό· δοκεῖ δὲ ἐπιλήψεως εἶναι τὸ γῆρας τὸ τοῦδε τοῦ ζῴου ἀντίπαλον. οἶδε δὲ καὶ ἔλαφος τὸ δεξιὸν κέρας ἔχων ἐς πολλὰ ἀγαθόν, καὶ μέντοι ⟨καὶ⟩[2] κατορύττει τε αὐτὸ καὶ ἀποκρύπτει φθόνῳ τοῦ τοσούτων[3] τινα ἀπολαῦσαι. ἴυγγας δὲ ἐρωτικὰς τῷ πώλῳ συντίκτουσα ἵππος οἶδε· ταῦτά τοι καὶ ἅμα τῷ τεχθῆναι τὸ βρέφος ἡ δὲ τὸ ἐπὶ τῷ μετώπῳ σαρκίον ἀπέτραγεν. ἱππόμανες ἄνθρωποι καλοῦσιν αὐτό. καὶ οἱ γόητες τὰ τοιαῦτά φασιν ὁρμάς τινας ἑλκτικὰς ἐς μίξιν ἀκατάσχετον καὶ οἶστρον ἀφροδίσιον παρέχειν καὶ ἐξάπτειν. οὔκουν τὴν ἵππον ἐθέλειν ἀνθρώπους μεταλαγχάνειν τοῦ γοητεύματος τοῦδε, ὥσπερ οὖν ἀγαθοῦ μεγίστου φθονοῦσαν. οὐ γάρ;

18. Ἐν τῇ Ἐρυθρᾷ θαλάττῃ[4] ἰχθὺν Λεωνίδης ὁ Βυζάντιος γίνεσθαί[5] φησι, κωβιοῦ τοῦ τελείου μείονα οὐδὲ ἕν· ἔχειν δὲ οὔτε[6] ὀφθαλμοὺς αὐτὸν οὔτε στόμα ἐν νόμῳ τῷ τῶν ἰχθύων. προσπέφυκε δέ οἱ βράγχια καὶ σχῆμα κεφαλῆς, ὡς εἰκάσαι, οὐ μὴν ἐκμεμόρφωται εἶδος· κάτω δὲ ἄρα ὑπὸ τῇ γαστρὶ αὐτῷ ἐντέθλασται τύπος κολπώδης ἡσυχῇ, καὶ ἐκπέμπει σμαράγδου χρόαν. τοῦτον οὖν εἶναι καὶ ὀφθαλμόν οἵ φησι καὶ στόμα.

---

[1] ὄντα τόν.
[2] ⟨καὶ⟩ add. H.
[3] Jac : τοσούτου.
[4] θαλάττῃ κόλπῳ δὲ τῷ Ἀραβίῳ.

17. Euripides says [*fr.* 403 N] that jealousy is an accursed thing. It seems that there are certain animals in which this quality resides. For instance, the Gecko, according to Theophrastus [*fr.* 175], when it has sloughed its skin, turns and makes away with it by swallowing it. It seems that the slough of this creature is a remedy for epilepsy. And the Deer too, knowing that its right horn serves many purposes, goes so far as to bury it and secrete it out of jealousy lest anyone should benefit thereby. The Mare also knows that with the birth of a foal she is producing love-spells; and that is why the moment the foal is born, the Mare bites off the piece of flesh on its forehead. Men call it ' mare's-frenzy.' *a* And wizards maintain that such things produce and excite impulses to unrestrained sexual intercourse and a lecherous passion. So the Mare does not wish men to have any of this spell, as though she grudged them a boon beyond compare. And is it not so?

Jealousy in certain animals

18. Leonidas of Byzantium asserts that there occurs in the Red Sea a fish *b* of exactly the same size as a full-grown goby: it has neither eyes nor mouth after the manner of fishes, but grows gills and a kind of head, so far as one can guess, though its form is not perfectly developed. But lower down beneath its stomach is a slightly indented depression which emits the colour of an emerald; and this, they say, is both its eye and its mouth. But anyone who

The ' Inflater' fish

---

a See 14. 18.
b Probably the *Tetrodon* or Globe-fish.

---

⁵ *Schn*: γενέσθαι.   οὐδέ.

ὅστις δὲ αὐτοῦ γεύεται,[1] σὺν τῷ κακῷ τῷ ἑαυτοῦ
ἐθήρασεν αὐτόν. καὶ τῆς διαφθορᾶς ὁ τρόπος, ὁ
γευσάμενος ᾤδησεν, εἶτα ἡ γαστὴρ κατέρραξε, καὶ
ὁ ἄνθρωπος ἀπόλωλε. δίδωσι δὲ καὶ αὐτὸς ἁλοὺς
δίκας. πρῶτον μὲν ἔξω τοῦ κύματος γενόμενος
οἰδαίνει, καὶ εἴ τις αὐτοῦ ψαύσειεν,[2] ὁ δὲ ἔτι καὶ
μᾶλλον πίμπραται. καὶ εἴ τις ἐπιμείνειε ψαλάτ-
των, γίνεται πᾶς ὑπὸ σήψεως διαυγέστατος, ὡς
ὑδεριῶν· εἶτα τελευτῶν διερράγη. εἰ δὲ αὐτὸν
ἐθέλοι τις ἔτι ζῶντα ἐς τὴν θάλατταν μεθεῖναι, ὁ
δὲ ἐπινήχεται δίκην κύστεως ἀρθείσης πνεύματι.
καί φησιν ὅτι ἐκ τοῦ πάθους φύσαλον ἐκάλουν
αὐτόν.

19. Φώκη δέ, ὡς ἀκούω, τὴν πυετίαν τὴν
ἑαυτῆς ἐξεμεῖ,[3] ἵνα μὴ τοῖς ἐπιλήπτοις ᾖ ἰᾶσθαι.
βάσκανον δὴ τὸ ζῷον ἡ φώκη, ναὶ μὰ τόν.

20. Οἱ πελεκᾶνες ⟨οἱ⟩[4] ἐν τοῖς ποταμοῖς ⟨τὰς⟩
κόγχας περιχαίνοντες εἶτα καταπίνουσιν, ἔνδον
δὲ καὶ ἐν ⟨τῷ⟩[5] μυχῷ τῆς γαστρὸς ὑποθαλ-
ψαντες ἀνεμοῦσι, καὶ τὰ μὲν ὀστράκια ἐκ τῆς
ἀλέας διέστη, ὥσπερ οὖν ⟨τὰ⟩[6] τῶν ἑφθῶν, οἱ δὲ
ἐξορύττουσι τὰ κρέα, καὶ ἔχουσι δεῖπνον. καὶ
μέντοι καὶ οἱ λάροι, ὡς Εὔδημός φησι, τοὺς
κοχλίας μετεωρίζοντες καὶ ὑψοῦ αἴροντες ταῖς
πέτραις βιαιότατα προσαράττουσιν.

21. Λέγει Εὔδημος, ἐν Παγγαίῳ τῷ Θρακίῳ
κοίτη λέοντος ἐρήμῳ φυλακῆς ἐπιστᾶσαν ἄρκτον

---

[1] γεύσεται.   [2] ψαύσοι.   [3] Ges: ἐκροφεῖ.

eats it has fished to his own undoing. And this is
how he is destroyed: the man who has eaten it
swells up; then his stomach bursts and he dies. But
the fish itself when caught pays for it, for first, when
it is out of the water, it swells, and if one touches it,
it swells even more; while if one continues to handle
it, it turns to corruption and becomes quite trans-
lucent, like a man with dropsy, and finally bursts. If
however one is prepared to return it still alive to the
sea, it swims on the surface like an inflated bladder.
Leonidas says that in consequence of this property
men call it the 'inflater.'

19. The Seal, I am told, vomits up the curdled milk The Seal
from its stomach so that epileptics may not be cured
thereby. Upon my word the Seal is indeed a
malignant creature.

20. Pelicans that live in rivers take in mussels and The Pelican
then swallow them, and when they have warmed
them deep within the recesses of their belly, they
disgorge them. Now the mussels open under the
influence of the heat, just like the shells of things
when cooked, and the Pelicans scoop out the flesh
and make a meal. So too Sea-mews, as Eudemus The
observes, lift snails into the air and carry them high Sea-mew
up and then dash them violently upon the rocks.

21. Eudemus records how on mount Pangaeus in A Bear and
Thrace a Bear came upon a Lion's lair which was two Lions

---

4 ⟨οἱ⟩ . . . ⟨τάς⟩ add. H, cp. Arist. HA 614 b 27.
5 ⟨τῷ⟩ add. H.
6 ⟨τά⟩ add. H.

AELIAN

⟨τοὺς⟩[1] σκύμνους τοῦ λέοντος διαφθεῖραι διὰ τὸ
μικρούς τε εἶναι ἔτι καὶ ἀμῦναί σφισιν ἀδυνάτους.
ἐπεὶ δὲ ἀφίκοντο[2] ἔκ τινος ἄγρας ὅ τε πατὴρ καὶ
ἡ μήτηρ, καὶ εἶδον τοὺς παῖδας ἐν ταῖς φοναῖς,
οἷα εἰκὸς ἤλγουν, καὶ ἐπὶ τὴν ἄρκτον ἵεντο· ἡ δὲ
δείσασα εἴς τι δένδρον ᾗ ποδῶν εἶχεν ἀνέθει, καὶ
καθῆστο τὴν ἐπιβουλὴν τὴν ἐξ ἐκείνων ἐκκλῖναι
πειρωμένη. ὡς δὲ ἐδόκουν τοῦ τιμωρήσασθαι τὸν
λυμεῶνα ἥκειν δεῦρο, ἐνταῦθα ἡ μὲν λέαινα οὐ
λείπει τὴν φυλακήν, ἀλλ᾽ ὑπὸ τῷ πρέμνῳ καθῆστο
ἐλλοχῶσα καὶ ὕφαιμον ἄνω βλέπουσα, ὁ δὲ λέων,
οἷα ἀδημονῶν καὶ ἀλύων ὑπὸ τοῦ ἄχους,[3] ἐν τοῖς
ὄρεσιν ἠλᾶτο, καὶ ἀνδρὶ ὑλουργῷ περιτυγχάνει· ὁ
δὲ ἔδεισε καὶ ἀφίησι τὸν πέλεκυν, τὸ δὲ θηρίον ὁ
λέων ἔσαινέ τε καὶ ἑαυτὸν ἀνατείνας ἠσπάζετο,
ὡς οἷός τε ἦν, καὶ τῇ γλώττῃ τὸ πρόσωπον
ἐφαίδρυνεν αὐτῷ. καὶ ἐκεῖνος ὑπεθάρρησεν, ὅ τε
λέων περιβαλὼν οἷ τὴν οὐρὰν ἦγεν αὐτόν, καὶ
ἀφέντα[4] τὸν πέλεκυν οὐκ εἴα, ἀλλὰ ἐσήμαινε τῷ
ποδὶ ἀνελέσθαι. ὡς δὲ οὐ συνίει, ὁ δὲ τῷ στόματι
ἐλάβετο, καὶ ὤρεξέν οἱ, καὶ εἵπετο ἐκεῖνος, ἄγει
τε αὐτὸν ἐπὶ τὸ αὔλιον. καὶ ἡ λέαινα ⟨ὡς⟩[5] εἶδε,
καὶ αὐτὴ προσελθοῦσα ὑπέσαινε,[6] καὶ ἑώρα
οἰκτρόν, καὶ ἀνέβλεπε πρὸς τὴν ἄρκτον. συνιδὼν
οὖν ὁ ἄνθρωπος καὶ συμβαλὼν ἠδικῆσθαί τι τούτους ἐξ ἐκείνης, ὡς εἶχε ῥώμης τε καὶ χειρῶν,
ἐξέκοψε τὸ δένδρον. καὶ τὸ μὲν ἀνετράπη, ἡ δὲ
κατηνέχθη· καὶ διεσπάσαντό γε[7] οἱ θῆρες αὐτήν.

---

1 ⟨τοὺς⟩ add. H.   2 ἀφίκετο.
3 ἄχους ὡς ἄνθρωπος εἶτα.
4 ἀφιέντα.
5 ⟨ὡς⟩ add. H.

unguarded and slew the Lion's cubs, they being small and unable to protect themselves. But when the father and mother returned from hunting somewhere and saw their young ones slaughtered, they were naturally filled with grief, and set upon the Bear. She in terror ran up a tree as fast as her legs could carry her and sat there trying to escape their fell design. But as they came there with the intention of wreaking vengeance upon the murderer, the Lioness did not relax her watch but sat down beneath the tree-trunk, lying in wait and gazing upward with a look that meant blood. Meantime the Lion in anguish and distraught with grief roamed the mountains and came upon a woodcutter. The man was terrified and dropped his axe, but the animal fawned upon him and reaching upwards greeted him as well as it could, stroking his face with its tongue. And the man took courage, while the Lion, wrapping its tail around him, led him on and would not permit him to leave the axe but signified with its paw that he was to pick it up. But since the man failed to understand, the Lion took it in its mouth and offered it to him; the man followed and the Lion led him to the lair. As soon as the Lioness saw him she too came up and began to fawn upon him with a piteous expression as she looked up at the Bear. So the man grasped their meaning and guessing that they had been somehow injured by the Bear, began to fell the tree with all the strength of his hands. And the tree was overturned and the Bear brought down and the Lions tore her to pieces. As for the man, the Lion

---

6 *Reiske* : ἐπεσήμαινεν MSS, ὑπέσηνε *Jac.*
7 τε.

τὸν δὲ [1] ἄνθρωπον ὁ λέων ἀπαθῆ τε καὶ ἀσινῆ
πάλιν ἐπανήγαγεν ἐς τὸν χῶρον, οὗ πρότερον
ἐνέτυχεν αὐτῷ, καὶ ἀπέδωκε τῇ ἐξ ἀρχῆς ὑλοτομίᾳ.

22. Αἰγυπτίων μάχη θηρίων ἀσπίδος καὶ ἰχνεύμονος.[2] καὶ ὁ μὲν ἰχνεύμων οὐκ ἀβούλως οὐδὲ
ἐκπλήκτως ἐπὶ τὸν ἀγῶνα ἀφικνεῖται τὸν πρὸς τὸν
ἀντίπαλον, ἀλλ' ὡς ἀνὴρ πανοπλίᾳ φραξάμενος,
οὕτως ἐκεῖνος τῷ πηλῷ ἐγκυλίσας [3] ἑαυτὸν καὶ
ἀναπλήσας τοῦ περιπαγέντος ἔοικεν ἔχειν ἀρκοῦν
πρόβλημα καὶ στεγανόν. εἰ δὲ ἀπορίᾳ εἴη πηλοῦ,
λούσας ἑαυτὸν ὕδατι καὶ ἐς ἄμμον βαθεῖαν ὑγρὸν
ἔτι ἐμβαλών, ἐκ τῆσδε τῆς ἐπινοίας τὸ ἀμυντήριον
ἐξ ἀπόρων σπάσας, ἐπὶ τὴν μάχην ἔρχεται. τῆς
τε ῥινὸς τὸ ἄκρον ἁπαλὸν ὂν καὶ [4] ἐγχρίσει τῇ τῆς
ἀσπίδος τρόπον τινὰ ἐκκείμενον φρουρεῖ τὴν
οὐρὰν [5] ἀνακλάσας καὶ ἀποφράξας δι' αὐτῆς αὐτό.[6]
καὶ ἐὰν μὲν ἡ ἀσπὶς τούτου τύχῃ, τὸν ἀνταγωνι-
στὴν καθεῖλεν· εἰ δὲ μή, μάτην τοὺς ὀδόντας τῷ
πηλῷ πονεῖται, πάλιν τε ὁ ἰχνεύμων προσερπύσας
ἀδοκήτως καὶ τοῦ τραχήλου λαβόμενος ἀπέπνιξε
τὴν ἀσπίδα. νικᾷ δὲ ὁ πρῶτος φθάσας.

23. Τρέφειν μὲν τοὺς πατέρας πελαργοὶ γεγηρα-
κότας καὶ ἐθέλουσι καὶ ἐμελέτησαν· κελεύει δὲ
αὐτοὺς νόμος ἀνθρωπικὸς οὐδὲ εἷς τοῦτο, ἀλλὰ
αἰτία τούτων φύσις.[7] οἱ αὐτοὶ δὲ καὶ τὰ ἑαυτῶν
ἔκγονα φιλοῦσι· καὶ τὸ [8] μαρτύριον, ὅταν ὁ

---

[1] τε.
[2] The sentence is incomplete: μάχη⟨ν⟩ . . . ἰχνεύμονος ⟨ἄξιον
ἀκοῦσαι⟩· ὁ μεν i., ex. gr. H.
[3] Schn : κυλίσας.

brought him back untouched and unscathed to the spot where it first met him and restored him to his original task of cutting wood.

22. A battle between two animals of Egypt, the Asp and the Ichneumon. . . . The Ichneumon does not attack his adversary without deliberation or rashly, but like a man fortifying himself with all his weapons, rolls in the mud and covers himself with a hard coating, thereby obtaining, it seems, an adequate and impenetrable defence. But if he is at a loss for mud, he washes himself in water and plunges still wet into deep sand—a device which secures his protection in difficult circumstances—and goes forth to battle. But the tip of his nose, which is sensitive and somewhat exposed to the bite of the Asp, he protects by bending back his tail, thereby blocking the approach to it. If however the Asp can reach it, the snake kills its adversary; otherwise it plies its fangs against the mud in vain, while the Ichneumon on the other hand makes a sudden dash, seizes the Asp by the neck, and strangles it. And the victory goes to the one that gets in first.

Ichneumon and Asp

23. When their parents have grown old, Storks tend them voluntarily and with studied care; yet there is no law of man that bids them do so; the cause of their actions is Nature. And the same birds love their offspring too. Here is the proof: when the full-

The Stork

---

⁴ ἁπαλὸν ὂν καί del. *H.*
⁵ οὐρὰν ὑποκάμψας μᾶλλον καί.
⁶ αὐτὸ οὕτως γὰρ ποιεῖν εἴωθεν.
⁷ φύσις ἀγαθή.   ⁸ τούτου.

τέλειος ἐνδεὴς ᾖ τροφῆς ἀπτῆσιν ἔτι καὶ ἁπαλοῖς
τοῖς νεοττοῖς ἐν τῇ καλιᾷ παραθεῖναι, γενομένης
αὐτῷ κατὰ τύχην ἀπορίας, ὁ δὲ τὴν ἑαυτοῦ
χθιζὴν ἀνεμέσας ἐκείνους τρέφει. καὶ τοὺς ἐρω-
διοὺς ἀκούω ποιεῖν ταὐτόν, καὶ τοὺς πελεκᾶνας
μέντοι. προσακούω δὲ τοὺς πελαργοὺς καὶ αὐτοὺς[1]
συμφεύγειν ταῖς γεράνοις καὶ συναποδιδράσκειν
τὸν χειμῶνα· τῆς ὥρας δὲ τῆς κρυμώδους διελθού-
σης, ὅταν ὑποστρέψωσιν[2] ἐς τὰ ἴδια καὶ οἶδε καὶ
αἶδε, τὴν ἑαυτῶν ἕκαστος καλιὰν ἀναγνωρίζουσιν,
ὡς τὴν οἰκίαν ἄνθρωποι. Ἀλέξανδρος δὲ ὁ
Μύνδιός φησιν,[3] ὅταν ἐς γῆρας ἀφίκωνται, παρελ-
θόντας[4] αὐτοὺς ἐς[5] τὰς Ὠκεανίτιδας νήσους
ἀμείβειν τὰ εἴδη ἐς ἀνθρώπου μορφήν, καὶ
εὐσεβείας γε τῆς ἐς τοὺς γειναμένους ἆθλον τοῦτο
ἴσχειν, ἄλλως τε, ⟨εἴ τι⟩[6] ἐγὼ νοῶ, καὶ ὑποθέσθαι
τῶν θεῶν βουλομένων τοῦτο γοῦν τῶν ἀνθρώπων
τῶν ἐκεῖθι τὸ γένος εὐσεβὲς καὶ ὅσιον, ἐπεὶ
⟨οὐχ⟩[7] οἷόν τε ἦν ἐν τῇ ἄλλῃ τῇ ὑφ᾽ ἡλίῳ[8] τοιοῦτον
διαβιοῦν. καὶ οὔ μοι δοκεῖ μῦθος εἶναι. ἢ τί καὶ
βουλόμενος ὁ Ἀλέξανδρος τοῦτο ἂν ἐτερατεύσατο
κερδαίνων μηδὲ ἕν; ἄλλως τε οὐδ᾽ ἂν ἔπρεπεν
ἀνδρὶ συνετῷ πρὸ τῆς ἀληθείας ποιήσασθαι τὸ
ψεῦδος, οὐδὲ ἐπὶ κέρδει τῷ μεγίστῳ, μή τι γοῦν
ἐς λαβὰς ἐμπεσουμένῳ τὰς ὑπὲρ τῶν τοιούτων
ἀκερδεστάτας.

24. Ἡ χελιδὼν ὅτε[9] εὐποροίη πηλοῦ, τοῖς
ὄνυξι φέρει καὶ συμπλάττει τὴν καλιάν· εἰ δὲ
ἀπορία εἴη, ὡς Ἀριστοτέλης λέγει, ἑαυτὴν βρέχει,

---

[1] αὑτοῖς.   [2] ὑποστρέφωσιν.

grown bird is in want of food to give to its still un
fledged and tender chicks, some accident having
occasioned a shortage, the Stork disgorges its food of
yesterday and feeds its young. And I am told that
Herons do the same, and Pelicans also.

I learn further that Storks migrate along with *its migra-*
Cranes and all together avoid the winter. But when *tions*
the season of frost is over and both Storks and Cranes
return to their own homes, each kind recognises its
own nests, as men do their own houses.

Alexander of Myndus asserts that when they reach *transformed*
old age they pass to the islands of Ocean and are *into a*
transformed into human shape, and that this is a re- *human*
ward for their filial piety towards their parents, since, *being*
if I am not mistaken, the gods especially desire to
hold up there if nowhere else a human model of piety
and uprightness, for in no other country under the
sun could such a race continue to exist. This is in
my opinion no fairy-tale, otherwise what was Alexander's design in relating such marvels when he had
nothing to gain from it? Anyhow it would have ill
become an intelligent man to sacrifice truth to false-
hood, be the gain never so great, still less if he was
going to fall into an opponent's grasp, from which
act nothing whatsoever was to be gained.

24. Whenever there is plenty of mud the Swallow *The Swallow*
brings it in her claws and builds her nest. If how- *and its nest*
ever mud is lacking, as Aristotle says [*HA* 612 b 23],

---

3 φησιν, τῶν πελαργῶν τοὺς ἅμα βιώσαντας.
4 περιελθόντας.   5 ὡς.
6 ⟨εἴ τι⟩ add. H.   7 ⟨οὐχ⟩ add. Ges.
8 Jac : ὑφηλίῳ.   9 Ges : ὅταν.

καὶ ἐς κόνιν ἐμπεσοῦσα [1] φύρει τὰ πτερά, καὶ τοῦ πηλοῦ περιπαγέντος, ἐντεῦθεν ὑπαποψήχουσα τῷ ῥάμφει τὴν προκειμένην οἰκοδομίαν χειρουργεῖ. ἀπαλά τε ὄντα τὰ νεόττια καὶ τῶν πτίλων γυμνὰ οἶδε καλῶς ἐπὶ ψιλῶν καρφῶν εἰ ἀναπαύοιτο ὅτι κολασθήσεται ἀλγοῦντα. οὐκοῦν ἐπὶ τὰ νῶτα τῶν προβάτων ἱζάνει, καὶ ἀποσπᾷ τοῦ μαλλοῦ, καὶ ἐντεῦθεν τοῖς ἑαυτῆς βρέφεσι τὸ λέχος μαλακὸν ἔστρωσεν.

25. Δικαίους ἡ μήτηρ ἡ χελιδὼν τοὺς ἑαυτῆς νεοττοὺς ἐργάζεται, τὸ ἰσότιμον αὐτοῖς διὰ τῆς τροφῆς τῆς ἴσης φυλάττουσα· μίαν δὲ ἄρα οὐ κομίζει πᾶσιν, ἐπεὶ μηδὲ δύναται· ἀλλὰ μικρὰ μὲν καὶ ὀλίγα ἐστὶν ὅσα ἄγει, τὸν πρῶτον δὲ τεχθέντα πρῶτον τρέφει, δεύτερον δὲ τὸν ἐπ᾽ ἐκείνῳ, καὶ τρίτον σιτίζει τὸν τῆς τρίτης ὠδῖνος, καὶ μέχρι τοῦ πέμπτου πρόεισι τὸν τρόπον τοῦτον· οὔτε γὰρ κύει χελιδὼν πλείονας οὔτε τίκτει. αὐτὴ δὲ τοσοῦτον κατασπᾷ τῆς τροφῆς, ὅσον ἂν ἐν τῇ καλιᾷ κερδᾶναι δυνηθῇ παραρρεῦσαν αὐτῇ. βραδέως δὲ ἐκβλέπει τὰ [2] ταύτης βρέφη, ὡς καὶ τὰ τῶν κυνῶν σκυλάκια· πόαν δὲ κομίζει καὶ προσάγει, τὰ δὲ ὑπαναβλέπει, εἶτα ἀτρεμήσαντα ὀλίγον ἐκπετήσιμα ὄντα πρόεισι τῆς καλιᾶς ἐπὶ τὴν νομήν.[3] ταύτης τῆς πόας ἄνθρωποι γενέσθαι ἐγκρατεῖς διψῶσι, καὶ οὐδέπω [4] τῆς σπουδῆς κατέτυχον.

26. Οἱ ἔποπές εἰσιν ὀρνίθων ἀπηνέστατοι, καί μοι δοκοῦσι τῶν προτέρων τῶν ἀνθρωπικῶν ἐν

[1] ἐμπεσοῦσα after πτερά.

she souses herself in water and plunging into dust befouls her feathers. Then when the mud has stuck to her all over, she scrapes it off by degrees with her beak and constructs her proposed dwelling. And as her young are tender and unfledged, she knows full well that if she lets them rest on bare twigs, they will suffer and be in pain. Accordingly she settles on the backs of sheep, plucks some wool, and with it makes their bed soft for her offspring.

25. The mother Swallow trains her young ones to be just by carefully distributing food in equal portions. So she does not bring one meal for all, because she is not able to do so, but brings small objects and a few at a time; she feeds the first-born first, after it the second, thirdly her third offspring, proceeding as far as the fifth in the same way; for the Swallow neither conceives nor hatches more than five. She herself only consumes as much food as she can obtain in the nest, that is, anything that is dropped beside it. Her young are slow to open their eyes, in the same way as puppies. But she collects and brings a herb,[a] and they by degrees gain their sight; then after remaining quiet for a while, when able to fly, they leave the nest to seek for food. Men long to possess this herb but have not yet obtained their desire. *The Swallow and its young*

26. Among birds Hoopoes are the most savage; and in my opinion it is due to the recollection of their *The Hoopoe*

[a] Pliny (*HN* 8. 27; 25. 8) calls it *chelidonia, i.e.* Greater celandine.

---

[2] καὶ τά.   [3] *Ges:* τῆς νομῆς.   [4] οὐδέπω νῦν.

μνήμῃ καὶ μέντοι καὶ μίσει τοῦ γένους τοῦ τῶν
γυναικῶν ὑποπλέκειν τὰς καλιὰς ἐν ταῖς ἐρήμοις
καὶ τοῖς πάγοις τοῖς ὑψηλοῖς· καὶ ὑπὲρ τοῦ μὴ
προσιέναι τοὺς ἀνθρώπους αὐτῶν τοῖς βρέφεσιν οἴ-
δε ἀντὶ τοῦ πηλοῦ χρίουσι τὰς καλιάς, ἀποπάτημα
ἀνθρώπου περιβαλόντες, τῇ δυσωδίᾳ τε καὶ κακοσμίᾳ ἀνείργοντες καὶ ἀναστέλλοντες τὸ ζῷον τὸ
ἑαυτοῖς πολέμιον. ἔτυχε δὲ καὶ ἐν τῷ τείχους [1]
ἐρημοτέρῳ ὅδε ὁ ὄρνις παιδοποιησάμενος ἔν τινι
ῥήγματι λίθου ὑπὸ τοῦ χρόνου διαστάντι. οὐκοῦν
ὁ τοῦ τείχους μελεδωνὸς ἰδὼν ἔνδον τὰ βρέφη
κατήλειψε τὸν χηραμὸν τῷ πηλῷ. καὶ ὑποστρέψας
ὁ ἔποψ, ὡς εἶδεν αὐτὸν ἀποκλεισθέντα, πόαν
ἐκόμισε, καὶ προσήνεγκε τῷ πηλῷ· ὁ δὲ κατερρύη,
καὶ προσῆλθε πρὸς τὰ αὑτοῦ ἐκεῖνος τέκνα, εἶτα
ἐπὶ ⟨τὴν⟩ [2] νομὴν ᾖξεν. αὖθις οὖν ὁ αὐτὸς ἐπήλει-
ψεν ἄνθρωπος, καὶ ὁ [3] ὄρνις τῇ αὐτῇ πόᾳ ἀνέῳξε
τὸν χηραμόν· καὶ τὸ τρίτον ἐπράχθη τὰ αὐτά. ὁ
τοίνυν τοῦ τείχους φύλαξ ἰδὼν τὸ πραττόμενον,
τὴν πόαν [4] ἀνελόμενος ἐχρῆτο οὐκ ἐς τὰ αὐτά,
ἀλλ᾽ ἀνέῳγεν [5] μηδέν οἱ προσήκοντας θησαυρούς.

27. Ἡ Πελοπόννησος λεόντων ἄγονός ἐστι· καὶ
οἷα [6] εἰκὸς Ὅμηρος πεπαιδευμένῃ φρενὶ συνιδὼν
τοῦτο τὴν Ἄρτεμιν ἐκεῖθι θηρῶσαν ᾄδων εἶπεν
ὅτι ἄρα ἔπεισι τόν τε Ταΰγετον καὶ τὸν Ἐρύμανθον

τερπομένη κάπροισι καὶ ὠκείης ἐλάφοισιν.

---

[1] τείχους A, τοῦ τ. most MSS.
[2] ⟨τὴν⟩ add. H.
[3] ἤ.
[4] συντεθείσης τῆς πόας.
[5] ἀνοίγων.
[6] ὅσα γε.

former existence as human beings and more especially from their hatred of the female sex,[a] that they build their nests in desolate regions and on high rocks; and to prevent human beings from getting near their young they smear their nests not with mud but with human excrement, and by dint of its disgusting and evil smell they repel and keep away the creature that is their enemy.

It happened that this bird had raised a family in the deserted part of a fortress, in the cleft of a stone that had split with age. So the guardian of the fortress, observing the young birds inside, smeared the hole over with mud. When the Hoopoe returned and saw itself excluded, it fetched a herb and applied it to the mud. The mud was dissolved; the bird reached its young, and then flew off to get food. So once again the man smeared the spot over, and the bird by means of the same herb opened the hole. And the same thing happened a third time. Therefore the guardian of the fortress, seeing what was done, himself gathered the herb and used it not for the same purpose; instead he laid open treasures that were none of his.

27. The Peloponnese does not breed Lions, and Homer (as you would expect) with his trained intelligence realising the fact, says in singing of Artemis and her hunting there that she passes over Taygetus [b] and Erymanthus

'delighting in boars and swift-footed stags'
[*Od.* 6. 104].

The Peloponnese devoid of Lions

[a] See 2. 3 n.
[b] Mountain range to the W and S of Sparta.—Erymanthus, mt on the borders of Achaia and Arcadia.

ἐπεὶ δὲ [1] ἔρημα λεόντων τάδε τὰ ὄρη, καὶ μάλα γε
εἰκότως οὐκ ἐμνήσθη αὐτῶν.

28. Γίνεται δὲ ἐν τῇ Ἐρυθρᾷ θαλάττῃ ἰχθύς,
καὶ ὅσα γε εἰδέναι ἐμέ, ἔθεντο Περσέα ⟨οἱ⟩[2]
ἐπιχώριοι ὄνομα αὐτῷ. καὶ οἱ μὲν Ἕλληνες
αὐτὸν οὕτω, καλοῦσι δὲ καὶ Ἄραβες ὁμοίως τοῖς
Ἕλλησι· Διὸς γὰρ υἱὸν καὶ ἐκεῖνοι ᾄδουσι τὸν
Περσέα, καὶ ἀπ᾽ αὐτοῦ γε τὸν ἰχθὺν ὑμνοῦσι
λέγεσθαι. μέγεθος μὲν οὖν ἐστι κατὰ τὸν ἀνθίαν
τὸν μέγιστον, ἰδεῖν δὲ ὅμοιος λάβρακι· γρυπός γε
μὴν ἡσυχῇ οὕτω, καὶ ζώναις πεποίκιλται χρυσῷ
προσεικασμέναις· ἄρχονται δὲ ἀπὸ τῆς κεφαλῆς
ἐπικάρσιοι αἱ ζῶναι, καὶ ἐς τὴν γαστέρα κατα-
λήγουσι. πέφρακται δὲ ὀδοῦσι μεγάλοις καὶ
πυκνοῖς. λέγεται δὲ ἰχθύων περιεῖναι ῥώμῃ τε
σώματος καὶ βίᾳ· ἀλλὰ οὐδὲ τόλμης οἱ ἐνδεῖ.
θήραν δὲ αὐτοῦ καὶ ἄγραν εἶπον ἀλλαχόθι.

29. Ἡ πίννη θαλάττιον ζῷον, καὶ ἔστι τῶν
ὀστρείων. κέχηνε δὲ τῇ διαστάσει τῶν περικει-
μένων ὀστράκων, καὶ προτείνει σαρκίον ἐξ ἑαυτῆς
οἱονεὶ δέλεαρ τοῖς παρανηχομένοις τῶν ἰχθύων.
καρκίνος δὲ αὐτῇ παραμένει σύντροφός τε καὶ
σύννομος. οὐκοῦν ὅταν τις τῶν ἰχθύων προσνέῃ,
ὁ δὲ ὑπένυξεν ἡσυχῇ αὐτήν· καὶ ἡ πίννη μᾶλλον
ἀνέῳξεν ἑαυτήν, καὶ ἐδέξατο ἔσω τοῦ ἐπιόντος
ἰχθύος τὴν κεφαλὴν (καθίησι γὰρ ὡς ἐπὶ τροφῇ)
καὶ ἐσθίει αὐτήν.

---

[1] ἐπειδή.   [2] ⟨οἱ⟩ add. Schn.

---

[a] Not in any surviving work.

And since these mountains are destitute of Lions he was quite right not to mention them.

28. There occurs in the Red Sea a fish, and, so far The Perseus fish as I know, the people there have given it the name of *Perseus*. And the Greeks call it so, and the Arabians in like manner with the Greeks. For they too call Perseus the son of Zeus, and it is after him that they declare the fish is named. Its size is that of the largest anthias; in appearance it is like a basse; its nose is somewhat hooked, and it is dappled with rings as it were of gold round its body, and these rings begin at the head at right angles to it and cease at the belly. It is armed with large teeth set close. It is said to surpass other fish in the strength and power of its body, neither is it wanting in courage. How to fish for it and how to catch it I have explained elsewhere.[a]

29. The Pinna is a marine creature and belongs to Pinna and Crab the class of bivalves. It opens by parting the shells that enclose it, and extends a small piece of its flesh like a bait to fish that swim by. The Crab however remains by its side, sharing its food and its feeding-ground. So when some fish comes swimming up, the Crab gives the Pinna a gentle prick, whereat the Pinna opens its shell wider and admits the head of the approaching fish—for it lowers its head to feed—and eats it.

30. Ἦν δὲ ἄρα οἰκεῖα τῷ πεπαιδευμένῳ καὶ ταῦτα εἰδέναι. σοφώτατος ὁ κόκκυξ καὶ πλέκειν εὐπόρους ἐξ ἀπόρων μηχανὰς δεινότατος. ἑαυτῷ μὲν γὰρ συνεπίσταται ἐπῳάζειν οὐ δυναμένῳ καὶ ἐκλέπειν διὰ ψυχρότητα τῆς ἐν τῷ σώματι συγκράσεως, ὥς φασιν. οὐκοῦν ὅταν τίκτῃ, οὔτε αὐτὸς νεοττιὰν ὑποπλέκει οὔτε τιθηνεῖται τὰ βρέφη, φυλάττει δὲ ἄρα τοὺς τῶν νεοττιῶν δεσπότας ἀφεστῶτας καὶ πλανωμένους, καὶ παρελθὼν ἐς καταγωγὴν ὀθνείαν ἐντίκτει. οὐ πάντων δὲ ὀρνίθων καλιαῖς ἐπιπηδᾷ οὗτός γε, ἀλλὰ κορύδου καὶ φάττης καὶ χλωρίδος καὶ πάππου· τούτοις γὰρ συνεπίσταται ὅμοια αὐτῷ ᾠὰ τίκτουσι. καὶ κενῶν μὲν αὐτῶν οὐσῶν, οὐκ ἂν παρέλθοι· ᾠῶν δὲ ἔνδον ὄντων εἶτα μέντοι τὰ ἑαυτοῦ παρενέμιξεν. ἐὰν δὲ ᾖ πολλὰ τὰ ἐκείνων, τὰ μὲν ἐκκυλίσας ἠφάνισε, τὰ δὲ ἑαυτοῦ κατέλιπε, διαγνωσθῆναί τε καὶ φωραθῆναι δι' ὁμοιότητα μὴ δυνάμενα. καὶ οἱ μὲν ὄρνιθες οἱ προειρημένοι τὰ μηδέν σφισι προσήκοντα ἐκγλύφουσιν, ὑποπηγνύμενα δὲ ἐκεῖνα ἑαυτοῖς συνεγνωκότα τὴν νοθείαν ἐκπέτεταί τε καὶ παρὰ τὸν γειναμένον στέλλεται· τῶν γὰρ πτερῶν αὐτοῖς περιχυθέντων γνωρίζεται ἀλλότρια ὄντα, καὶ αἰκίζεται πικρότατα. ὁρᾶται [1] δὲ μίαν ὥραν τοῦ ἔτους τὴν ἀρίστην ὁ κόκκυξ· ἦρος γὰρ ὑπαρχομένου καὶ αὐτὸς ἐμφανής ἐστιν ἐς ἀνατολὰς Σειρίου, εἶτα τῆς τῶν πολλῶν ὄψεως ἀνεχώρησεν.

31. Ἀλεκτρυόνα φοβεῖται λέων. καὶ βασιλίσκος δὲ τὸν αὐτὸν ὄρνιν, ὥς φασιν, ὀρρωδεῖ, καὶ κατιδὼν τρέμει, καὶ ἀκούων ᾄδοντος σπᾶταί τε καὶ ἀποθνή-

---

[1] καὶ ὁρᾶται.

30. It seems after all fitting that an educated man The Cuckoo
should be acquainted with these facts as well. The
Cuckoo is extremely clever and most adroit at devising ingenious solutions to difficulties. For the bird is
conscious that it cannot brood and hatch eggs because
of the cold nature of its bodily constitution, so they
say. Therefore, when it lays its eggs, it neither builds
itself a nest nor nurses its young, but watches until
birds that have nestlings are flown and abroad, enters
the strange lodging, and there lays its eggs. The
rascal does not however assail the nests of all birds,
only those of the lark, the ring-dove, the greenfinch,
and the pappus,[a] knowing as it does that these birds
lay eggs resembling its own. And if the nests are
empty, it will not go near them, but if they contain
eggs, then it mixes its own with them. But if the
eggs of the other bird are numerous, it rolls them out
and destroys them and leaves its own behind, their
resemblance making it impossible to know them
apart and detect them. And the aforesaid birds
hatch the eggs which are none of theirs. But when
the Cuckoo's young have grown strong and are conscious of their bastardy, they fly away and resort to
their parent. For directly they are fledged they are
recognised as alien and are grievously ill-treated.

The Cuckoo is seen only at one season, and that the
best, of the year. For it is actually visible from the
beginning of spring until the rising of the Dog-star; [b]
after that it withdraws from the sight of man.

31. The Lion dreads a Cock, and the Basilisk too, The Cock,
they say, goes in fear of the same bird: at the sight feared by
of one it shudders, and at the sound of its crowing it Basilisk

---

[a] Unknown bird.   [b] About mid-July.

σκει. ταῦτα ἄρα καὶ οἱ τὴν Λιβύην ὁδοιποροῦντες
τὴν τῶν τοιούτων τροφὸν δέει τοῦ προειρημένου
βασιλίσκου εἶτα μέντοι συνέμπορον καὶ κοινωνὸν
τῆς ὁδοῦ τὸν ἀλεκτρυόνα ἐπάγονται, ὅσπερ οὖν τὸ
τηλικοῦτον κακὸν ἀπαλλάξει αὐτοῖς.

32. Ἡ Κρήτη καὶ τοῖς λύκοις καὶ τοῖς ἑρπετοῖς
θηρίοις ἐχθίστη ἐστίν. ἀκούω ⟨δὲ⟩[1] Θεοφράστου
λέγοντος καὶ ἐν τῷ Μακεδονικῷ Ὀλύμπῳ τοῖς
λύκοις ἄβατα εἶναι. αἶγες δὲ ἄρα αἱ Κεφαλληνίδες
οὐ πίνουσι μηνῶν ἕξ. οἷς δὲ Βουδινὰς[2] οὐκ ὄψει
λευκάς, ὥς φασι, μελαίνας δὲ πάσας. διαφορότης
δὲ ἄρα τῶν ζῴων καὶ ἰδιότης εἴη ἂν καὶ ταύτῃ·
τὰ μὲν γὰρ αὐτῶν ἐστι δακετὰ καὶ ἐνίησιν ἀπὸ
τοῦ ὀδόντος φάρμακον, βλητικὰ[3] δὲ ὅσα παίσαντα
εἶτα μέντοι καὶ ἐκεῖνα τὸ[4] τοιοῦτον κακὸν
ἐνίησιν.

33. Ἡ Λίβυσσα δ' ἀσπίς, ἀκούω, τὸν πρὸς τὸ
φύσημα αὐτῆς ἀντιβλέψαντα[5] τυφλοῖ τὴν ὄψιν·
ἡ δὲ ἄλλη οὐ τυφλοῖ μέν, ἀποκτείνει δὲ ῥᾷστα.
Λέγονται δὲ βόες Ἠπειρωτικαὶ πλεῖστον ὅσον
ἀμέλγεσθαι καὶ αἶγες αἱ Σκύριαι γάλα ἀφθονώτα-
τον παρέχειν, ὅσον οὐκ ἄλλαι αἶγες. αἱ δὲ
Αἰγύπτιαι ἔστιν αἷ[6] πέντε ἀποτίκτουσι,[7] καὶ αἱ
πλεῖσται δίδυμα. λέγεται δὲ αἴτιος ὁ Νεῖλος
εἶναι, εὐτεκνότατον παρέχων ὕδωρ. ἔνθεν τοι καὶ
τῶν νομέων τοὺς ἄγαν φιλοκάλους καὶ τῆς ποίμνης
τῆς σφετέρας ἔχοντας πεφροντισμένως ὕδωρ ἐκ
τοῦ Νείλου ταῖς ἑαυτῶν ἀγέλαις ἄγειν μηχανῇ

---

[1] ⟨δὲ⟩ add. H. [2] Ἀβυδηνάς.
[3] Schn : βλητά. [4] τι.

is seized with convulsions and dies. This is why travellers in Libya, which is the nurse of such monsters, in fear of the aforesaid Basilisk take with them a Cock as companion and partner of their journey to protect themselves from so terrible an infliction.

32. Crete is exceedingly hostile to wolves and reptiles; and I learn from Theophrastus *a* that there are places on Macedonian Olympus where wolves do not go. Goats in Cephallenia go without drinking for six months. Among the Budini,*b* they say, you will not see a white sheep: they are all black. *Local peculiarities*

It seems that one peculiarity that distinguishes animals consists in this: some bite and inject poison from a fang, while others are given to striking, and having struck also inject a like deadly substance.

33. The Libyan Asp, I am told, blinds the sight of the man who faces its breath. But the other kind does not indeed blind but kills at once. *The Asp in Libya*

It is said that the Cows of Epirus give a most copious supply of milk, and the Goats of Scyros a far more generous yield than any other goats. And there are Goats in Egypt that produce quintuplets, while most produce twins. The Nile is said to be the cause of this, as the water it provides is extremely progenitive. For that reason shepherds who like fine flocks and devote much care to them have a device for drawing as much water as is possible from the Nile *Goats in Scyros, in Egypt*

---

a There is no such statement in his extant remains.
b The Budini were a tribe living N of the Sea of Azov.

---

5 ἀντιβλέψαντα ὅταν πρησθῇ τὸν τράχηλον.
6 ἑκάστη.   7 ἀποτίκτει.

ὅσον δυνατόν ἐστι, καὶ ταῖς γε στερίφαις [1] ἔτι καὶ
μᾶλλον.

34. Πτολεμαίῳ τῷ δευτέρῳ φασὶν ἐξ Ἰνδῶν
κέρας ἐκομίσθη, καὶ τρεῖς ἀμφορέας ἐχώρησεν.
οἷος [2] ἄρα ὁ βοῦς ἦν, ὡς ἐκπεφυκέναι οἱ τηλικοῦτον
κέρας.

35. Περδίκων φθέγμα ἓν οὐδέποτ' ἂν ἀκούσειας [3]
ἁπάντων, ἀλλὰ ἔστι διάφορα. καὶ Ἀθήνησί γε
οἱ ἐπέκεινα τοῦ Κορυδαλλέων δήμου ἄλλο [4]
ἠχοῦσι, καὶ οἱ ἐπίταδε ἄλλο. τίνα δέ ἐστι τοῖς
φθέγμασι τὰ ὀνόματα, ἐρεῖ Θεόφραστος. ἐν δὲ
τῇ Βοιωτίᾳ καὶ τῇ ἀντιπέρας Εὐβοίᾳ ὁμόφωνοί τέ
εἰσι καὶ ὡς ἂν εἴποι τις ὁμόγλωττοι. ἄφωνα δέ
ἐστι τὸ παράπαν ἐν Κυρήνῃ μὲν οἱ βάτραχοι, ἐν
Μακεδονίᾳ δὲ ὗς. καὶ τεττίγων τι γένος, ἄφωνοι
καὶ οὗτοι.

36. Γένος φαλαγγίου φασὶν εἶναι, καλοῦσι δὲ
ῥᾶγα τὸ φαλάγγιον, εἴτε ὅτι μέλαν ἐστὶ καὶ τῷ
ὄντι προσέοικε σταφυλῆς ῥαγὶ καὶ πως ὁρᾶται καὶ
περιφερές, εἴτε δι' αἰτίαν ἑτέραν.[5] γίνεται δὲ ἐν
τῇ Λιβύῃ, καὶ ἔχει πόδας μικρούς [6]· στόμα δὲ
εἴληχεν ἐν μέσῃ τῇ γαστρί, καὶ ἔστιν ἀποκτεῖναι
τάχιστον.

37. Ἐν Σερίφῳ βάτραχοι, τὸ παράπαν οὐκ ἂν
αὐτῶν ἀκούσειας [7] φθεγγομένων· εἰ δὲ αὐτοὺς
κομίσειας [8] ἀλλαχόθι, διάτορόν τε [9] καὶ τραχύτα-

[1] ταῖς στερίφαις γε.   [2] ὅσος conj. H, οὗτος AL.
[3] οὐδέποτε ἀκούσαις.

196

for their herds, especially for animals that are barren.

34. They say that a horn was brought from the Indies to Ptolemy II, and it held three *amphorae.*[a] Imagine an ox that could produce a horn of that size.

35. You would never hear the same note from all Partridges, but they vary. At Athens for instance those on the far side of the deme Corydallus emit one note, those on this side another. What names these notes have Theophrastus will tell us [*fr.* 181]. But in Boeotia and on the opposite shore of Euboea they have the same note and, as it were, the same language. In Cyrene the Frogs are completely dumb; in Macedonia, the Pigs; and there is also a kind of Cicada that is dumb.

36. There is a kind of Spider which they call the 'Grape-spider,' either because it is dark and does in fact resemble a grape in a bunch—it has a somewhat spherical appearance—or for some other reason. It occurs in Libya and has short legs; it has a mouth in the middle of its belly, and can kill in a twinkling.

37. In Seriphus you will never hear the Frogs croaking at all. If however you transport them elsewhere, they emit a piercing and most harsh sound.

---

[a] About 26 gallons.

---

[4] ἄλλο γε.
[5] ἑτέραν, καταγνῶναι τοῦτο ῥᾷον οὐκ ἐστι.
[6] *Ges* : μακρούς.   [7] ἀκούσαις.
[8] κομίσαις.   [9] τι.

τον ἠχοῦσιν. ἐν Πιέρῳ δὲ τῆς Θετταλίας λίμνη [1]
ἐστίν, οὐκ ἀέναος, ἀλλὰ χειμῶνος ἐκ τῶν συρ-
ρεόντων ἐς αὐτὴν ὑδάτων τίκτεται. οὐκοῦν ἐὰν
ἐμβάλῃ τις βατράχους ἐς αὐτήν, σιωπῶσιν,
ἀλλαχοῦ φθεγγόμενοι. ὑπὲρ δὲ τῶν Σερίφων
βατράχων κομπάζουσι Σερίφιοι ἐλθεῖν ἐκ τοῦ
κατὰ τῆς Γοργόνος ἄθλου τὸν Περσέα πολλὴν
περιελθόντα γῆν, καὶ οἷα εἰκὸς καμόντα ἀνα-
παύσασθαι τῆς λίμνης πλησίον καὶ κατακλιθῆναι
ὕπνου δεόμενον. τοὺς δὲ βατράχους βοᾶν καὶ
ἐρεσχελεῖν τὸν ἥρωα καὶ τὸν ὕπνον αὐτῷ διακόπτειν [2]· τὸν Περσέα δὲ εὔξασθαι τῷ πατρὶ τοὺς
βατράχους κατασιγάσαι. τὸν δὲ ὑπακοῦσαι καὶ
χαριζόμενον τῷ υἱεῖ τῶν ἐκεῖθι βατράχων αἰώνιον
σιγὴν καταψηφίσασθαι. λέγει δὲ Θεόφραστος
ἐκβάλλων τὸν μῦθον καὶ Σερίφιους τῆς ἀλαζονείας
παραλύων τὴν τοῦ ὕδατος ψυχρότητα αἰτίαν εἶναι
τῆς ἀφωνίας τῶν προειρημένων.

38. Ἐν τοῖς ὑγροῖς χωρίοις καὶ ἔνθα νοτιώτα-
τος [3] ὁ ἀὴρ ὑπεράγαν, οἱ ἀλεκτρυόνες οὐκ ᾄδουσι,
φησὶ Θεόφραστος. ἡ δὲ ἐν Φενεῷ λίμνη ἰχθύων
ἄγονός ἐστι. ψυχροὶ δὲ ἄρα ὄντες τὴν σύγκρασιν
οἱ τέττιγες εἶτα μέντοι πυρούμενοι τῷ ἡλίῳ [4]
ᾄδουσι, ἐκεῖνος λέγει.

39. Τολμηρότατος [5] ⟨δὲ⟩ [6] ἄρα ζώων ὁ αἰγιθήλας
ἦν· τῶν μὲν γὰρ ὀρνίθων ὑπερφρονεῖ τῶν μικρῶν,
ἐπιτίθεται δὲ ταῖς αἰξὶ κατὰ τὸ καρτερόν, καὶ
μέντοι ⟨καὶ⟩ [7] τοῖς οὔθασιν αὐτῶν προσπετόμενος

---

[1] Ges: λίμνη ἥ.
[2] διακόπτειν καὶ λυπεῖν δηλονότι.

On mount Pierus in Thessaly there is a lake; it is not perennial but is created in winter by the waters which flow together into it. Now if one throws Frogs into it they become silent, though vocal elsewhere. Touching the Seriphian Frogs the people of Seriphus boast that Perseus arrived from his contest with the and Perseus Gorgon after covering an immense distance, and being naturally fatigued rested by the lake side and lay down wishing to sleep. The Frogs however worried the hero with their croaking and interrupted his slumbers. But Perseus prayed to his father to silence the Frogs. His father gave ear and to gratify his son condemned the Frogs there to everlasting silence. Theophrastus however upsets the story [*fr*. 186] and relieves the Seriphians of their imposture by asserting that it is the coldness of the water that causes the aforesaid Frogs to be dumb.

38. In moist places and where the air is excessively Local pecu- damp Cocks do not crow, according to Theophrastus liarities [*fr*. 187]. And the lake at Pheneus produces no fish. It is because Cicadas are constitutionally cold that, when warmed by the sun, they sing, says the same writer.

39. It seems that the Goatsucker is the most The Goat- audacious of creatures, for it despises small birds but sucker assails goats with the utmost violence, and more than that, it flies to their udders and sucks out the milk

---

³ νοτιώτερος.

⁴ *Jac*: πυρουμένου τοῦ ἡλίου.

⁵ τολμηρότατον *Ges*, τολμηρότερον.

⁶ ⟨δέ⟩ *add. H*.

⁷ ⟨καί⟩ *add. H*.

εἶτα ἐκμυζᾷ τὸ γάλα,[1] καὶ τὴν τιμωρίαν τὴν ἐκ
τοῦ αἰπόλου οὐ δέδοικε, καίτοι πονηρότατον
αὐταῖς μισθὸν ὑπὲρ τῆς πλησμονῆς ἀποδιδούς·
τυφλοῖ γὰρ τὸν μαστόν,[2] καὶ ἀποσβέννυσι τὴν
ἐκεῖθεν ἐπιρροήν.

40. Μητροδίδακτον μὲν τὸν τῆς Ἀρήτης[3] υἱὸν
τὸν τῆς ἀδελφῆς τῆς Ἀριστίππου ὑμνοῦσιν οἱ
πολλοί· λέγει δὲ Ἀριστοτέλης ἰδεῖν αὐτὸς τὰ
νεόττια τῆς ἀηδόνος ὑπὸ τῆς μητρὸς διδασκόμενα
ᾄδειν. ἦν δὲ ἄρα ὀρνίθων ἡ ἀηδὼν ἐλευθερίας
ἐράστρια ἰσχυρῶς, καὶ διὰ ταῦτα ἡ ἐντελὴς τὴν
ἡλικίαν ὅταν θηραθῇ καὶ καθειργμένη ᾖ,[4] ᾠδῆς[5]
ἀπέχεται, καὶ ἀμύνεται τὸν ὀρνιθοθήραν ὑπὲρ τῆς
δουλείας τῇ σιωπῇ. οὗπερ οὖν οἱ ἄνθρωποι
πεπειραμένοι, τὰς μὲν ἤδη πρεσβυτέρας[6] μεθιᾶσι,
σπουδάζουσι δὲ θηρᾶν τὰ νεόττια.

41. Ἵππους μονόκερως γῆ Ἰνδικὴ τίκτει, φασί,
καὶ ὄνους μονόκερως ἡ αὐτὴ τρέφει, καὶ γίνεταί
γε ἐκ τῶν κεράτων τῶνδε ἐκπώματα. καὶ εἴ τις
ἐς αὐτὰ ἐμβάλοι φάρμακον θανατηφόρον, ὁ πιών,
οὐδὲν ἡ ἐπιβουλὴ λυπήσει αὐτόν· ἔοικε γὰρ
ἀμυντήριον τοῦ κακοῦ τὸ κέρας καὶ τοῦ ἵππου καὶ
τοῦ ὄνου εἶναι.

42. Ὁ πορφυρίων ὡραιότατός τε ἅμα καὶ
φερωνυμώτατός ἐστι ζῴων, καὶ χαίρει κονιόμενος,

---

[1] ἐκ τοῦ γάλακτος.
[2] μαστὸν ὅταν σπάσῃ MSS, ὃν ἂν σ. Jac.
[3] Cas : Ἀρίστης.
[4] ᾖ ἐν τῷ οἰκίσκῳ φυλάττεται.

without any fear of vengeance from the goatherd,
although it makes the basest return for being filled
with milk, for it makes the dug 'blind' and staunches
its flow.

40. Many people sing the praises of the son of The
Arete, the sister *a* of Aristippus, as being taught by Nightingale
his mother. Aristotle says [*HA* 536 b 17] that he
has with his own eyes seen the young of the Nightingale being instructed by their mother how to sing.
It seems that the Nightingale passionately loves its
freedom, and for that reason when a mature bird is
caught and confined in a cage, it refrains from song
and takes vengeance on the birdcatcher for its enslavement by silence. Consequently men who have
had this experience let them go when they are older
and do their best to catch the young.

41. India produces horses with one horn, they say, The Horn
and the same country fosters asses with a single horn. of the
And from these horns they make drinking-vessels, Unicorn
and if anyone puts a deadly poison in them and a man
drinks, the plot will do him no harm. For it seems
that the horn both of the horse and of the ass is an
antidote to the poison.

42. The Purple Coot is the most beautiful and the The Purple
most appropriately named of creatures, and it de- Coot

---

a Arete was the daughter, not the sister, of Aristippus, and
her son was called after his grandfather.

---

⁵ καὶ τροφῶν καὶ ᾠδῆς.
⁶ πρεσβυτέρας καὶ ἀλούσας.

ἤδη δὲ καὶ λοῦται[1] τὸ τῶν περιστερῶν λουτρόν·
οὐ πρότερον δὲ ἑαυτὸν ἐπιδίδωσι ταῖς κονίστραις
καὶ τοῖς λουτροῖς, πρὶν ἂν βαδίσῃ τινὰ ἀριθμὸν
βαδίσεων[2] ἀρκοῦντά οἱ. σιτούμενος δὲ ἐπὶ μαρτύ-
ρων ἄχθεται, καὶ διὰ ταῦτα ἀναχωρεῖ, καὶ
ὑπολανθάνων ἐσθίει. ζηλότυπος δέ ἐστιν ἰσχυρῶς,
καὶ τὰς ὑπάνδρους τῶν γυναικῶν παραφυλάττει,
καὶ ἐὰν καταγνῷ μοιχεύεσθαι τῆς οἰκίας τὴν
δέσποιναν, ἀπάγχει ἑαυτόν. οὐ πέτεται δὲ ὑψηλός.
χαίρουσί γε μὴν οἱ ἄνθρωποι αὐτῷ, καὶ τρέφουσι
πεφεισμένως καὶ προμηθῶς αὐτόν. καὶ ἔοικεν ἢ
σοβαρᾶς οἰκίας καὶ μέγα πλουσίας ἄθυρμα εἶναι,
ἢ ὑποδέχεται νεὼς αὐτόν, καὶ ἄφετος ἀλᾶται καὶ
ἱερὸς περίεισιν ἔσω περιβόλου. τὸν ταῶν μὲν
οὖν ὡραῖον ὄντα καὶ καταθύουσι καὶ σιτοῦνται οἱ
ἄσωτοι· τοῦ γὰρ ὄρνιθος τὰ μὲν πτερὰ κόσμος
ἐστί, τὸ δὲ σῶμα ἤ τι ἢ οὐδέν.[3] πορφυρίωνα δὲ
οὐκ οἶδα καταθύσαντα οὐδένα ἐπὶ δείπνῳ, οὐ
Καλλίαν οὐ Κτήσιππον τοὺς Ἀθηναίους, οὐ
Λεύκολλον[4] οὐχ Ὁρτήσιον τοὺς Ῥωμαίους·
εἶπον δὲ ὀλίγους ἐκ πολλῶν ἀσώτους καὶ ἀκρατε-
στάτους τῇ τε ἄλλῃ καὶ μέντοι καὶ περὶ γαστέρα.

43. Ὁ κόραξ ὁ ἤδη γέρων ὅταν μὴ δύνηται
τρέφειν τοὺς νεοττούς, ἑαυτὸν αὐτοῖς προτείνει
τροφήν· οἱ δὲ ἐσθίουσι τὸν πατέρα. καὶ τὴν

---

[1] λούεται. [2] βαδίσεως.
[3] Jac : ἤν τι οὐδέν.
[4] λεύκουλλον most MSS, εὔκολον A.

---

[a] Callias : end of 5th cent. B.C., a wealthy and frivolous
Athenian. Both Xenophon and Plato lay the scene of their

lights to dust itself, and it also bathes just as pigeons do. But it does not devote itself to the dusting-place or to the bath until it has walked a certain number of paces to satisfy itself. It cannot bear being seen feeding, and for that reason it retires and eats in concealment. It is violent in its jealousy and keeps a close watch on the mated female birds, and if it discovers the mistress of its house to be adulterous, it strangles itself. It does not fly high. Yet men take pleasure in it and tend it with care and consideration. And apparently it is either a pet in a sumptuous and opulent household, or else it is admitted into a temple and roams unconfined, moving about as a sacred creature within the precinct.

The Peacock on the contrary, which is a beautiful bird, is killed and eaten by voluptuaries. The feathers of this bird are a decoration, though its body is of little or no account. But I never heard of anyone killing a Purple Coot for a meal, not Callias [a] nor Ctesippus the Athenians, not Lucullus nor Hortensius the Romans. I have named but a few out of many who were luxurious and insatiate in other ways but especially where their bellies were concerned. *The Peacock*

43. When the Raven on reaching old age can no longer feed its young, it offers itself as their food; and they eat their father. And this is alleged to be *The Raven in old age*

Symposia at his house.—Ctesippus, pleasure-loving Athenian, defended by Demosthenes in his speech against Leptines; became a butt for Comic poets.—Lucullus: 1st cent. B.C., conqueror of Mithridates; his name became proverbial for wealth.—Hortensius: 1st cent. B.C., famous as an orator, the rival of Cicero, and possessor of immense wealth.

παροιμίαν ἐντεῦθέν φασι τὴν γένεσιν λαβεῖν τὴν λέγουσαν 'κακοῦ κόρακος κακὸν ᾠόν.'

44. Σωφρονέσταται ὀρνίθων αἱ φάτται ᾄδονται. ὁ γοῦν ἄρρην καὶ ὁ θῆλυς συνδυασθέντες καὶ οἱονεὶ συμπνεύσαντες ἐς γάμον ἀλλήλων ἔχονται καὶ σωφρονοῦσι, καὶ οὐκ ἂν ὀθνείου λέχους οὐδέτερος ἅψαιτο τῶν ὀρνίθων τῶνδε. ἐὰν δὲ ἐποφθαλμιάσωσιν ἑτέροις, περιέρχονται αὐτοὺς οἱ λοιποί, καὶ τὸν μὲν ἄρρενα οἱ ὁμογενεῖς διασπῶσιν, αἱ θήλειαι δὲ τὸν θῆλυν. οὗτος ἄρα ὁ τῆς σωφροσύνης νόμος καὶ ἐς τὰς τρυγόνας ἀφικνεῖται καὶ ἄτρεπτος μένει,[1] πλὴν τοῦ ⟨μὴ⟩[2] θανατοῦσθαι ἑκάτερον τὸν ὄρνιν· ἐπεὶ τὸν μὲν ἄρρενα ἀναιροῦσι, τὸν δὲ θῆλυν ᾤκτειραν καὶ εἴασαν ἀπαθῆ, καὶ περίεισι χῆρος.

45. Ἀριστοτέλης λέγει τῶν περιστερῶν τοὺς[3] ἄρρενας ταῖς θηλείαις ταῖς τικτούσαις συνωδίνειν καὶ ἀλωμένας τῆς καλιᾶς ἔξω συνωθεῖν τε καὶ συνελαύνειν, καὶ ὅταν τέκωσιν, ἐπῳάζειν ἐκβιάζεσθαι. θάλπειν δὲ[4] καὶ τοὺς ἄρρενας τὰ νεόττια καὶ συνεκτρέφειν[5] ταῖς θηλείαις ὁ αὐτός φησι, καὶ ὑπὲρ τοῦ μὴ κακοσίτους εἶναι τοὺς νεοττοὺς πρώτην τροφὴν διδόναι[6] τοῖς βρέφεσι τοὺς γειναμένους ἁλμυρίδα γῆν, ἧσπερ οὖν γευσάμενα εἶτα μέντοι καὶ τῶν λοιπῶν σιτεῖσθαι ἑτοίμως τὸ ἐντεῦθεν αὐτά. δοκεῖ δέ πως ταῖς περιστεραῖς[7] πρὸς μὲν τοὺς ἄλλους ὄρνιθας τοὺς ἁρπακτικοὺς ἔνσπονδα εἶναι, τοὺς μέντοι ἁλιαέτους καὶ τοὺς

---

[1] μένει καὶ ἐς τὰς περιστερὰς τὰς λευκάς.
[2] ⟨μὴ⟩ add. H.

the origin of the proverb which says ' A bad egg of a bad raven.'

44. Ringdoves are celebrated as the most con- The
tinent of birds. For instance, when once the male Ringdove
and the female have paired and are, so to say, of one
mind to wed, they cling to one another and are
continent, and neither bird would touch a strange
bed. If however they cast amorous glances at other
birds, the rest gather round them and the male is
torn to pieces by those of his own sex, the female by
the females. This then is the law of continence
which extends to doves and remains unchanged,
except that they do not put to death both birds:
when they kill the male they take compassion on the
female and leave her unharmed; and she goes about,
a widow.

45. Aristotle says [*HA* 613 a 1] that male Pigeons The Pigeon
share the birth-pangs of the females, and if they
wander from the nest the males will push and drive
them in; and when they have laid their eggs the
males will force them to brood them. But the male
birds also keep the chicks warm and help the females
to feed them, according to the same writer. And to
prevent the chicks from being underfed the parents
begin by giving them saline earth, so that when they
have tasted it, they then readily eat the rest of their
food. It would seem that there is a treaty of peace
between Pigeons and such others as are birds of prey,
but they are said to live in fear of sea-eagles and

---

³ καὶ τούς.   ⁴ τε.
⁵ συνδιατρέφειν.   ⁶ ἐνδιδόναι.
⁷ *Jac*: περιστεραῖς τὸ ἐντεῦθεν.

κίρκους ὡς πεφρίκασί φασι. πρὸς δὲ τοὺς ἱέρακας
οἷα παλαμῶνται ἀκοῦσαι ἄξιον. ὅταν μὲν αὐτὰς
διώκῃ ὁ μετάρσιός τε καὶ ἐς ὕψος πεφυκὼς πέ-
τεσθαι, αἱ δὲ ὑπολισθάνουσι [1] καὶ κατωτέρω
ἑαυτὰς καθέλκουσι καὶ τὸ πτερὸν [2] πειρῶνται
πιέζειν· ὅταν δὲ ὁ κατωτέρω λαχὼν ἐκ τῆς
φύσεως τὴν πτῆσιν, αἱ δὲ αἴρονταί τε καὶ μετεωρο-
ποροῦσι, καὶ ὑπὲρ αὐτοῦ πετόμεναι θαρροῦσιν,
ἀνωτέρω ἆξαι μὴ δυναμένου.

46. Ἐλέφαντος πωλίῳ περιτυγχάνει λευκῷ πω-
λευτὴς Ἰνδός, καὶ παραλαβὼν ἔτρεφεν ἔτι νεαρόν,
καὶ κατὰ μικρὰ ἀπέφηνε χειροήθη, καὶ ἐπωχεῖτο
αὐτῷ, καὶ ἤρα τοῦ κτήματος καὶ ἀντηρᾶτο, ἀνθ'
ὧν ἔθρεψε τὴν ἀμοιβὴν κομιζόμενος ἐκεῖνος. ὁ
τοίνυν βασιλεὺς τῶν Ἰνδῶν πυθόμενος ᾔτει λαβεῖν
τὸν ἐλέφαντα. ὁ δὲ ὡς ἐρώμενος ζηλοτυπῶν καὶ
μέντοι ⟨καὶ⟩[3] περιαλγῶν εἰ ἔμελλε δεσπόσειν
αὐτοῦ ἄλλος, οὐκ ἔφατο δώσειν, καὶ ᾤχετο ἀπιὼν
ἐς τὴν ἔρημον, ἀναβὰς τὸν ἐλέφαντα. ἀγανακτεῖ
ὁ βασιλεύς, καὶ πέμπει κατ' αὐτοῦ τοὺς ἀφαιρησο-
μένους καὶ ἅμα καὶ τὸν Ἰνδὸν ἐπὶ τὴν δίκην ἄξον-
τας. ἐπεὶ δὲ ἧκον, ἐπειρῶντο βίαν [4] προσφέρειν.
οὐκοῦν καὶ ὁ ἄνθρωπος ἔβαλλεν αὐτοὺς ἄνωθεν,
καὶ τὸ θηρίον ὡς ἀδικούμενον συνημύνετο. καὶ τὰ
μὲν πρῶτα ἦν τοιαῦτα· ἐπεὶ δὲ βληθεὶς ὁ Ἰνδὸς
κατώλισθε, περιβαίνει μὲν τὸν τροφέα ὁ ἐλέφας
κατὰ τοὺς ὑπερασπίζοντας ἐν τοῖς ὅπλοις, καὶ τῶν
ἐπιόντων πολλοὺς ἀπέκτεινε, τοὺς δὲ ἄλλους
ἐτρέψατο· περιβαλὼν δὲ τῷ τροφεῖ τὴν προβοσ-

---

[1] ὑπολισθάνουσι τὴν πτῆσιν.   [2] Reiske : πτερὸν δέ.

falcons. But their method of dealing with hawks is a and Hawks tale worth hearing. When the hawk, which is accustomed to soar high in the air, gives chase, the Pigeons glide and sink lower and attempt to reduce their flight. When attacked however by some bird which by nature flies at a lower level than they, the Pigeons mount up and travel through the sky, and flying overhead they have no fear, because the other cannot harry them from above.

46. An Indian trainer finding a young white A white Elephant Elephant took and reared it during its early years; he gradually tamed it and used to ride upon it and grew fond of his chattel, which returned his affection and recompensed him for his fostering care. Now the king of the Indies hearing of this, asked to be given the animal. But the trainer in his affection was jealous and even overcome with grief at the thought of another man being its master, and declined to give it up; and so, mounting the Elephant, he went off into the desert. The king in his indignation despatched men to take the Elephant away and at the same time to bring the Indian to judgment. When they arrived they attempted to apply force. So the man struck at them from his mount, and the beast helped to defend its master as he was being injured. Such was the beginning of the affair. But when the Indian was wounded and fell, the Elephant bestrode its keeper after the manner of armed men covering a comrade with their shields, slew many of the attackers, and put the remainder to flight. Then, winding its trunk round its keeper, it raised

---

³ ⟨καί⟩ add. H. ⁴ Ges: πεῖραν.

κίδα, αἴρει τε αὐτὸν καὶ ἐπὶ τὰ αὔλια κομίζει,
καὶ παρέμεινεν ὡς φίλῳ φίλος πιστός, καὶ τὴν
εὔνοιαν ἐπεδείκνυτο. ὦ ἄνθρωποι πονηροὶ καὶ
περὶ τράπεζαν μὲν καὶ ταγήνου ψόφον †ἀεί, ἐπ'
ἄριστά τε χορεύοντες,† [1] ἐν δὲ τοῖς κινδύνοις
προδόται, καὶ μάτην καὶ ἐς οὐδὲν τὸ τῆς φιλίας
ὄνομα χαίνοντες. [2]

47. Δότε μοι τοὺς τραγῳδοὺς πρὸς τοῦ πατρῴου
Διὸς καὶ πρό γε ἐκείνων τοὺς μυθοποιοὺς ἐρέσθαι
τί βουλόμενοι τοσαύτην ἄγνοιαν τοῦ παιδὸς τοῦ
Λαΐου καταχέουσι τοῦ συνελθόντος τῇ μητρὶ τὴν
δυστυχῆ σύνοδον, καὶ τοῦ Τηλέφου [3] τοῦ μὴ
πειραθέντος μὲν τῆς ὁμιλίας, συγκατακλινέντος δὲ
τῇ γειναμένῃ καὶ πράξαντος ἂν τὰ αὐτά, εἰ μὴ [4]
θεία πομπῇ διεῖρξεν ὁ δράκων· εἴ γε ἡ φύσις τοῖς
ἀλόγοις ζῴοις τὴν τοιαύτην μίξιν καὶ ἐκ τοῦ
χρωτὸς [5] δίδωσι κατανοῆσαι, καὶ οὐ δεῖται
γνωρισμάτων οὐδὲ τοῦ ἐκθέντος ἐς τὸν Κιθαιρῶνα. [6]
οὐκ ἂν γοῦν ποτε τῇ τεκούσῃ ὁμιλήσειε [7] κάμηλος.
ὁ δέ τοι νομεὺς τῆς ἀγέλης κατακαλύψας τὸν
θῆλυν ὡς οἶόν τε ἦν καὶ ἀποκρύψας πάντα πλὴν
τῶν ἄρθρων, τὸν παῖδα ἐπάγει τῇ μητρί, καὶ
ἐκεῖνος λάθριος ὑπὸ ὁρμῆς τῆς πρὸς μίξιν ἔδρασε
τὸ ἔργον καὶ συνῆκε. καὶ τὸν μὲν αἴτιον τῆς
ὁμιλίας οἱ τῆς ἐκθέσμου δάκνων καὶ πατῶν καὶ

---

[1] ἀεί . . . χορεύοντες corrupt, ἐπὶ ῥαστώνης Grasberger.
[2] Jac : χραίνοντες.
[3] καὶ τοῦ Τηλέφου after καταχέουσι MSS, transposed by H.
[4] Jac : εἰ μὴ πολλάκις.
[5] χρωτὸς προσαψαμένοις.
[6] Κιθαιρῶνα ὡς ὁ Οἰδίπους ὁ τοῦ Σοφοκλέους.
[7] ὁμιλῆσαι.

him and brought him to its stable and stayed by his side, as one trusty friend might do to another, thus showing its kindly nature.

O wicked men, for ever busy (?) about the table and the clash of frying-pans and dancing to your lunch, but traitors in the hour of danger, in whose mouth the word ' Friendship ' is vain and of no effect.

47. In the name of Zeus our father, permit me to ask the tragic dramatists and their predecessors, the inventors of fables, what they mean by showering such a flood of ignorance upon the son of Laïus[a] who consummated that disastrous union with his mother; and upon Telephus[b] who, without indeed attempting union, lay with his mother and would have done the same as Oedipus, had not a serpent sent by the gods kept them apart, when Nature allows unreasoning animals to perceive by mere contact the nature of this union, with no need for tokens nor for the presence of the man who exposed Oedipus on Cithaeron.

The Camel, for instance, would never couple with its mother. Now the keeper of a herd of camels covered up a female as far as possible, hiding all but its parts, and then drove the son to its mother. The beast, all unwitting, in its eagerness to copulate, did the deed, then realised what it had done. It bit and trampled on the man who was the cause of its un-

Examples of incest

---

[a] Oedipus, after having unwittingly slain his father Laïus, married his widow Iocasta.

[b] Telephus, son of Heracles and Auge. According to one story Teuthras king of Mysia, unaware of their relationship, gave his daughter Auge in marriage to Telephus who was equally unaware.

τοῖς γόνασι παίων ἀπέκτεινεν ἀλγεινότατα, ἑαυτὸν
δὲ κατεκρήμνισεν. ἀμαθὴς δὲ καὶ κατὰ τοῦτο
Οἰδίπους, οὐκ ἀποκτείνας,[1] ἀλλὰ πηρώσας τὴν
ὄψιν, καὶ τὴν τῶν κακῶν λύσιν μὴ γνοὺς ἐξὸν
ἀπηλλάχθαι καὶ μὴ τῷ οἴκῳ καὶ τῷ γένει κατα-
ρώμενον εἶτα μέντοι κακῷ ἀνηκέστῳ ἰᾶσθαι κακὰ
τὰ ἤδη παρελθόντα.

[1] ἀποκτείνας ⟨ἑαυτόν⟩ Schn.

lawful union, and kneeling on him put him to an agonising death, and then threw itself over a precipice.

And here Oedipus was ill-advised in not killing himself but blinding his eyes; in not realising how to escape from his calamities when he might have made away with himself instead of cursing his house and his family; and finally in seeking by an irremediable calamity to remedy calamities already past.

BOOK IV

Δ

1. Ἀκολαστότατοι ὀρνίθων οἱ πέρδικές εἰσι. ταῦτά τοι καὶ τῶν θηλειῶν ἐρῶσι δριμύτατα, καὶ τῆς λαγνείας ἡττώμενοι συνεχέστατά εἰσιν οἵδε. οὐκοῦν οἱ τρέφοντες τοὺς ἀθλητὰς πέρδικας, ὅταν αὐτοὺς ἐς τὴν μάχην τὴν κατὰ ἀλλήλων ὑποθήγωσι, τὴν θήλειαν παρεστάναι ποιοῦσιν ἑκάστῳ τὴν σύννομον, σόφισμα τοῦτο δειλίας καὶ κάκης τῆς κατὰ τὴν ἀγωνίαν ἀντίπαλον αὐτοῖς εὑρόντες. οὐ γάρ τί που ἡττώμενος φανῆναι ἢ τῇ ἐρωμένῃ ἢ τῇ γαμετῇ ὁ πέρδιξ ὑπομένει· τεθνήξεται δὲ μᾶλλον παιόμενος ἢ ὁμόσε χωροῦντος ἀποστραφεὶς ἰδεῖν τολμήσει ταύτην ἀσχημόνως, παρ᾽ ᾗ βούλεται εὐδοκιμεῖν. τοῦτό τοι καὶ Κρῆτες ὑπὲρ τῶν ἐρωμένων ἐνενόουν. ἀκούω γὰρ Κρῆτα ἐραστὴν ἀγαθὸν τά τε ἄλλα καὶ τὰ πολέμια ἔχειν μὲν παιδικὰ εὐγενὲς μειράκιον ὥρᾳ διαπρεπὲς καὶ τὴν ψυχὴν ἀνδρεῖον καὶ πρὸς τὰ κάλλιστα τῶν μαθημάτων πεφυκὸς εὖ καὶ καλῶς, καλούμενον δὲ δι᾽ ἡλικίαν ἐς ὅπλα μηδέπω (εἶπόν γε μὴν ἀλλαχόθι καὶ τοῦ ἐραστοῦ καὶ τοῦ καλοῦ τὸ ὄνομα). ἀρετὰς μὲν οὖν ἐν τῇ μάχῃ τὸν νεανίαν ἀποδείξασθαί [1] φασιν οἱ Κρῆτες, ἀθρόας δὲ ἐς αὐτὸν ὠθουμένης τῆς τῶν ἐχθρῶν φάλαγγος προσπταῖσαι νεκρῷ κειμένῳ, καὶ περιτραπῆναι λέγουσιν αὐτόν. τῶν οὖν τις πολεμίων, ὁ μάλιστα πλησίον, ἀνατει

[1] Schn: ἀποδίδοσθαι.

BOOK IV

1. Partridges are the most incontinent of birds; that is the reason for their passionate love of the female birds and for their constant enslavement to lust. So those that rear fighting Partridges, when they egg them on to battle with one another, make the female stand each by her mate, as they have found this to be a device for countering any cowardice or reluctance to fight. For the Partridge that is defeated cannot endure to show himself either to his loved one or to his spouse. He will sooner die under the blows than turn away from his adversary and dare in his disgrace to look upon her whose good opinion he courts.

The Cretans also have taken this view regarding lovers. For I have heard that a Cretan lover, who had beside other qualities that of a fine soldier, had as his favourite a boy of good birth, conspicuous for his beauty, of manly spirit, excellently fitted by nature to imbibe the noblest principles, though on account of his youth he was not yet called to arms. (I have elsewhere [a] given the name of the lover and of the beautiful boy.) Now the Cretans say that the young man did acts of valour in the fight, but when the enemy's massed line pressed him hard, he stumbled over a dead body that lay there and was thrown down. Whereupon one of the enemy

---

[a] Not in any surviving work of Aelian's.

AELIAN

νάμενος παίειν ἔμελλε κατὰ τῶν μεταφρένων
τὸν ἄνδρα· ὁ δὲ ἐπιστραφεὶς 'μηδαμῶς' εἶπεν
'αἰσχρὰν καὶ ἀναλκῆ [1] πληγὴν ἐπαγάγῃς, ἀλλὰ
κατὰ τῶν στέρνων ἀντίαν παῖσον, ἵνα μή μου
δειλίαν ὁ ἐρώμενος καταψηφίσηται, καὶ φυλάξηται
περιστεῖλαί με νεκρόν, καὶ μάλα γε ἀσχημονοῦντι
προσελθεῖν οὐ τολμῶν.' αἰδεσθῆναι μὲν οὖν ἄνθρωπον ὄντα φανῆναι κακὸν οὔπω θαυμαστόν·
πέρδικι δὲ μετεῖναι αἰδοῦς ὑπέρσεμνον τοῦτο ἐκ
τῆς φύσεως τὸ δῶρον. Ἀριστόδημος δὲ ὁ τρέσας
καὶ Κλεώνυμος ὁ ῥίψας τὴν ἀσπίδα καὶ ὁ δειλὸς
Πείσανδρος οὔτε τὰς πατρίδας ᾐδοῦντο οὔτε τὰς
γαμετὰς οὔτε τὰ παιδία.

2. Ἐν Ἔρυκι τῆς Σικελίας ἑορτή ἐστιν, ἣν
καλοῦσιν Ἀναγώγια Ἐρυκῖνοί τε αὐτοὶ καὶ
μέντοι καὶ ὅσοι ἐν τῇ Σικελίᾳ πάσῃ. ἡ δὲ αἰτία
τοῦ τῆς ἑορτῆς ὀνόματος, τὴν Ἀφροδίτην λέγουσιν ἐντεῦθεν ἐς Λιβύην ἀπαίρειν ἐν ταῖσδε ταῖς
ἡμέραις. δοξάζουσι δὲ ἄρα ταῦτα ταύτῃ [2] τεκμαι-
ρόμενοι. περιστερῶν πλῆθός ἐστιν ἐνταῦθα πάμ-
πλειστον. οὐκοῦν αἱ μὲν οὐχ ὁρῶνται, λέγουσι δὲ
Ἐρυκῖνοι τὴν θεὸν δορυφορούσας ἀπελθεῖν· ἀθύρ-
ματα γὰρ Ἀφροδίτης περιστερὰς εἶναι ᾄδουσί τε
ἐκεῖνοι καὶ πεπιστεύκασι πάντες ἄνθρωποι. διελ-
θουσῶν δὲ ἡμερῶν ἐννέα μίαν μὲν διαπρεπῆ τὴν
ὥραν ἔκ γε τοῦ πελάγους τοῦ κομίζοντος ἐκ τῆς

---

[1] Jac : ἀνάλκῃ, ἄναλκιν.   [2] ταύτῃ ἐκεῖθεν.

---

[a] A Spartan who owing to sickness was absent from the
battle of Thermopylae. Later, at Plataea, he wiped out his
'disgrace.' See Hdt. 7. 229–32; 9. 71.

216

who was nearest, in his eagerness was about to strike him in the back. But the man turned and exclaimed ' Do not deal me a shameful and cowardly blow, but strike me in front, in the breast, in order that my loved one may not judge me guilty of cowardice and refrain from laying out my dead body : he could not bear to go near one who so disgraces himself.'

There is nothing wonderful in a man being ashamed to appear a coward, but that a Partridge should have some feeling of shame, this is a truly impressive gift of Nature. But Aristodemus the timid,[a] and Cleonymus who threw away his shield,[b] and Pisander the craven,[c] had no reverence for their country or for their wives or for their children.

2. At Eryx in Sicily there is a festival which not only the people of Eryx but everybody throughout the whole of Sicily as well call the ' Festival of the Embarkation.' And the reason why the festival is so called is this: they say that during these days Aphrodite sets out thence for Libya. They adduce in support of their belief the following circumstance. There is there an immense multitude of Pigeons. Now these disappear, and the people of Eryx assert that they have gone as an escort to the goddess, for they speak of Pigeons as ' pets of Aphrodite,' and so everybody believes them to be. But after nine days one bird of conspicuous beauty is seen flying in from the sea which brings it

[b] A frequent butt of Aristophanes.
[c] Athenian demagogue, end of 5th cent., lampooned by Comic poets for his bulk, his rapacity, and his cowardice. Helped to establish the rule of the Four Hundred.

Λιβύης ὁρᾶσθαι ἐσπετομένην, οὐχ οἵαν κατὰ τὰς
ἀγελαίας πελειάδας τὰς λοιπὰς εἶναι, πορφυρᾶν
δέ, ὥσπερ οὖν τὴν Ἀφροδίτην ὁ Τήιος ἡμῖν
Ἀνακρέων ᾄδει, 'πορφυρέην'[1] που λέγων. καὶ
χρυσῷ δὲ εἰκασμένη φανείη ἄν, καὶ τοῦτό γε κατὰ
τὴν Ὁμήρου θεὸν τὴν αὐτήν, ἣν ἐκεῖνος ἀναμέλπει
'χρυσῆν'. ἕπεται δὲ αὐτῇ τῶν περιστερῶν τὰ νέφη
τῶν λοιπῶν, καὶ ἑορτὴ πάλιν Ἐρυκίνοις καὶ
πανήγυρις τὰ Καταγώγια,[2] ἐκ τοῦ ἔργου καὶ
τοῦτο τὸ ὄνομα.

3. Λύκω συννόμω καὶ ἵππω, λέοντέ γε μὴν
οὐκέτι· λέαινα γὰρ καὶ λέων οὐ τὴν αὐτὴν ἴασιν
οὔτε ἐπὶ θήραν[3] οὔτε πιόμενοι. τὸ δὲ αἴτιον, τῇ
τοῦ σώματος ῥώμῃ θαρροῦντε[4] ἄμφω εἶτα οὐ
δεῖται θατέρου ὁ ἕτερος, ὥς φασιν οἱ πρεσβύτεροι.

4. Οὐ ῥᾳδίως οἱ λύκοι τὴν ὠδῖνα ἀπολύουσιν,
ἀλλὰ ἐν ἡμέραις δώδεκα καὶ νυξὶ τοσαύταις, ἐπεὶ
τοσούτῳ χρόνῳ τὴν Λητὼ ἐς Δῆλον ἐξ Ὑπερβο-
ρέων ἐλθεῖν Δήλιοί φασιν.

5. Ζῷα[5] πολέμια χελώνη τε καὶ πέρδιξ, καὶ
πελαργὸς καὶ κρὲξ πρὸς αἴθυιαν ⟨καὶ⟩[6] ἄρπη καὶ
ἐρωδιὸς πρὸς λάρον· κορυδαλλὸς δὲ ἀκανθυλλίδι
νοεῖ πολέμια, τρυγόνι ⟨δὲ⟩[7] πρὸς πυραλλίδα[8]
διαφορά, ἰκτῖνός γε μὴν καὶ κόραξ ἐχθροί· σειρὴν

---

[1] πορφυρῆν.
[2] Reiske : τὰ καταγώγια πανήγυρις.
[3] θήρας.
[4] θαρροῦ τε most MSS, θαρροῦσιν A.
[5] ζῷα ἀλλήλοις.
[6] ⟨καί⟩ add. H.

from Libya: it is not like the other Pigeons in a flock but is rose-coloured, just as Anacreon of Teos describes Aphrodite, styling her somewhere [*fr.* 2. 3 D] 'roseate.' And the bird might also be compared to gold, for this too is like the same goddess of whom Homer sings as ' golden ' [*Il.* 5. 427]. And after the bird follow the other Pigeons in clouds, and again there is a festal gathering for the people of Eryx, the ' Festival of the Return '; the name is derived from the event.

3. The Wolf and the she-Wolf feed together, like- Lion and wise the Horse and the Mare; the Lion and the Lioness Lioness however do not, for the Lioness and the Lion do not follow the same track either hunting or when drinking. And the reason is that both derive confidence from their bodily strength, so that neither has need of the other, as older writers assert.

4. Wolves are not easily delivered of their young, The Wolf only after twelve days and twelve nights, for the people of Delos maintain that this was the length of time that it took Leto to travel from the Hyperboreans to Delos.

5. Animals hostile to one another: the Tortoise Animal and the Partridge; the Stork and the Corncrake to enmities the Sea-gull; the Shearwater and the Heron to the Sea-mew. The Crested Lark feels enmity towards the Goldfinch; the Turtle-dove disagrees with the Pyrallis; [a] the Kite too and the Raven are enemies;

---

[a] Perhaps a kind of pigeon.

---

[7] ⟨δέ⟩ add. H.   [8] πύρραν.

δὲ [1] πρὸς κίρκην, κίρκη δὲ πρὸς κίρκον οὐ τῷ γένει μόνον, ἀλλὰ καὶ τῇ φύσει διαφέροντε πεφώρασθον.

Χάννη δὲ ἰχθὺς λαγνίστατος. λευκοὺς δὲ μύρμηκας ἐν Φενεῷ [2] τῆς Λακωνικῆς ἀκούειν πάρεστιν.

6. Τοὺς ἵππους ἕλεσί τε καὶ λειμῶσι καὶ τοῖς κατηνέμοις χωρίοις ἥδεσθαι μᾶλλον ἱπποτροφίας τε καὶ πωλοτροφικῆς ἄνθρωποι σοφισταὶ ὁμολογοῦσιν. ἔνθεν τοι καὶ Ὅμηρος ἐμοὶ δοκεῖν δεινὸς ὢν καὶ τὰ τοιαῦτα συνιδεῖν ἔφη που

τῷ τρισχίλιαι ἵπποι ἕλος κάτα βουκολέοντο.

ἐξηνεμῶσθαι δὲ ἵππους πολλάκις ἱπποφορβοὶ τεκμηριοῦσι καὶ κατὰ τὸν νότον ἢ τὸν βορρᾶν φεύγειν. εἰδότα οὖν τὸν αὐτὸν ποιητὴν εἰπεῖν

τάων καὶ Βορέης ἠράσσατο βοσκομενάων.

καὶ Ἀριστοτέλης δέ, ὡς ἐμὲ νοεῖν, λαβὼν ἐντεῦθεν εὐθὺ τῶν προειρημένων ἀνέμων οἰστρηθείσας διδράσκειν [3] ἔφατο αὐτάς.

7. Ἀκούω τὸν Σκυθῶν βασιλέα (τὸ δὲ ὄνομα εἰδὼς ἐῶ· τί γάρ μοι καὶ λυσιτελές ἐστιν;) ἵππον σπουδαίαν ἔχειν πᾶσαν ἀρετήν, ὅσην ἵπποι καὶ ἀπαιτοῦνται καὶ ἀποδείκνυνται, ἔχειν δὲ καὶ υἱὸν αὐτῆς ἐκείνης τῶν ἄλλων ἀρετῇ διαπρέποντα.

---

[1] σειρὴν μελίσσης ὄνομα.   [2] Πέφνῳ Venmans.
[3] ἀποδιδράσκειν.

---

[a] Probably the Serin-finch.
[b] The Circe has not been identified.

the Siren [a] and the Circe [b]; the Circe and the Falcon have been found to be at variance not only in the matter of sex but in their nature.

The Sea-perch is the most lecherous of fishes. In Pheneus in Laconia [c] one may hear tell of white Ants.

The Sea-perch

6. Men skilled in the breeding and care of Horses agree that Horses are most fond of marshy ground, meadows, and wind-swept spots. Hence we find Homer, who in my opinion had a remarkable knowledge of such matters, saying somewhere [*Il.* 20. 221]

The Horse

' For him three thousand mares grazed along the
    water-meadow.'

And horse-keepers frequently testify to Mares being impregnated by the wind, and to their galloping against the south or the north wind. And the same poet knew this when he said [*Il.* 20. 223]

Mares impregnated by the wind

' Of them was Boreas enamoured as they pastured.'

Aristotle too, borrowing (as I think) from him, said [*HA* 572 a 16] that they rush away in frenzy straight in the face of the aforesaid winds.

7. I am told that the King of the Scythians (his name I know but suppress, for I have nothing to gain by it) possessed a mare remarkable for every excellence which is expected of horses and for which they are displayed; and that he possessed also a foal of

Example of animal incest

---

[c] Pheneus was in Arcadia. Venmans, citing Paus. 3. 26. 2, 3, conjectures *Pephnus*, a place in Laconia at the NE corner of the Messenian Gulf. It was also the name of a rocky islet at the mouth of the Pamisus; see Frazer on Paus. *loc. cit.* The ' white ants ' are fabulous.

οὔκουν εὑρίσκοντα οὔτε ἐκείνην ἄλλῳ παραβα-
λεῖν ἀξίῳ, οὔτε ἐκεῖνον ἄλλῃ ἐπαγαγεῖν τὸ ἐξ
αὐτοῦ λαβεῖν σπέρμα ἀγαθῇ, διὰ ταῦτα ἄμφω
συναγαγεῖν ἐς τὸ ἔργον· τοὺς δὲ τὰ μὲν ἕτερα
ἀσπάζεσθαι σφᾶς καὶ φιλοφρονεῖσθαι, οὐ μὴν
ἐγχρίμπτεσθαι ἀλλήλοις. οὐκοῦν ἐπεὶ τῆς ἐπι-
βουλῆς τοῦ Σκύθου σοφώτερα ἦν τὰ ζῷα, ἐπηλύγασεν ἱματίοις καὶ τὸν καὶ τήν, καὶ ἐξειργάσαντο
τὸ ἔκνομόν τε καὶ ἔκδικον ἐκεῖνο ἔργον. ὡς δὲ
ἄμφω συνεῖδον τὸ πραχθέν, εἶτα μέντοι τὸ ἀσέβημα
διελύσαντο θανάτῳ, πηδήσαντε κατὰ κρημνοῦ.

8. Λέγει Εὔδημος ἵππου νέας καὶ τῶν νεμομένων τῆς ἀρίστης ἐρασθῆναι τὸν ἱπποκόμον, ὥσπερ
οὖν καλῆς μείρακος καὶ τῶν ἐν τῷ χωρίῳ ὡρικωτέ-
ρας πασῶν· καὶ τὰ μὲν πρῶτα ἐγκαρτερεῖν,
τελευτῶντα δὲ ἐπιτολμῆσαι τῷ λέχει τῷ ξένῳ καὶ
ὁμιλεῖν αὐτῇ. τῇ δὲ εἶναι πῶλον καὶ τοῦτον
καλόν, θεασάμενόν γε μὴν τὸ πραττόμενον ἀλγῆσαι,
ὥσπερ οὖν τυραννουμένης τῆς μητρὸς ὑπὸ τοῦ
δεσπότου, καὶ ἐμπηδῆσαι καὶ ἀποκτεῖναι τὸν
ἄνδρα, εἶτα μέντοι καὶ φυλάξαι ἔνθα ἐτάφη, καὶ
φοιτῶντα ἀνορύττειν αὐτόν, καὶ ἐνυβρίζειν τῷ
νεκρῷ καὶ λυμαίνεσθαι λύμην ποικίλην.[1]

9. Τῶν ἰχθύων διὰ τοῦ ἦρος οἱ πλεῖστοι ἐς[2]
ἀφροδίτην[3] πρόθυμοί εἰσι, καὶ ἀποκρίνουσί γε
αὐτοὺς ἐς τὸν Πόντον μᾶλλον· ἔχει γάρ πως
θαλάμας τε καὶ κοίτας, φύσεως ταῦτα ἰχθύσι[4] τὰ
δῶρα· ἀλλὰ καὶ θηρίων ἐλεύθερός ἐστιν ὅσα

---

[1] ποικίλην οὐκ αἰσθανομένῳ ἀλγοῦντα αὐτόν.

this same mare which surpassed all others in its excellence.   Being unable to find either another worthy mate for the mare or another mare fit to be impregnated by the foal, he therefore put the two together for that purpose.   They caressed each other in various ways and were friendly disposed, but refused to couple.   So as the animals were too clever for the Scythian's scheme, he blindfolded both mare and foal with cloths, and they accomplished the act so contrary to law and morality.   But when the pair realised what they had done, they atoned for their impious deed by death and threw themselves over a precipice.

8. Eudemus records how a groom fell in love with a young mare, the finest of the herd, as it might have been a beautiful girl, the loveliest of all thereabouts. And at first he restrained himself, but finally dared to consummate a strange union.   Now the mare had a foal, and a fine one, and when it saw what was happening it was pained, just as though its mother were being tyrannically treated by her master, and it leaped upon the man and killed him.   And it even went so far as to watch where he was buried, went to the place, dug up the corpse, and outraged it by inflicting every kind of injury. *Groom in love with Mare*

9. The majority of Fishes are eager for sexual intercourse throughout the springtime, and withdraw for choice to the Black Sea, for it contains caverns and resting-places which are Nature's gift to Fishes. Besides, its waters are free from the savage creatures *Fish in the mating season*

---

² ἐς (εἰς) om. AL.   ³ τὴν ἀφροδίτην.
⁴ Jac: ἰχθύσιν ὁ Πόντος.

223

βόσκει θάλαττα. δελφῖνες δὲ ἀλῶνται μόνοι,
λεπτοί τε καὶ ἀσθενικοί· καὶ μὴν καὶ πολύπου
χηρός ἐστι καὶ παγούρου ἄγονος, καὶ ἀστακὸν οὐ
τρέφει· μικρῶν δὲ ἰχθύων οἶδε ὄλεθρός εἰσιν.[1]

10. Πυνθάνομαι σελήνης ὑποφαινομένης νέας
τοὺς ἐλέφαντας κατά τινα φυσικὴν καὶ ἀπόρρητον
ἔννοιαν ἐκ τῆς ὕλης ἐν ᾗ νέμονται νεοδρεπεῖς
ἀφελόντας κλάδους εἶτα μέντοι μετεώρους ἀνατεί-
νειν, καὶ πρὸς τὴν θεὸν ἀναβλέπειν, καὶ ἡσυχῇ
τοὺς κλάδους ὑποκινεῖν, οἷον ἱκετηρίαν τινὰ ταύτην
τῇ θεῷ προτείνοντας ὑπὲρ τοῦ ἵλεώς τε καὶ εὐμενῆ
τὴν θεόν γε εἶναι αὐτοῖς.

11. Μόνας ἀκούω τῶν ζῴων τὰς ἵππους καὶ
κυούσας ὑπομένειν τὴν τῶν ἀρρένων μίξιν· εἶναι
γὰρ λαγνιστάτας. διὰ ταῦτά τοι καὶ τῶν γυναικῶν
τὰς ἀκολάστους ὑπὸ τῶν σεμνοτέρως αὐτὰς
εὐθυνόντων καλεῖσθαι ἵππους.

12. Οἱ πέρδικες ἐν τοῖς ᾠοῖς οἰκοῦντες ἔτι καὶ
κατειλημμένοι τοῖς περιπεφυκόσι σφίσιν ὀστρά-
κοις οὐκ ἀναμένουσι τὴν ἐκ τῶν γειναμένων
ἐκγλυφήν, ἀλλ᾽ αὐτοὶ δι᾽ ἑαυτῶν ὥσπερ θυροκο-
ποῦντες διακρούουσι[2] τὰ ᾠά, καὶ ἐκκύψαντες εἶτα
σφᾶς αὐτοὺς[3] ἀνωθοῦσι, καὶ τὸ τοῦ ᾠοῦ λέμμα
περιρρήξαντες ἤδη θέουσι, καὶ τὸ πρὸς τῷ οὐραίῳ
ἡμίτομον, εἰ προσέχοιτο, διασεισάμενοι ἐκβάλ-
λουσιν αὐτό, καὶ τροφὴν μαστεύουσι, καὶ πηδῶσιν
ὤκιστα.

---

[1] *Gron* : ἐστιν.   [2] *Meïn* : ἐκκρούουσι MSS, H.

which the sea breeds. Only dolphins roam there, and they are small and feeble. Moreover it is devoid of octopuses; it produces no crabs and does not breed lobsters: these are the bane of small fishes.

10. I am informed that when the new moon begins to appear, Elephants by some natural and un-explained act of intelligence pluck fresh branches from the forest where they feed and then raise them aloft and look upwards at the goddess, waving the branches gently to and fro, as though they were offering her in a sense a suppliant's olive-branch in the hope that she will prove kindly and benevolent to them. Elephants worship the Moon

11. I have heard that Mares are the only animals which when pregnant allow the male to have inter-course with them. For Mares are exceedingly lust-ful, and that is why strict censors call lecherous women 'mares.' The Mare

12. Partridges while still in the egg and confined by the shell that has formed around them do not wait for their parents to hatch them out, but alone and unaided, like house-breakers, peck through the eggs, peep out, and then lever themselves up, and then after cracking the egg-shell begin at once to run. And if half the shell is clinging to their tail they shake it off and cast it from them; and they hunt for food and dart about at great speed. The Partridge, its young

---

[3] ἑαυτούς.

13. Τῶν περδίκων οἱ τοροί τε καὶ ᾠδικοὶ τῇ σφετέρᾳ θαρροῦσιν εὐγλωττίᾳ· καὶ οἱ μαχητικοὶ δὲ καὶ ἀγωνιστικοὶ καὶ ἐκεῖνοι πεπιστεύκασιν ὅτι μή εἰσιν ἄξιοι παρανάλωμα γενέσθαι τεθηραμένοι· καὶ διὰ ταῦτα ἁλισκόμενοι ἧττον πρὸς τοὺς θηρῶντας διαμάχονται ὑπὲρ τοῦ μὴ ἁλῶναι.[1] οἱ δὲ ἄλλοι, καὶ ἔτι μᾶλλον οἱ Κιρραῖοι, συνεγνωκότες ἑαυτοῖς οὔτε ἀλκὴν ἀγαθοῖς οὔτε ᾄδειν, καλῶς δὲ διεγνωκότες ὅτι ἄρα ἁλόντες ἔσονται δεῖπνον τοῖς ᾑρηκόσι, παλαμῶνταί τινι σοφίᾳ φυσικῇ ἑαυτοὺς ἀβρώτους παρασκευάσαι· καὶ τῆς μὲν ἄλλης τροφῆς, ἥτις αὐτοὺς εὐφραίνει τε καὶ πιαίνει, ἀπέχονται, σκόροδα δὲ σιτοῦνται προθυμότατα. οἱ τοίνυν ταῦτα προμαθόντες ἐσπείσαντο πρὸς αὐτοὺς ἑκόντες ἀθηρίαν· ὅστις δὲ τῇ τούτων ἄγρᾳ οὐ προενέτυχε, συλλαβὼν καὶ καθεψήσας ἀπώλεσε καὶ τὸν χρόνον καὶ τὴν ἐπ᾽ αὐτοῖς σπουδήν, πονηροῦ κρέως πειραθείς.

14. Κακὸν θηρίον ἡ γαλῆ, κακὸν δὲ καὶ ὁ ὄφις. οὐκοῦν ὅταν μέλλῃ γαλῆ ὄφει μάχεσθαι, πήγανον διατραγοῦσα πρότερον εἶτα μέντοι ἐπὶ τὴν μάχην θαρροῦσα [2] ὥσπερ οὖν πεφραγμένη τε καὶ ὡπλισμένη παραγίνεται. τὸ δὲ αἴτιον, τὸ πήγανον πρὸς ὄφιν ἔχθιστόν ἐστιν.

15. Ὁ λύκος ἐμπλησθεὶς ἐς κόρον οὐδ᾽ ἂν τοῦ βραχίστου τὸ λοιπὸν ἀπογεύσαιτο· παρατείνεται [3] μὲν γὰρ ἡ γαστὴρ τῷδε, οἰδαίνει [4] δὲ ἡ γλῶττα, καὶ τὸ στόμα ἐμφράγνυται, πρᾴότατος δὲ ἐντυχεῖν

---

[1] ἁλῶναι ὅτι γὰρ σπουδασθήσονται καὶ οἶδε πιστεύουσι καὶ τῇ μάχῃ καὶ τῇ ᾠδῇ.

13. Partridges that utter clear, musical tones are confident in their vocal skill. So too the fighting birds which compete feel certain that when captured they will not be regarded as merely fit for sacrifice. And that is why when caught they struggle less against their pursuers in order to avoid capture. But the rest, and especially the Partridges of Cirrha, conscious that they possess neither strength nor ability to sing, and knowing full well that if caught they will furnish a meal for their captors, do their utmost, prompted by some natural intelligence, to render themselves unfit for eating. And they abstain from other food which delights and fattens them and feed most eagerly upon garlic. Hence those who are already aware of these facts have willingly agreed that they should be immune from pursuit. Whereas a man who has not previously chanced to hunt them, if he catches and cooks them, has wasted his time and his pains over them, when he finds their flesh disgusting.

The Partridge: three kinds

14. The Marten is an evil creature, and an evil creature is the Snake. And so when a Marten means to fight with a Snake, it chews some rue beforehand and then goes out boldly to battle, as though fortified and armed. The reason is that to a Snake rue is utterly abhorrent.

Marten and Snake

15. The Wolf when gorged to satiety will not thereafter taste the least morsel. For his belly is distended, his tongue swells, his mouth is blocked, and he is gentle as a lamb to meet, and would have no

The Wolf, when full-fed

---

² θαρροῦσα V, *del.* H, διαθαρροῦσα ἐπὶ τὴν μ. *most* MSS.
³ περι-.   ⁴ οἰδάνει H.

ἐστιν ἀμνοῦ δίκην, καὶ οὐκ ἂν ἐπιβουλεύσειεν [1] ἢ
ἀνθρώπῳ ἢ θρέμματι, οὐδὲ εἰ τῆς ἀγέλης βαδίζοι
μέσος. μειοῦται δὲ ἡσυχῇ καὶ κατ' ὀλίγον ἡ
γλῶττα αὐτῷ, εἶτα ἐς τὸ ἀρχαῖον σχῆμα ἐπάνεισι,
καὶ λύκος γίνεται αὖθις.

16. Ἀλεκτρυόνες ἐν ἀγέλῃ τὸν νέηλυν [2] ἀνα-
βαίνουσι πάντες. καὶ οἱ τιθασοὶ δὲ πέρδικες τὸν
ἥκοντα πρῶτον καὶ οὔπω πεπραϋσμένον τὰ αὐτὰ
δρῶσιν. ἀμειβόμενοι δὲ οἱ πέρδικες τοὺς τρέφοντας καὶ αὐτοὶ παλεύουσι τοὺς ἀφέτους καὶ ἀγρίους,
κατὰ τὰς περιστερὰς δρῶντες καὶ οὗτοι τοῦτο.
προσάγεται δὲ ἄρα ὁ πέρδιξ καὶ σειρῆνας ἐς τὸ
ἐφολκὸν προτείνει τὸ τῶν ἄλλων τὸν τρόπον
τοῦτον. ἕστηκεν ᾄδων [3] καὶ ἔστιν οἷ τὸ μέλος
προκλητικόν, ἐς μάχην ὑποθῆγον τὸν ἄγριον,
ἕστηκε δὲ ἐλλοχῶν πρὸς τῇ πάγῃ· ὁ δὲ [4] τῶν
ἀγρίων κορυφαῖος ἀντάσας πρὸ τῆς ἀγέλης μαχούμενος ἔρχεται. ὁ τοίνυν τιθασὸς ἐπὶ πόδα ἀναχωρεῖ, δεδιέναι σκηπτόμενος· ὁ δὲ ἔπεισι γαῦρος,
οἷα [5] δήπου κρατῶν ἤδη, καὶ ἑάλωκεν ἐνσχεθεὶς
τῇ πάγῃ. ἐὰν μὲν οὖν ᾖ ἄρρην ὁ τοῖς θηράτροις
περιπεσών, [6] πειρῶνται ἐπικουρεῖν οἱ σύννομοι τῷ
ἑαλωκότι· ἐὰν δὲ ᾖ θῆλυς, παίουσι τὸν ἐνσχεθέντα
ἄλλος ἀλλαχόθεν, ὡς διὰ τὴν λαγνείαν ἐς δουλείαν
ἐμπεσόντα. καὶ ἐκεῖνο δὲ οὐ παρήσω, ἐπεὶ καὶ
ἄξιον ἀκοῦσαι αὐτό. ἐὰν ᾖ θῆλυς ὁ παλεύων, ἵνα
μὴ ἐμπέσῃ ὁ ἄρρην, αἱ ἔξω θήλειαι μέλος ἀντῳδὸν
ἠχοῦσι, καὶ ῥύονται τὸν ἐμπεσούμενον ἐς τὴν πά-
γην ταῖς συννόμοις καὶ πλείοσιν ἀσμένως συμπα-

---

[1] ἐπιβουλεῦσαι.
[2] νέηλυν οὔσης θηλειῶν ἀπορίας.

designs on man or beast, even were he to walk through the middle of a flock. Gradually however and little by little his tongue shrinks and resumes its former shape, and he becomes once more a wolf.

16. Cockerels all tread a newcomer to the flock, The and tame Partridges do the same to the latest Partridge as arrival as yet untamed. And Partridges even requite their own parents by decoying those that are free and wild, acting in this respect just like pigeons. Now this is the way in which the Partridge draws them to him and displays the arts of a Siren to allure others. He stands uttering his cry, and his tune conveys a challenge, provoking the wild bird to fight; and he stands in ambush by the springe. Then the cock of the wild birds answers back and advances to do battle on behalf of his covey. So the tame bird withdraws, pretending to be afraid, while the other advances vaunting as though he were already victorious, is caught in the snare, and is captured. Now if it is a cock bird that falls into the trap, his companions attempt to bring help to the captive; but if it is a hen, one here and another there beats the captive for allowing her lust to bring her into slavery.

And here is a point that I will not omit, for it deserves attention. If the decoy-bird is a hen, the wild hens, in order to prevent the cock from falling into the trap, counter the challenge with their cries and rescue the cock that is about to be trapped, for he is glad to stay with those who are his mates and

---

³ ᾄδων ὁ πρᾶος.   ⁴ δή.
⁵ Reiske: ὡς οἷα.   ⁶ Reiske: παραμένων.

ραμένοντα,¹ ὡς ἂν ἴυγγί τινι ἐλχθέντα ναὶ μὰ Δι'
ἐρωτικῇ.

17. Ἔν τῶν βασκάνων ζῴων μέντοι καὶ ἐχῖνος
ὁ χερσαῖος εἶναι πεπίστευται. ὅταν γοῦν ἁλίσκη-
ται, παραχρῆμα ἐνεούρησε ² τῷ δέρματι, καὶ
ἀχρεῖον ἀπέφηνεν αὐτό· δοκεῖ δὲ ἐς πολλὰ
ἐπιτήδειον. καὶ ἡ λύγξ δὲ ἀποκρύπτει τὸ οὖρον·
ὅταν γὰρ παγῇ, λίθος γίνεται, καὶ γλυφαῖς ἐπιτή-
δειός ἐστι, καὶ τοῖς γυναικείοις κόσμοις συμμάχε-
ται, φασίν.³

18. Λεοντοφόνου φαγὼν ὁ λέων ἀποτέθνηκε.
τὰ δὲ ἔντομα φθείρεται, εἰ ἐλαίῳ τις ἐγχρίσειεν
αὐτά. γυπῶν γε μὴν τὸ μύρον ὄλεθρός ἐστι.
κάνθαρον δὲ ἀπολεῖς, εἰ ἐπιβάλοις τῶν ῥόδων
αὐτῷ.

19. Κύνες Ἰνδικοί, θηρία καὶ οἵδε εἰσὶ καὶ
ἀλκὴν ἄλκιμα καὶ ψυχὴν θυμοειδέστατα καὶ τῶν
πανταχόθεν κυνῶν μέγιστοι. καὶ τῶν μὲν ἄλλων
ζῴων ὑπερφρονοῦσι, λέοντι δὲ ὁμόσε χωρεῖ κύων
Ἰνδικός, καὶ ἐγκείμενον ὑπομένει, καὶ βρυχωμένῳ
ἀνθυλακτεῖ, καὶ ἀντιδάκνει δάκνοντα· καὶ πολλὰ
αὐτὸν λυπήσας καὶ κατατρώσας, τελευτῶν ἡττᾶται
ὁ κύων. εἴη δ' ἂν καὶ λέων ἡττηθεὶς ὑπὸ κυνὸς
Ἰνδοῦ, καὶ μέντοι καὶ δακὼν ὁ κύων ἔχεται καὶ
μάλα ἐγκρατῶς. κἂν προσελθὼν μαχαίρᾳ τὸ
σκέλος ἀποκόπτῃς τοῦ κυνός, ὁ δὲ οὐκ ἄγει
σχολὴν ἀλγήσας ἀνεῖναι τὸ δῆγμα, ἀλλὰ ἀπεκόπη

---

¹ Reiske : συνδραμόντα.　　　² ἐνούρησε.
³ φασίν διὰ τῆς γλυφῆς.

more numerous, seeming to be drawn by some spell that is in truth love.

17. The Hedgehog too is believed to be one of the animals that show spite. Thus, when it is caught it immediately makes water on its skin, so rendering it unfit for use, though it is thought to serve many purposes. The Lynx too hides its urine, for when it hardens it turns to stone *a* and is suitable for engraving, and is one of the aids to female adornment, so they say.

The Hedgehog

The Lynx

18. If a Lion eats a Lion's-bane,*b* it dies. And insects are destroyed if one drops oil on them. And perfumes are the death of Vultures. Beetles you will extirpate if you scatter roses on them.

Objects poisonous to certain animals

19. The Hounds of India are reckoned as wild animals; they are exceedingly strong and fierce-tempered, and are the largest dogs in the world. All other animals they despise; but an Indian Hound will engage with a lion and resist its onslaught, barking against its roar and giving bite for bite. Only after much worrying and wounding of the lion is the Hound finally overcome; and even a lion might be overcome by an Indian Hound, for once it has bitten, the Hound holds fast with might and main. And even if you take a sword and cut off a Hound's leg, it has no thought, in spite of its pain, of relaxing its

The Indian Hound

---

a The stone known as λυγγούριον was perhaps *amber*. The word was derived from λύγξ and οὖρον.

b In [Arist.] *Mir.* 845 a 28 it appears as a Syrian animal that was supposed to poison lions; to hunters who killed, cooked, and ate it it was equally fatal; cp. Plin. *NH* 8. 38. But L-S⁹ regard it as an insect.

μὲν πρότερον τὸ σκέλος, νεκρὸς δὲ ἀνῆκε τὸ στόμα,
καὶ κεῖται βιασθεὶς ἀποστῆναι τῷ θανάτῳ. ἃ δὲ
προσήκουσα,[1] ἐρῶ ἀλλαχόθι.

20. Ἀνθρώπου μόνου καὶ κυνὸς κορεσθέντων
ἀναπλεῖ ἡ τροφή. καὶ τοῦ μὲν ἀνθρώπου ἡ
καρδία τῷ μαζῷ τῷ λαιῷ προσήρτηται, τοῖς γε
μὴν ἄλλοις ζῴοις ἐν μέσῳ τῷ στήθει προσπέπλα-
σται. γαμψώνυχον δὲ ἄρα οὐδὲ ἓν οὔτε πίνει
οὔτε οὐρεῖ οὔτε μὴν συναγελάζεται ἑτέροις.

21. Θηρίον Ἰνδικὸν βίαιον τὴν ἀλκήν,[2] μέγεθος
κατὰ τὸν λέοντα τὸν μέγιστον, τὴν δὲ χρόαν
ἐρυθρόν, ὡς κινναβάρινον[3] εἶναι δοκεῖν, δασὺ δὲ
ὡς κύνες, φωνῇ τῇ Ἰνδῶν μαρτιχόρας ὠνόμασται.
τὸ πρόσωπον δὲ κέκτηται τοιοῦτον, ὡς δοκεῖν οὐ
θηρίου τοῦτό γε, ἀλλὰ ἀνθρώπου ἔχειν.[4] ὀδόντες
δὲ[5] τρίστοιχοι ἐμπεπήγασιν οἱ ἄνω αὐτῷ, τρίστοιχοι δὲ οἱ κάτω, τὴν ἀκμὴν ὀξύτατοι, τῶν
κυνείων ἐκεῖνοι μείζους· τὰ δὲ ὦτα ἔοικεν ἀνθρώπῳ καὶ ταῦτα,[6] μείζω δὲ καὶ δασέα· τοὺς δὲ
ὀφθαλμοὺς γλαυκός ἐστι, καὶ ἐοίκασιν ἀνθρωπίνοις
καὶ οὗτοι. πόδας δέ μοι νόει καὶ ὄνυχας οἵους εἶναι
λέοντος. τῇ δὲ οὐρᾷ ἄκρᾳ προσήρτηται σκορπίου
κέντρον, καὶ εἴη ἂν ὑπὲρ πῆχυν τοῦτο, καὶ παρ'
ἑκάτερα αὐτῷ ἡ οὐρὰ κέντροις διείληπται· τὸ δὲ
οὐραῖον τὸ ἄκρον ἐς θάνατον ἐκέντησε τὸν περιτυ-

---

[1] προσήκουσα ἑτέρως.
[2] τὴν ἀκοὴν καὶ ἀλκήν L.
[3] κιννάβαριν.
[4] θηρίον . . . ἄνθρωπον ὁρᾶν.
[5] μέν. [6] ταῦτα τήν γε ἑαυτῶν πλάσιν.

bite, but though its leg has been cut off, only when dead does it let go and lie still, forced by death to desist.

What more I have learned I will recount elsewhere.[a]

20. Men and Dogs are the only creatures that belch after they have eaten their fill. A man's heart is attached to his left breast, but in other creatures it is fixed in the centre of the thorax. Among birds of prey there is not one that drinks or makes water, or even gathers in flocks with others of its kind.

Peculiarities of various creatures

21. There is in India a wild beast, powerful, daring, as big as the largest lion, of a red colour like cinnabar, shaggy like a dog, and in the language of India it is called *Martichoras.*[b] Its face however is not that of a wild beast but of a man, and it has three rows of teeth set in its upper jaw and three in the lower; these are exceedingly sharp and larger than the fangs of a hound. Its ears also resemble a man's, except that they are larger and shaggy; its eyes are blue-grey and they too are like a man's, but its feet and claws, you must know, are those of a lion. To the end of its tail is attached the sting of a scorpion, and this might be over a cubit in length; and the tail has stings at intervals on either side. But the tip of the tail gives a fatal sting to anyone who encounters

The Mantichore

[a] See 8. 1.
[b] The English form is *mantichore*. The word is derived from the Persian *mardkhora* = 'man-slayer'; perhaps a man-eating tiger.

χόντα, καὶ διέφθειρε παραχρῆμα. ἐὰν δέ τις αὐτὸν [1]
διώκῃ, ὁ δὲ ἀφίησι τὰ κέντρα πλάγια ὡς βέλη, καὶ
ἔστι τὸ ζῷον ἐκηβόλον. καὶ ἐς τοὔμπροσθεν μὲν
ὅταν ἀπολύῃ τὰ κέντρα, ἀνακλᾷ τὴν οὐράν· ἐὰν δὲ
ἐς τοὐπίσω κατὰ τοὺς Σάκας, ᾗ δὲ ἀποτάδην αὐτὴν
ἐξαρτᾷ. ὅτου δ' ἂν τὸ βληθὲν τύχῃ, ἀποκτείνει·
ἐλέφαντα δὲ οὐκ ἀναιρεῖ μόνον. τὰ δὲ ἀκοντιζό-
μενα κέντρα ποδιαῖα τὸ μῆκός ἐστι, σχοίνου δὲ τὸ
πάχος. λέγει δὲ ἄρα Κτησίας καὶ φησιν ὁμολο-
γεῖν αὐτῷ τοὺς Ἰνδούς, ἐν ταῖς χώραις τῶν
ἀπολυομένων ἐκείνων κέντρων ὑπαναφύεσθαι ἄλλα,
ὡς εἶναι τοῦ κακοῦ τοῦδε ἐπιγονήν. φιληδεῖ δέ,
ὡς ὁ αὐτὸς λέγει, μάλιστα ἀνθρώπους ἐσθίων,
καὶ ἀναιρεῖ γε [2] ἀνθρώπους πολλούς, καὶ οὐ καθ'
ἕνα ἐλλοχᾷ, δύο [3] δ' ἂν ἐπίθοιτο καὶ τρισί, καὶ
κρατεῖ τῶν τοσούτων μόνος. καταγωνίζεται δὲ
καὶ τῶν ζῴων τὰ λοιπά, λέοντα δὲ οὐκ ἂν καθέλοι
ποτέ. ὅτι δὲ κρεῶν ἀνθρωπείων ἐμπιπλάμενον
τόδε τὸ ζῷον ὑπερήδεται, κατηγορεῖ καὶ τὸ
ὄνομα· νοεῖ [4] γὰρ τῇ Ἑλλήνων φωνῇ [5] ἀνθρω-
ποφάγον αὐτὸ εἶναι. ἐκ δὲ τοῦ ἔργου καὶ κέκληται. πέφυκε δὲ κατὰ τὴν ἔλαφον ὤκιστος. τὰ
βρέφη δὲ τῶνδε τῶν ζῴων Ἰνδοὶ θηρῶσιν ἀκέντρους τὰς οὐρὰς ἔχοντα, καὶ λίθῳ γε [6] διαθλῶσιν
αὐτάς, ἵνα ἀδυνατῶσι τὰ κέντρα ἀναφύειν. φωνὴν
δὲ σάλπιγγος ὡς ὅτι ἐγγυτάτω προΐεται. λέγει
δὲ καὶ ἑορακέναι [7] τόδε τὸ ζῷον ἐν Πέρσαις
Κτησίας ἐξ Ἰνδῶν κομισθὲν δῶρον τῷ Περσῶν
βασιλεῖ, εἰ δή τῳ ἱκανὸς τεκμηριῶσαι ὑπὲρ τῶν

---

[1] αὐτό. [2] δέ. [3] καὶ δύο.
[4] Reiske: νοεῖται. [5] φωνῇ ἡ Ἰνδῶν.
[6] γε ἔτι. [7] ἑωρακέναι.

it, and death is immediate. If one pursues the beast it lets fly its stings, like arrows, sideways, and it can shoot a great distance; and when it discharges its stings straight ahead it bends its tail back; if however it shoots in a backward direction, as the Sacae [a] do, then it stretches its tail to its full extent. Any creature that the missile hits it kills; the elephant alone it does not kill. These stings which it shoots are a foot long and the thickness of a bulrush. Now Ctesias asserts (and he says that the Indians confirm his words) that in the places where those stings have been let fly others spring up, so that this evil produces a crop. And according to the same writer the Mantichore for choice devours human beings; indeed it will slaughter a great number; and it lies in wait not for a single man but would set upon two or even three men, and alone overcomes even that number. All other animals it defeats: the lion alone it can never bring down. That this creature takes special delight in gorging human flesh its very name testifies, for in the Greek language it means *man-eater*, and its name is derived from its activities. Like the stag it is extremely swift.

Now the Indians hunt the young of these animals while they are still without stings in their tails, which they then crush with a stone to prevent them from growing stings. The sound of their voice is as near as possible that of a trumpet.

Ctesias declares that he has actually seen this animal in Persia (it had been brought from India as a present to the Persian King)—if Ctesias is to be

[a] Iranian nomads inhabiting the country SE of the Sea of Aral between the rivers Jaxartes and Oxus. They contributed a contingent to the Persian army.

τοιούτων Κτησίας. ἀκούσας γε μὴν τὰ ἴδιά τις
τοῦδε τοῦ ζῴου εἶτα μέντοι τῷ συγγραφεῖ τῷ
Κνιδίῳ προσεχέτω.

22. Σκολόπενδρα θαλαττία διαρρήγνυται, ὥς
φασιν, ἀνθρώπου διαπτύσαντος αὐτῆς.[1]

23. Καρπὸν δὲ ἰτέας εἴ τις θλιβέντα δοίη πιεῖν
τοῖς ἀλόγοις, λυπεῖται ἐκεῖνα οὐδὲ ἕν, μᾶλλον δὲ
καὶ τρέφεται· πιὼν δὲ ἄνθρωπος τὴν σπορὰν τὴν
παιδοποιόν τε καὶ ἔγκαρπον ἀπώλεσε. καί μοι
δοκεῖ Ὅμηρος καὶ τὰ τῆς φύσεως ἀπόρρητα
ἀνιχνεύσας εἶτα μέντοι ʽ καὶ ἰτέαι ὠλεσίκαρποιʼ ἐν
τοῖς ἑαυτοῦ μέτροις εἰπεῖν τοῦτο αἰνιττόμενος.
κωνείου δὲ ἄνθρωπος πιὼν κατὰ τὴν τοῦ αἵματος
πῆξίν τε καὶ ψῦξιν ἀποθνήσκει, ὗς δὲ κωνείου
ἐμπίπλαται καὶ ὑγιαίνει.

24. Οἱ Ἰνδοὶ τέλειον μὲν ἐλέφαντα συλλαβεῖν
ῥᾳδίως ἀδυνατοῦσιν,[2] ἐς δὲ τὰ ἕλη φοιτῶντες τὰ
γειτνιῶντα τῷ ποταμῷ εἶτα μέντοι λαμβάνουσιν
αὐτῶν τὰ βρέφη. ἀσπάζεται γὰρ ὁ ἐλέφας τὰ
ἔνδροσα χωρία καὶ μαλακά, καὶ φιλεῖ τὸ ὕδωρ,
καὶ ἐν τοῖσδε τοῖς ἤθεσι διαιτᾶσθαι ἐθέλει, καὶ ὡς
ἂν εἴποις ἕλειός ἐστι. λαβόντες οὖν ἀπαλὰ καὶ
εὐπειθῆ τρέφουσι κολακείᾳ τε τῇ κατὰ γαστέρα
καὶ θεραπείᾳ τῇ περὶ τὸ σῶμα καὶ φωνῇ θωπευ-
τικῇ (συνιᾶσι γὰρ ἐλέφαντες καὶ γλώττης ἀνθρωπίνης τῆς ἐπιχωρίου), καὶ συνελόντι εἰπεῖν ὡς
παῖδας αὐτοὺς ἐκτρέφουσι, καὶ κομιδὴν προσά-

---

[1] προσπτύσαντος αὐτῇ H.

regarded as a sufficient authority on such matters.
At any rate after hearing of the peculiarities of this
animal, one must pay heed to the historian of Cnidos.

22. The Sea-scolopendra bursts, they say, when a The power
man spits in its face. of human
spittle

23. If one crushes the fruit of a Willow-tree and The Willow
gives it to animals to drink, they suffer no injury at
all, rather they thrive on it.   But if a man drinks it,
his semen loses its procreative strength.   And I fancy
that Homer had explored the secrets of nature when
he wrote in his verses [*Od.* 10. 510] ' and willows that
lose their fruit,' and that he was making a cryptic
allusion to this.   And if a man drink Hemlock, he dies The
from the congealing and chilling of his blood, whereas Hemlock
a hog can gorge itself with Hemlock and remain in
good health.

24. The Indians have difficulty in capturing a full- The taming
grown Elephant.   So they resort to the swamps by a of Elephants
river and then capture the young ones.   For the
Elephant delights in moist places where the ground is
soft, and loves the water, and prefers to pass his time
in these haunts: he is, so to say, a creature of the
swamps.   So having caught them while tender and
docile, they look after them, pandering to their
appetites, grooming their bodies, and using soothing
words—for the Elephants understand the speech of
the natives—and, in a word, they foster them like
children and bestow care upon them, instructing

---

² ἀδυνατοῦσιν, οὔτε γὰρ τοσαῦτα δράσουσιν οὔτε τοσοίδε
παρέσονται.

γουσιν αὐτοῖς καὶ παιδεύματα ποικίλα. οἱ δὲ
πείθονται.

25. Ὅταν ἀλοητὸς ᾖ, καὶ στρέφωνται περὶ τὸν
δῖνον οἱ βόες, καὶ πεπληρωμένη τῶν δραγμάτων ἡ
ἅλως ᾖ, ὑπὲρ τοῦ τοὺς βοῦς μὴ ἀπογεύσασθαι
τῶν σταχύων βολίτῳ τὰς ῥῖνας ἐπιχρίουσιν αὐτῶν,
σόφισμα ἐπινοήσαντες τοῦτο καὶ μάλα γε ἐπιτήδειον. τοῦτο γὰρ τὸ ζῷον μυσαττόμενον τὴν
προειρημένην χρῖσιν οὐκ ἄν τινος ἀπογεύσαιτο,
οὐδ' εἰ τῷ βαρυτάτῳ λιμῷ πιέζοιτο.

26. Τοὺς λαγὼς καὶ τὰς ἀλώπεκας θηρῶσιν οἱ
Ἰνδοὶ τὸν τρόπον τοῦτον. κυνῶν ἐς τὴν ἄγραν
οὐ δέονται, ἀλλὰ νεοττοὺς συλλαβόντες ἀετῶν καὶ
κοράκων καὶ ἰκτίνων προσέτι τρέφουσι καὶ
ἐκπαιδεύουσι τὴν θήραν. καὶ ἔστι τὸ μάθημα,
πράῳ λαγῷ καὶ ἀλώπεκι τιθασῷ κρέας προσαρ-
τῶσι, καὶ μεθιᾶσι θεῖν, καὶ τοὺς ὄρνιθας αὐτοῖς
κατὰ πόδας ἐπιπέμψαντες τὸ κρέας ἀφελέσθαι
συγχωροῦσιν. οἱ δὲ ἀνὰ κράτος διώκουσι, καὶ
ἑλόντες ἢ τὸν ἢ τὴν ἔχουσιν ὑπὲρ τοῦ καταλαβεῖν
ἆθλον τὸ κρέας. καὶ τοῦτο μὲν αὐτοῖς δέλεάρ
ἐστι καὶ μάλα ἐφολκόν. οὐκοῦν ὅταν ἀκριβώσωσι
τὴν σοφίαν τὴν θηρατικήν, ἐπὶ τοὺς ὀρείους λαγὼς
μεθιᾶσιν αὐτοὺς καὶ ἐπὶ τὰς ἀλώπεκας τὰς ἀγρίας.
οἱ δὲ ἐλπίδι τοῦ δείπνου τοῦ συνήθους, ὅταν τι
τούτων φανῇ, μεταθέουσι, καὶ αἱροῦσιν ὤκιστα,
καὶ τοῖς δεσπόταις ἀποφέρουσιν, ὡς λέγει Κτησίας.
καὶ ὅτι ὑπὲρ τοῦ τέως προσηρτημένου κρέως
αὐτοῖς τὰ σπλάγχνα τῶν ᾑρημένων δεῖπνόν [1]
ἐστιν, ἐκεῖθεν καὶ τοῦτο ἴσμεν.

them in various ways. And the baby Elephants learn to obey.

25. In the threshing season when the oxen move round the threshing-floor and the space is filled with sheaves, in order to prevent the oxen from eating the ears, the men smear their nostrils with dung—a device which they have hit upon and which serves them well. For this animal is so disgusted at the aforesaid smearing that it would not touch any food, even though it were assailed with the fiercest hunger.

Oxen treading out the corn

26. This is the way in which the Indians hunt Hares and Foxes: they have no need of hounds for the chase, but they catch the young of Eagles, Ravens, and Kites also, rear them, and teach them how to hunt. This is their method of instruction: to a tame Hare or to a domesticated Fox they attach a piece of meat, and then let them run; and having sent the birds in pursuit, they allow them to pick off the meat. The birds give chase at full speed, and if they catch the Hare or the Fox, they have the meat as a reward for the capture: it is for them a highly attractive bait. When therefore they have perfected the birds' skill at hunting, the Indians let them loose after mountain Hares and wild Foxes. And the birds, in expectation of their accustomed feed, whenever one of these animals appears, fly after it, seize it in a trice, and bring it back to their masters, as Ctesias tells us. And from the same source we learn also that in place of the meat which has hitherto been attached, the entrails of the animals they have caught provide a meal.

Falconry in India

---

[1] τὸ δεῖπνον.

27. Τὸν γρῦπα ἀκούω τὸ ζῷον τὸ Ἰνδικὸν τετράπουν εἶναι κατὰ τοὺς λέοντας, καὶ ἔχειν ὄνυχας καρτεροὺς ὡς ὅτι μάλιστα, καὶ τούτους μέντοι τοῖς τῶν λεόντων παραπλησίους· κατάπτερον δὲ εἶναι, καὶ τῶν μὲν νωτιαίων [1] πτερῶν τὴν χρόαν μέλαιναν ᾄδουσι, τὰ δὲ πρόσθια ἐρυθρά φασι, τάς γε μὴν πτέρυγας αὐτὰς οὐκέτι τοιαύτας, ἀλλὰ λευκάς. τὴν δέρην δὲ αὐτῶν κυανοῖς διηνθίσθαι τοῖς πτεροῖς Κτησίας ἱστορεῖ, στόμα δὲ ἔχειν ἀετῶδες καὶ τὴν κεφαλὴν ὁποίαν οἱ χειρουργοῦντες γράφουσί τε καὶ πλάττουσι. φλογώδεις δὲ τοὺς ὀφθαλμούς φησιν αὐτοῦ. νεοττιὰς δὲ ἐπὶ τῶν ὀρῶν ποιεῖται, καὶ τέλειον μὲν λαβεῖν ἀδύνατόν ἐστι, νεοττοὺς δὲ αἱροῦσι. καὶ Βάκτριοι μὲν γειτνιῶντες Ἰνδοῖς λέγουσιν αὐτοὺς φύλακας εἶναι τοῦ χρυσοῦ ⟨τοῦ⟩[2] αὐτόθι, καὶ ὀρύττειν τε αὐτόν φασιν αὐτοὺς καὶ ἐκ τούτου τὰς καλιὰς ὑποπλέκειν, τὸ δὲ ἀπορρέον Ἰνδοὺς λαμβάνειν. Ἰνδοὶ δὲ οὔ φασιν αὐτοὺς φρουροὺς εἶναι τοῦ προειρημένου· μηδὲ γὰρ δεῖσθαι χρυσίου γρῦπας (καὶ ταῦτα εἰ λέγουσι, πιστὰ ἔμοιγε δοκοῦσι λέγειν)· ἀλλὰ αὐτοὺς μὲν ἐπὶ τὴν τοῦ χρυσίου ἄθροισιν ἀφικνεῖσθαι, τοὺς δὲ ὑπέρ τε τῶν σφετέρων βρεφῶν δεδιέναι καὶ τοῖς ἐπιοῦσι μάχεσθαι. καὶ διαγωνίζεσθαι μὲν πρὸς τὰ ἄλλα ζῷα καὶ κρατεῖν ῥᾷστα, λέοντι δὲ μὴ ἀνθίστασθαι μηδὲ ἐλέφαντι. δεδιότες δὲ ἄρα τὴν τῶνδε τῶν θηρίων ἀλκὴν οἱ ἐπιχώριοι, μεθ᾽ ἡμέραν ἐπὶ τὸν χρυσὸν οὐ στέλλονται, νύκτωρ δὲ ἔρχονται· ἐοίκασι γὰρ τηνικάδε τοῦ καιροῦ λανθάνειν μᾶλλον. ὁ δὲ χῶρος οὗτος, ἔνθα

---

[1] εἶναι . . . νωτιαίων] τὰ νῶτα εἶναι καὶ τούτων τῶν.
[2] ⟨τοῦ⟩ add. Reiske.

27. I have heard that the Indian animal the Gryphon is a quadruped like a lion; that it has claws of enormous strength and that they resemble those of a lion. Men commonly report that it is winged and that the feathers along its back are black, and those on its front are red, while the actual wings are neither but are white. And Ctesias records that its neck is variegated with feathers of a dark blue; that it has a beak like an eagle's, and a head too, just as artists portray it in pictures and sculpture. Its eyes, he says, are like fire. It builds its lair among the mountains, and although it is not possible to capture the full-grown animal, they do take the young ones. And the people of Bactria, who are neighbours of the Indians, say that the Gryphons guard the gold in those parts; that they dig it up and build their nests with it, and that the Indians carry off any that falls from them. The Indians however deny that they guard the aforesaid gold, for the Gryphons have no need of it (and if that is what they say, then I at any rate think that they speak the truth), but that they themselves come to collect the gold, while the Gryphons fearing for their young ones fight with the invaders. They engage too with other beasts and overcome them without difficulty, but they will not face the lion or the elephant. Accordingly the natives, dreading the strength of these animals, do not set out in quest of the gold by day, but arrive by night, for at that season they are less likely to be detected. Now the region where the Gryphons live

οἵ τε γρῦπες διαιτῶνται καὶ τὰ χρυσεῖά [1] ἐστιν,
ἔρημος πέφυκε δεινῶς. ἀφικνοῦνται δὲ οἱ τῆς
ὕλης τῆς προειρημένης θηραταὶ κατὰ χιλίους τε
καὶ δὶς τοσούτους ὡπλισμένοι, καὶ ἅμας κομίζουσι
σάκκους τε, καὶ ὀρύττουσιν ἀσέληνον ἐπιτηροῦντες
νύκτα. ἐὰν μὲν οὖν λάθωσι τοὺς γρῦπας, ὤνηνται
διπλῆν τὴν ὄνησιν· καὶ γὰρ σῴζονται καὶ μέντοι
καὶ οἴκαδε τὸν φόρτον κομίζουσι, καὶ ἐκκαθήραν-
τες [2] οἱ μαθόντες χρυσοχοεῖν [3] σοφίᾳ τινὶ σφετέρᾳ
πάμπολυν πλοῦτον ὑπὲρ τῶν κινδύνων ἔχουσι τῶν
προειρημένων· ἐὰν δὲ κατάφωροι γένωνται, ἀπο-
λώλασιν. ἐπανέρχονται δὲ ἐς τὰ οἰκεῖα ὡς
πυνθάνομαι δι' ἔτους τρίτου καὶ τετάρτου.

28. Χελώνης θαλαττίας ἀποτμηθεῖσα ἡ κεφαλὴ [4]
βλέπει καὶ καταμύει τὴν χεῖρα προσάγοντος· ἤδη
δ' ἂν καὶ δάκοι, εἰ περαιτέρω προσαγάγοις τὴν
χεῖρα. καὶ ἐπὶ μακρὸν ἐκλάμποντας ἔχει τοὺς
ὀφθαλμούς· αἱ γάρ τοι κόραι λευκόταταί τε καὶ
περιφανέσταταί εἰσι, καὶ ἐξαιρεθεῖσαι χρυσίῳ καὶ
ὅρμοις ἐντίθενται. ἔνθεν τοι καὶ δοκοῦσι ταῖς
γυναιξὶ θαυμασταί. γίνονται δὲ ὡς πυνθάνομαι αἱ
χελῶναι αἵδε ἐν τῇ θαλάττῃ, ἣν ᾄδουσιν Ἐρυθράν.

29. Ὁ ἀλεκτρυὼν τῆς σελήνης ἀνισχούσης
ἐνθουσιᾷ φασι καὶ σκιρτᾷ. ἥλιος δὲ ἀνίσχων οὐκ
ἄν ποτε αὐτὸν διαλάθοι, ᾠδικώτατος δὲ ἑαυτοῦ [5]
ἐστι τηνικάδε. πυνθάνομαι δὲ ὅτι ἄρα καὶ τῇ

---

[1] Reiske : τὰ χωρία τὰ χρυσεῖα.
[2] ἐκκαθάραντες.
[3] Ges : χρυσωρυχεῖν.
[4] κεφαλὴ οὔποτε θνήσκει ἀλλά.

and where the gold is mined is a dreary wilderness.
And the seekers after the aforesaid substance arrive,
a thousand or two strong, armed and bringing spades
and sacks; and watching for a moonless night they
begin to dig. Now if they contrive to elude the
Gryphons they reap a double advantage, for they not
only escape with their lives but they also take home
their freight, and when those who have acquired a
special skill in the smelting of gold have refined it,
they possess immense wealth to requite them for the
dangers described above. If however they are
caught in the act, they are lost. And they return
home, I am told, after an interval of three or four
years.

28. The head of a Turtle, after it has been cut off, The Turtle
sees and closes its eyes if one brings one's hand near; and its eyes
and it would still bite if you brought your hand too
near. It has eyes that flash a long way off, for the
pupils are the purest white and very conspicuous, and
when removed are set in gold and necklaces.[a] For
that reason they are greatly admired by women.
These Turtles, I learn, are natives of what is commonly called the ' Red Sea.'

29. The Cock, they say, at moonrise becomes pos- The Cock
sessed and jumps about. Never would a sunrise pass and its
unnoticed by him, but at that hour he excels himself crowing
in crowing. And I learn that the Cock is the

---

[a] χελωνία, *tortoise-stone*; an unknown gem. Cp. Plin. *HN*
37. 10.

---

[5] ᾠδικώτερος δὲ ἑαυτοῦ μᾶλλον.

AELIAN

Λητοῖ φίλον ἐστὶν ὁ ἀλεκτρυὼν [1] τὸ ὄρνεον. τὸ
δὲ αἴτιον, παρέστη φασὶν αὐτῇ τὴν διπλῆν τε καὶ
μακαρίαν ὠδῖνα ὠδινούσῃ. ταῦτά τοι καὶ νῦν
ταῖς τικτούσαις ἀλεκτρυὼν πάρεστι, καὶ δοκεῖ
πως εὐώδινας ἀποφαίνειν. τῆς δὲ ὄρνιθος ἀπο-
λωλυίας, ἐπῴάζει αὐτός, καὶ ἐκλέπει τὰ ἐξ ἑαυτοῦ
νεόττια σιωπῶν· οὐ γὰρ ᾄδει τότε θαυμαστῇ τινι
καὶ ἀπορρήτῳ αἰτίᾳ, ναὶ μὰ τόν· δοκεῖ γάρ μοι
συγγινώσκειν ἑαυτῷ θηλείας ἔργα καὶ οὐκ ἄρρενος
δρῶντι τηνικάδε. μάχῃ [2] ⟨δὲ⟩ [3] ἀλεκτρυὼν καὶ
τῇ πρὸς ἄλλον ἡττηθεὶς ἀγωνίᾳ οὐκ ἂν ᾄσειε [4]· τὸ
γάρ τοι φρόνημα αὐτῷ κατέσταλται, [5] καὶ καταδύε-
ταί γε ὑπὸ τῆς αἰδοῦς. κρατήσας δὲ γαῦρός ἐστι,
καὶ ὑψαυχενεῖ, καὶ κυδρουμένῳ ἔοικε. θαυμάσαι
δὲ τοῦ ζῴου ὑπεράξιον καὶ ἐκεῖνο δήπου· θύραν
γὰρ ὑπιὼν καὶ τὴν ἄγαν ὑψηλήν, ὁ δὲ ἐπικύπτει,
ἀλαζονέστατα δρῶν ἐκεῖνος τοῦτο· φειδοῖ γὰρ
τοῦ λόφου πράττειν ἔοικε τὸ εἰρημένον.

30. Οἱ κολοιοὶ δεινῶς φιλοῦσι τὸ ὁμόφυλον.
τοῦτό τοι καὶ διαφθείρει αὐτοὺς πολλάκις, καὶ τό
γε δρώμενον τοιοῦτόν ἐστιν. ὅτῳ μέλει θηρᾶσαι
κολοιούς, τοιαῦτα παλαμᾶται. ἔνθα οἶδεν αὐτῶν
νομὰς καὶ τροφὰς καὶ ἀθροιζομένους ὁρᾷ κατ᾽
ἀγέλας, ἐνταῦθα λεκανίδας ἐλαίου μεστὰς διατί-
θησιν. οὐκοῦν διειδὲς μὲν τὸ ἔλαιον, περίεργον δὲ
τὸ ὀρνίθιον, καὶ ἀφικνεῖται καὶ ἐπὶ τὸ χεῖλος τοῦ
σκεύους κάθηται, καὶ κύπτει κάτω καὶ ὁρᾷ τὴν
ἑαυτοῦ σκιάν, καὶ οἴεται κολοιὸν βλέπειν ἄλλον,
καὶ κατελθεῖν πρὸς αὐτὸν σπεύδει. κάτεισί τε

---
[1] ὁ ἀλεκτρυων del. Cobet.   [2] ἐν μάχῃ.

favourite bird of Leto. The reason is, they say, that
he was at her side when she was so happily brought to
bed of twins. That is why to this very day a Cock
is at hand when women are in travail, and is believed
somehow to promote an easy delivery.

If the Hen dies the Cock himself sits on the eggs
and hatches his own eggs in silence, for then for some
strange and inexplicable reason, I must say, he does
not crow. I fancy that he is conscious that he is then
doing the work of a female and not of a male.

A Cock that has been defeated in battle and in a
struggle with another will not crow, for his spirit is
depressed and he hides himself in shame. On the
other hand if he is victorious, he is proud and holds his
head high and appears exultant. Here too is a most
astonishing trait, I think. As he passes beneath a
doorway, no matter how high, the Cock lowers his
head—a most pretentious action, done apparently
to protect his comb.

30. Jackdaws are devoted to their own species; The
and this it is that often causes their destruction. And Jackdaw
it happens in this way. The man who intends to
hunt Jackdaws adopts the following plan. In the how caught
place where he knows that they feed and where he
sees them gathering in flocks he arranges basins full
of oil. Now the oil is transparent and the bird is in-
quisitive, and it comes and perches on the rim of the
vessel, bends down, and sees its own reflexion, and
supposing it to be another Jackdaw, makes haste to
go down to it. So it descends, flaps its wings, and

---

³ ⟨δέ⟩ add. Reiske.   ⁴ ᾆσαι.
⁵ κατέσταλται καὶ μεμείωται.

οὖν καὶ πτερύσσεται [1] καὶ περιβάλλει τὸ ἔλαιον αὐτῷ,[2] καὶ ἀναπτερυγίσαι [3] ἥκιστός ἐστι, καὶ χωρὶς δικτύων καὶ πάγης καὶ ἀρπεδόνων τὸ ζῷον μένει ὡς ἂν εἴποις πεπεδημένον.

31. Ὁ ἐλέφας, οἱ μὲν αὐτοῦ προκύπτειν χαυλιόδοντάς φασιν, οἱ δὲ κέρατα. ἔχει δὲ καὶ καθ᾽ ἕκαστον πόδα δακτύλους πέντε, ὑποφαίνοντας μὲν τὰς ἐκφύσεις, οὐ μὴν διεστῶτας. ταῦτά τοι καὶ νηκτικός ἐστιν ἥκιστα. σκέλη δὲ τὰ κατόπιν τῶν προσθίων [4] βραχύτερά ἐστι· μαζοὶ δὲ αὐτῷ πρὸς ταῖς μασχάλαις εἰσί· μυκτῆρα δὲ κέκτηται χειρὸς παγχρηστότερον καὶ γλῶτταν βραχεῖαν· χολὴν δὲ αὐτὸν ἔχειν οὐ κατὰ τὸ ἧπαρ ἀλλὰ πρὸς τῷ ἐντέρῳ [5] φασί. κύειν δὲ πυνθάνομαι δύο ἐτῶν τὸν ἐλέφαντα. οἱ δὲ οὐ τοσοῦτον χρόνον, ἀλλὰ ὀκτωκαίδεκα μηνῶν ὁμολογοῦσιν. ἀποτίκτει δὲ ἰσήλικα τὸ μέγεθος μόσχῳ ἐνιαυσίῳ, σπᾷ δὲ τῆς θηλῆς τῷ στόματι. ἐνθουσιῶν δὲ ἐς μίξιν οἴστρῳ τε φλεγόμενος ἐμπίπτει τοίχῳ καὶ ἀνατρέπει, καὶ φοίνικας κλίνει, τὸ μέτωπον προσαράττων κατὰ τοὺς κριούς. πίνει δὲ ὕδωρ οὐ διειδὲς οὐδὲ καθαρόν, ἀλλ᾽ ὅταν ὑποθολώσῃ τε καὶ ὑποταράξῃ. καθεύδει γε μὴν ὀρθοστάδην· κατακλιθῆναι γὰρ καὶ ἐξαναστῆναι ἐργῶδες αὐτῷ. ἀκμὴ δὲ ἐλέφαντι ἑξήκοντα ἔτη,[6] διατείνει δὲ τὸν βίον καὶ ἐς διπλῆν ἑκατοντάδα. κρυμῷ δὲ ὁμιλεῖν ἥκιστός ἐστι.[7]

---

[1] Jac : περιπτύσσεται.

[2] Ges : αὐτό.

[3] καὶ ἀναπτερυγίσαι] ὃν γλίσχρον καὶ συνδεῖται· τὸ δὲ αἴτιον ἀναπτερυγίσαι.

[4] Ges : τὰ πρόσθια τῶν κατόπιν.

scatters the oil all over itself. Being quite unable to fly up again the bird remains, so to speak, fettered, though neither net nor trap nor snare is there.

31. The Elephant has what some call protruding tusks, what others call horns. On each foot he has five toes; their growth is just visible although they are not separate; and that is why he is ill-adapted for swimming. His hind legs are shorter than his forelegs; his paps are close to his armpits: he has a proboscis which is far more serviceable than a hand, and his tongue is short; his gall-bladder is said to be not near the liver but close to the intestines. I am informed that the duration of the Elephant's pregnancy is two years, although others maintain that it is not so long, but only eighteen months. It bears a young one as big as a one-year-old calf, which pulls at the dug with its mouth. When it is possessed with a desire to copulate and is burning with passion, it will dash at a wall and overturn it, will bend palm-trees by butting its forehead against them, as rams do. It drinks water not when clear and pure but when it has dirtied and stirred it up a little. But it sleeps standing upright, for it finds the act of lying down and of rising troublesome. The Elephant reaches its prime at the age of sixty, though its life extends to two hundred years. But it cannot endure cold.

The Elephant, its anatomy and habits

---

[5] *Camper* : στέρνῳ.

[6] ἐλέφαντος ἐξήκοντα ἔτη γεγονέναι.

[7] *The sentence* κρυμῷ . . . ἐστι *appears in the* MSS *between* ἔτη *and* διατείνει; *transposed by* H (*Hermes* 11. 233).

32. Προβατεῖαι δὲ Ἰνδῶν ὁποῖαι μαθεῖν ἄξιον. τὰς αἶγας καὶ τὰς οἶς ὄνων τῶν μεγίστων μείζονας ἀκούω καὶ ἀποκύειν τέτταρα ἑκάστην· μείω γε μὴν τῶν τριῶν οὔτ᾽ αἴξ Ἰνδικὴ οὔτ᾽ ἂν οἶς ποτε τέκοι. καὶ τοῖς μὲν προβάτοις αἱ οὐραὶ πρὸς τὸν πόδα τέτανται, αἱ δὲ αἶγες μηκίστας ἔχουσιν, ὥστ᾽ ἐπιψαύειν γῆς ὀλίγου. τῶν μὲν οὖν οἰῶν τῶν τίκτειν ἀγαθῶν ἀποκόπτουσι τὰς οὐρὰς οἱ νομεῖς, ἵνα ἀναβαίνωνται, ἐκ δὲ τῆς πιμελῆς τῆς τούτων καὶ ἔλαιον ἀποθλίβουσι· τῶν δὲ ἀρρένων διατέμνουσι τὰς οὐράς, καὶ ἐξαιροῦσι τὸ στέαρ καὶ ἐπιρράπτουσι, καὶ ἐνοῦται πάλιν ἡ τομή, καὶ ἀφανίζεται τὰ ἴχνη αὐτῆς.

33. Ἀλέξανδρος ὁ Μύνδιος τὸν χαμαιλέοντα λυπεῖν τοὺς ὄφεις καὶ ἀσιτίᾳ περιβάλλειν τὸν τρόπον τοῦτόν φησι. κάρφος πλατὺ καὶ στερεὸν ἐνδακὼν ἑαυτὸν ἐπιστρέφει, καὶ ἀντιπρόσωπος [ὁμόσε] [1] χωρεῖ τῷ πολεμίῳ. ὁ δὲ αὐτοῦ λαβέσθαι ἀδυνατεῖ, τοῦ κάρφους τὸ πλάτος οὐκ ἔχων περιχανεῖν. οὐκοῦν ἄδειπνος τό γε ἐπ᾽ ἐκείνῳ μένει ὁ ὄφις· δάκνων γάρ τοι τὰ λοιπὰ τῶν μελῶν αὐτοῦ οὐδὲν ἀνύτει· στερεὰν γὰρ τὴν φορίνην ἔχει, καὶ ἐπαΐει τῶν ἐκείνου ὀδόντων ὁ χαμαιλέων οὐδὲ ἕν.

34. Ὁ αὐχὴν ὁ τοῦ λέοντος ἐξ ὀστέου [2] συνέστηκεν, οὐ μὴν ἐκ σφονδύλων πολλῶν. εἰ δέ τις τὰ ὀστᾶ τοῦ λέοντος διακόπτοι, πῦρ αὐτῶν ἐξάλλεται. μυελοὺς δὲ οὐκ ἔχει· οὐδὲ γάρ ἐστι κοῖλα αὐλῶν δίκην. μίξεως δὲ αὐτὸν οὐδεμία ἔτους

---

[1] ὁμόσε del. H (1876).

32. It is worth while learning the nature of the flocks that belong to the Indians. I have heard that their Goats and their Sheep are larger than the largest asses, and that each one gives birth to quadruplets; anyhow no Goat or Sheep in India would ever give birth to less than three at a time. The Sheep have tails reaching down to their feet, while the Goats have tails of such length as all but touch the ground. The shepherds cut off the tails of the ewes which are good for breeding so that the rams may mount them, and they press oil out of the fat contained in them. In the rams' tails also they make an incision and extract the fat and sew them up again. And the cut joins up once more and all traces of it disappear.

33. Alexander of Myndus declares that the Chameleon annoys snakes and makes them go hungry in this way. Taking in its teeth a piece of wood, broad and solid, it turns about and goes to face its enemy. But the Snake is unable to seize it as its jaws cannot compass the width of the wood; and so the Snake goes without a meal as far as the Chameleon is concerned, for although it may bite the rest of its body it gains nothing, since the Chameleon has a solid hide and cares not at all for the fangs of the Snake.

34. The neck of a Lion consists of a single bone and not of a number of vertebrae. And if a man cuts through the bones of a Lion fire leaps forth. But they are devoid of marrow, nor are they hollow like tubes. There is no season of the year in which it

---

² *Jac* : ὀστέων.

ἀναστέλλει ὥρα. κύει δὲ ἄρα[1] μηνῶν δύο.
τίκτει δὲ[2] πεντάκις, καὶ τῇ μὲν ὠδῖνι τῇ πρώτῃ
πέντε, τῇ δὲ δευτέρᾳ τέτταρα, τρία τε ⟨τῇ⟩ ἐπὶ
ταύτῃ, καὶ δύο ⟨τῇ⟩[3] ἐπ' ἐκείνῃ, καὶ ἕν ἐπὶ
πάσαις. οἱ δὲ σκύμνοι ἀρτιγενεῖς μικροί τέ εἰσι
καὶ τυφλοὶ κατὰ τὰ σκυλάκια· βαδίσεως δὲ
ὑπάρχονται, ὅταν δύο μῆνας ἀπὸ γενεᾶς διαβιῶ-
σιν.[4] ὁ λόγος δέ, ὅστις λέγει διαξαίνειν αὐτοὺς
τὰς μήτρας, μῦθός ἐστι. λιμώττων μὲν οὖν λέων
ἐντυχεῖν χαλεπός ἐστι, κορεσθεὶς δὲ πραότατος·
φασὶ δὲ καὶ φιλοπαίστην εἶναι τηνικάδε αὐτόν.
φύγοι[5] δὲ οὐκ ἄν ποτε τὰ νῶτα τρέψας λέων,
ἡσυχῇ δὲ ἐπὶ πόδα ἀναχωρεῖ βλέπων ἀντίος.[6]
τοῦ γήρως δὲ ὑπαρχομένου ἐπὶ τὰ αὔλια ἔρχεται
καὶ ἐπὶ τὰς καλύβας καὶ ἐπὶ τὰς οἰκήσεις τὰς τῶν
νομέων τὰς ὑπάντρους, καὶ εἰκότως· ταῖς γὰρ
ὀρείοις ἔτι θήραις ἐπιθαρρεῖν ἀδύνατός ἐστι. πῦρ
δὲ ὀρρωδεῖ. ὅστις μὲν οὖν ἐστιν αὐτῶν γυρότερος
καὶ συνεστραμμένος καὶ τὴν χαίτην λασιώτερος,
ἀθυμότερός τε καὶ ἀτολμότερος δοκεῖ μᾶλλον· ὁ
δὲ μήκους[7] εὖ ἥκων καὶ εὐθυτενὴς τὴν τρίχα
ἀνδρειότερος πεπίστευται καὶ θυμοειδέστερος.
ἀδηφάγος δὲ ὢν καὶ ὅλα φασὶ μέλη βρύκων ἂν
καταπίοι. τούτων οὖν πεπληρωμένος καὶ τριῶν
ἡμερῶν οὐκ ἐσθίει πολλάκις, ἔστ' ἂν ὑπαναλωθῇ
τὰ πρῶτά οἱ καὶ πεφθῇ. πίνει δὲ ὀλίγα.

35. Ὁ βοῦς ὁ πρᾶος τοῦ πλήττοντος καὶ κολά-
ζοντος οὐκ ἄν ποτε λήθην λάβοι, ἀλλ' ἀπομνησθεὶς[8]

---

[1] Jac: ἀνά.   [2] δὲ καί.
[3] ⟨τῇ⟩ . . . ⟨τῇ⟩ add. H.
[4] διαβιώσῃ τὰ τοῦ λέοντος βρέφη.

abstains from coupling, and the Lioness is pregnant
for two months. Five times does she give birth, at
the first birth to five cubs, at the second to four, after
that to three, after that to two, and finally to one.
The cubs when new-born are small and, like puppies,
blind,[a] and they begin to walk when they have completed two months from birth. But the account which
says that they scratch through the womb is a fable.
To encounter a Lion when famished is dangerous,
but when he has eaten his fill he is extremely gentle;
they even say that at that time he is playful. A Lion
will never turn his back and flee, but withdraws,
looking you straight in the face, and by degrees.
But when he begins to age he visits folds and huts
and spots where shepherds lodge in caves; which is
to be expected, because he no longer has the spirit
for hunting on the mountains. He has a horror of fire.
Any Lion that inclines to roundness and a compact
figure, and that has too shaggy a mane, appears to be
lacking in spirit and daring; whereas the beast that
attains a good length and has a straight mane is regarded as bolder and fiercer. Possessing a ravenous
appetite he will, they say, devour and swallow whole
limbs. So when he has taken his fill of them he will
often not eat for the space of three days until his
former meal has been gradually absorbed and
digested. He drinks but little.

35. A domesticated Ox will never forget the man *The Ox and*
who strikes and chastises him, but he remembers and *its memory*

[a] See 5. 39.

---

⁵ καὶ φύγοι.   ⁶ ἀντίος καὶ ἐπιβραχύ.
⁷ εἰς μῆκος.   ⁸ ὑπομνησθείς.

τιμωρεῖται καὶ διαστήματος ἐγγενομένου. ὧν μὲν
γὰρ ὑπὸ ζεύγλην καὶ τρόπον τινὰ καθειργμένος,
ἔοικε δεσμώτῃ καὶ ἡσυχάζει· ὅταν δὲ ἀφεθῇ,
πολλάκις ⟨μὲν⟩[1] τῷ σκέλει παίσας συνέτριψε
μέλος[2] τι τοῦ βουκόλου, πολλάκις δὲ καὶ θυμωθεὶς
ἐς κέρας εἶτα ἐμπεσὼν ἀπέκτεινεν αὐτόν. ἐντεῦθεν
πρὸς τοὺς ἄλλους πρᾷός ἐστι, καὶ πάρεισιν ἐς τὸ
αὔλιον ἡσυχῇ· οὐ γάρ ἐστιν ἀνήμερος πρὸς οὓς
οὐκ ἔχει τοῦ θυμοῦ τὴν ὑπόθεσιν.

36. Ἡ τῶν Ἰνδῶν γῆ, φασὶν αὐτὴν οἱ συγγραφεῖς πολυφάρμακόν τε καὶ τῶν βλαστημάτων
τῶνδε δεινῶς πολύγονον εἶναι. καὶ τὰ μὲν σώζειν
αὐτῶν καὶ ἐκ τῶν κινδύνων ῥύεσθαι τοὺς ὑπὸ τῶν
δακετῶν ὁμοῦ τῷ θανάτῳ ὄντας (πολλὰ δὲ ἐκεῖθι
τοιαῦτα), τὰ δὲ ἀπολλύναι καὶ διαφθείρειν ὀξύτατα,
ὧνπερ οὖν[3] καὶ τὸ ἐκ τοῦ ὄφεως ⟨τοῦ πορφυροῦ⟩[4]
γινόμενον εἴη ἄν. ἔστι δὲ ἄρα οὗτος ὁ ὄφις κατὰ
σπιθαμὴν τὸ μῆκος ὅσα ἰδεῖν· χρόαν δὲ ἔοικε
πορφύρᾳ τῇ βαθυτάτῃ. λευκὴν δὲ κεφαλὴν καὶ
οὐκέτι πορφυρᾶν περιηγοῦνται αὐτοῦ, λευκὴν δὲ
οὐχ ὡς εἰπεῖν ἔπος, ἀλλὰ καὶ χιόνος ἐπέκεινα καὶ
γάλακτος,[5] ὀδόντων δὲ ἄγονός ἐστιν ὁ ὄφις
οὗτος· εὑρίσκεται δ' ἐν τοῖς πυρωδεστάτοις τῆς
Ἰνδικῆς χωρίοις. καὶ δάκνειν μὲν ἥκιστός ἐστι,
καὶ κατά γε τοῦτο φαίης ἂν τιθασὸν αὐτὸν εἶναι
καὶ πρᾶον· οὗ δ' ἂν κατεμέσῃ, ὡς ἀκούω, ἢ
ἀνθρώπου τινὸς ἢ θηρίου, τοῦδε τὸ μέλος διασα-
πῆναι ἀνάγκη πᾶν. οὐκοῦν θηραθέντα αὐτὸν ἐκ
τοῦ οὐραίου μέρους ἐξαρτῶσι, καὶ οἷα εἰκὸς κάτω

---

[1] ⟨μέν⟩ add. H.   [2] Wytt : μέρος.

takes his revenge even after a long interval. For being under the yoke and in a certain degree confined, he is like a prisoner and keeps still; but when he is let out he has often kicked and broken some limb of his herdsman; often too he has put passion into his horns and has fallen upon a man and killed him. After that he is gentle to others and goes quietly to the fold, for he is not savage towards those against whom he has no ground for anger.

36. Historians say that India is rich in drugs and remarkably prolific of medicinal plants, of which some save life and rescue from danger men who have been brought to death's door through the bites of noxious creatures (and there are many such in India); while other drugs are swift to kill and destroy; and to this class might be assigned the drug which comes from the Purple Snake. Now this snake appears to be a span long; its colour is like the deepest purple, but its head they describe as white and not purple, and not just white, but whiter even than snow or milk. But this snake has no fangs and is found in the hottest regions of India, and though it is quite incapable of biting—for which reason you might pronounce it to be tame and gentle—yet if it vomits upon anyone (so I am told), be it man or animal, the entire limb inevitably putrefies. Therefore when caught men hang it up by the tail, and naturally it has its head hanging down, looking at the ground. And below the creature's mouth they place a bronze vessel, into

The Purple Snake of India

---

[3] ὧν οὖν (or ἕν)περ.
[4] ⟨τοῦ πορφυροῦ⟩ add. Jac.
[5] γάλακτος πλέον λευκήν.

τὴν κεφαλὴν ἔχει, καὶ ἐς γῆν ὁρᾷ· ὑπ' αὐτὸ δὲ τὸ
στόμα [1] τοῦ θηρὸς ἀγγεῖόν τι τιθέασι πεποιημένον
χαλκοῦ. καὶ [2] διὰ τοῦ στόματος σταγόνες ἐκείνῳ [3]
λείβονται ἐς τοῦτο, καὶ τὸ καταρρεῦσαν συνίσταταί
τε καὶ πήγνυται, καὶ ἐρεῖς ἰδὼν ἀμυγδαλῆς δάκρυον
εἶναι. καὶ ὁ μὲν ἀποθνήσκει ὁ ὄφις, ὑφαιροῦσι
δὲ τὸ σκεῦος, καὶ προστιθέασιν [4] ἄλλο, χαλκοῦν
καὶ ἐκεῖνο· νεκροῦ δὲ ἐκρεῖ πάλιν ὑγρὸς ἰχώρ,[5]
καὶ ἔοικεν ὕδατι. τριῶν δὲ ἡμερῶν ἐῶσι, καὶ
συνίσταται μέντοι καὶ οὗτος. εἴη δ' [6] ἂν ἀμφοῖν [7] διαφορὰ κατὰ τὴν χρόαν· ἡ μὲν γὰρ δεινῶς
ἐστι μέλαινα, ἡ δὲ ἠλέκτρῳ εἴκασται. οὐκοῦν
τούτου μὲν εἰ δοίης τινὶ ὅσον σησάμου μέγεθος
ἐμβαλὼν [8] ἐς οἶνον ἢ ἐς σιτίον, πρῶτον μὲν αὐτὸν
σπασμὸς περιλήψεται καὶ μάλα ἰσχυρός, εἶτα
διαστρέφονταί οἱ τὼ ὀφθαλμώ, ὁ δὲ ἐγκέφαλος
διὰ τῶν ῥινῶν κατολισθάνει [9] λειβόμενος,[10] καὶ
ἀποθνήσκει καὶ μάλα οἴκτιστα·[11] ἐὰν δὲ ἔλαττον
λάβῃ τοῦ φαρμάκου, ἄφυκτα μὲν αὐτῷ τὸ [12]
ἐντεῦθέν ἐστι, χρόνῳ δὲ ἀπόλλυται. ἐὰν δὲ τοῦ
μέλανος ὀρέξῃς, ὅπερ οὖν κατέρρευσε τεθνεῶτος,
ὅσον [13] σησάμου καὶ τοῦτο μέγεθος, ὑπόπνους
γίνεται, καὶ φθόη καταλαμβάνει τὸν λαβόντα, καὶ
ἐνιαυτοῦ ἀναλίσκεται τηκεδόνι· πολλοὶ δὲ καὶ ἐς
ἔτη δύο προῆλθον, κατὰ μικρὰ ἀποθνήσκοντες.

37. Ἡ στρουθὸς ἡ μεγάλη ᾠὰ μὲν ἀποτίκτει
πολλά, οὐ πάντα δὲ ἐκγλύφει,[14] ἀλλὰ ἀποκρίνει τὰ
ἄγονα, τοῖς δὲ ἐγκάρποις ἐπῳάζει. καὶ ἐκ μὲν

---

[1] αὐτῷ δὲ τῷ στόματι.   [2] καὶ αἱ.
[3] ἐκεῖναι.   [4] τιθεῖσιν.
[5] ἰχὼρ οὗτος.   [6] Jac: ἡ δ'.

which there ooze drops from its mouth; and the
liquid sets and congeals, and if you saw it you would
say that it was gum from an almond-tree. So when
the snake is dead they remove the vessel and substitute another, also of bronze; and again from the
dead body there flows a liquid serum which looks like
water. This they leave for three days, and it too
sets; but there will be a difference in colour between
the two, for the latter is a deep black and the former
the colour of amber. Now if you give a man a piece
of this no bigger than a sesame seed, dropping it into
his wine or his food, first he will be seized with convulsions of the utmost violence; next, his eyes squint
and his brain dissolves and drips through his nostrils,
and he dies a most pitiable death. And if he takes a
smaller dose of the poison, there is no escape for him
hereafter, for in time he dies. If however you
administer some of the black matter which has flowed
from the snake when dead, again a piece the size of a
sesame seed, the man's body begins to suppurate, a
wasting sickness overtakes him, and within a year he
is carried off by consumption. But there are many
whose lives have been prolonged for as much as two
years, while little by little they died.

37. Although the Ostrich lays a number of eggs it   The Ostrich
does not hatch all of them but sets aside the sterile
ones and sits upon those that are fertile; and from

---

⁷ ἐπ᾽ ἀμφοῖν.
⁸ *Schn* : ἀφελὼν καὶ ἐμβαλών.
⁹ κατολισθαίνει.
¹⁰ *Reiske* : θλιβόμενος.
¹¹ καὶ οἴκτιστα μὲν ἀλλὰ ᾤκιστα.
¹² καί.   ¹³ ὡς εἶναι.   ¹⁴ τρέφει.

255

τούτων τοὺς νεοττοὺς ἐξέλεψεν, ἐκεῖνα δὲ τὰ
ἐκφαυλισθέντα τούτοις τροφὴν παρατίθησιν. εἰ
δὲ αὐτὴν διώκοι τις, ἡ δὲ οὐκ ἐπιτολμᾷ τῇ πτήσει,
θεῖ δὲ τὰς πτέρυγας ἁπλώσασα· εἰ δὲ ἁλίσκεσθαι
μέλλοι, τοὺς παραπίπτοντας λίθους ἐς τοὐπίσω
σφενδονᾷ τοῖς ποσίν.

38. Οἱ στρουθοὶ οἱ σμικροὶ συνειδότες ἑαυτοῖς
ἀσθένειαν διὰ σμικρότητα τοῦ σώματος, ἐπὶ τοῖς
ἀκρεμόσι τῶν κλάδων τοῖς φέρειν αὐτοὺς δυναμέ-
νοις τὰς νεοττιὰς συμπλάσαντες εἶτα μέντοι τὴν ἐκ
τῶν θηρατῶν ἐπιβουλὴν ὡς τὰ πολλὰ διαφεύγουσιν
ἐπιβῆναι τῷ [1] κλαδὶ μὴ δυναμένων· οὐ γὰρ
αὐτοὺς φέρει διὰ λεπτότητα.

39. Αἱ δὲ ἀλώπεκες ἐς ὑπερβολὴν προήκουσαι
πανουργίας καὶ τρόπου δολεροῦ ὅταν θεάσωνται
σφηκιὰν εὐθενουμένην,[2] αὐταὶ[3] μὲν ἀποστρέφονται
τὸν χηραμὸν ἐκνεύουσαι καὶ τὰς ἐκ τῶν κέντρων
τρώσεις φυλαττόμεναι· καθιᾶσι δὲ τὴν οὐρὰν
δασυτάτην τε οὖσαν καὶ μηκίστην τὴν αὐτὴν καὶ
διασείουσι τοὺς σφῆκας· οἱ δὲ προσέχονται τῷ
τῶν τριχῶν δάσει. ὅταν δὲ ἐμπαλαχθῶσιν[4] αὐτῷ,
προσαράττουσι τὴν οὐρὰν ἢ δένδρῳ ἢ τειχίῳ[5]
ἢ αἱμασιᾷ· παιόμενοι δὲ οἱ σφῆκες ἀποθνήσκουσιν.
εἶτα ἦλθον ἐπὶ τὸν αὐτὸν τόπον, καὶ τοὺς λοιποὺς
προσαναλέξασαι καὶ ἀποκτείνασαι κατὰ τοὺς
πρώτους, ὅταν ἐννοήσωσι λοιπὸν εἰρήνην εἶναι καὶ
ἀπὸ τῶν κέντρων ἐλευθερίαν, καθῆκαν τὸ στόμα
καὶ τὰ σφηκία ἐσθίουσι, μήτε θορυβούμεναι μήτε
μὴν τὰ κέντρα ὑφορώμεναι.

these it hatches its young, giving them the other, rejected eggs to eat. And if one chases the Ostrich it does not venture to fly but spreads its wings and runs. And if it is in danger of being captured it slings the stones that come in its way backwards with its feet.

38. Sparrows, conscious that their weakness is due to the small size of their bodies, build their nests upon those twigs of branches which are strong enough to support them, and so generally escape the machinations of bird-catchers who cannot climb the branch : it is too slender to bear them. The Sparrow

39. Foxes pass all bounds in their mischievousness and trickery. When they observe a thriving Wasps' nest they turn their back upon it and avoid the hole so as to protect themselves from being stung. But their tail, which is very bushy and long, they let down into the hole and shake up the Wasps. And these fasten on the thick hairs. But when they are entangled in them the Foxes beat their tail against a tree or fence or stone wall, and the Wasps are killed by the blows. Then the Foxes return to the same spot, collect the remaining Wasps, and kill them as they did the first lot. When they know that they will have peace and be free from stings they put down their heads and eat up the combs, with nothing to disturb them and no need to look out for stings. The Fox and Wasps

---

1 *Schn* : τῇ.
2 εὐθην- MSS *always.*
3 *Reiske* : αὗται.
4 ἀναπλασθῶσιν MSS, ἐμπλασ- *Jac.*
5 τειχίῳ H (1875). τοίχῳ.

40. Κυνὸς κρανίον ῥαφὴν οὐκ ἔχει. δραμὼν δὲ
ἐπὶ πλέον λάγνης γίνεται, φασί.[1] κυνὸς δὲ γηρῶν-
τος ἀμβλεῖς οἱ ὀδόντες καὶ μελαίνονται. εὔρινος
δέ ἐστιν οὕτως ὡς μήποτ' ἂν ὀπτοῦ κυνείου
κρέως μηδ'[2] ἂν καρυκείᾳ τῇ ποικιλωτάτῃ καὶ
δολερωτάτῃ καταγοητευθέντος γεύσασθαι. τρεῖς
δὲ ἄρα νόσοι κυνὶ ἀποκεκλήρωνται καὶ οὐ πλείους,
κυνάγχη λύττα ποδάγρα· ἀνθρώποις γε μὴν
μυρίαι. πᾶν δὲ ὅ τι ἂν ὑπὸ κυνὸς λυττῶντος
δηχθῇ, τοῦτο ἀποθνήσκει. κύων δὲ ποδαγρήσας,
σπανίως ἀναρρωσθέντα ὄψει αὐτόν. κυνὶ δὲ βίος
ὁ μήκιστος τεσσαρεσκαίδεκα ἔτη. Ἄργος δὲ ὁ
Ὀδυσσέως καὶ ἡ περὶ αὐτὸν ἱστορία ἔοικε παιδιὰ
Ὁμήρου εἶναι.

41. Γένος ὀρνίθων Ἰνδικῶν βραχυτάτων καὶ
τοῦτο εἴη ἄν. ἐν τοῖς πάγοις τοῖς ὑψηλοῖς νεοτ-
τεύει καὶ ταῖς πέτραις ταῖς καλουμέναις λεπραῖς,[3]
καὶ ἔστι τὸ μέγεθος τὰ ὀρνύφια ὅσονπερ ᾠὸν
πέρδικος· σανδαρακίνην δέ μοι νόει τὴν χρόαν
αὐτῶν. καὶ Ἰνδοὶ μὲν αὐτὸ φωνῇ τῇ σφετέρᾳ
δίκαιρον φιλοῦσιν ὀνομάζειν, Ἕλληνες δὲ ὡς
ἀκούω δίκαιον. τούτου τὸ ἀποπάτημα εἴ τις
λάβοι ὅσον κέγχρου μέγεθος λυθὲν[4] ἐν τῷ
πώματι, ὁ δὲ[5] ἐς ἑσπέραν ἀπέθανεν. ἔοικε δὲ ὁ
θάνατος ὕπνῳ καὶ μάλα γε ἡδεῖ καὶ ἀνωδύνῳ καὶ
οἷον οἱ ποιηταὶ λυσιμελῆ φιλοῦσιν ὀνομάζειν ἢ
ἀβληχρόν· εἴη γὰρ ἂν καὶ οὗτος ἐλεύθερος ὀδύνης
καὶ τοῖς δεομένοις διὰ ταῦτα ἥδιστος. σπουδὴν

---

[1] φασὶ μᾶλλον.   [2] μήτ'.
[3] λιπταῖς MSS, λισσ- Schn.
[4] ἔωθεν conj. Jac; cp. Ctes. ap. Phot. Bibl. 47ᵃ. 30.

40. A Dog's skull has no suture. Running, they say, makes a Dog more lustful. In old age a Dog's teeth are blunt and turn black. He is so keen-scented that he will never touch the roasted flesh of a dog, be it bewitched by the subtlest and craftiest of rich sauces. Now there are three diseases which fall to the lot of a Dog and no more, viz. dog-quinsy, rabies, and gout, while mankind has an infinite number. Everything that is bitten by a mad Dog dies. If a Dog once gets gout you will hardly see him recover his strength. The life of a Dog at its longest is fourteen years; so Argus, the dog of Odysseus, and the story about him [*Od.* 7. 291] look like a playful tale of Homer's.

The Dog

41. The following species of bird belongs to the very smallest of those in India. They build their nests on high mountains and among what are called ' rugged ' rocks. These tiny birds are the size of a partridge's egg, and you must know that they are orange-coloured. The Indians are accustomed to call the bird in their language *dikairon,*[a] but the Greeks, so I am informed, *dikaion.* If a man take of its droppings a quantity the size of a millet-seed dissolved in his drink, he is dead by the evening. But his death is like a very pleasant and painless sleep, and such as poets are fond of describing as ' limb-relaxing ' and ' gentle.' For death too may be free from pain, and for that reason most welcome to those

The ' Dikairon ' (dung-beetle)

[a] ' The " bird " was the Dung-beetle, *Scarabaeus sacer* . . . the " dung " was probably . . . a resinous preparation of Indian hemp ' (Thompson, *Gk. birds,* s.v.).

[5] εἶτα.

δὲ ἄρα τὴν ἀνωτάτω τίθενται Ἰνδοὶ ἐς τὴν
κτῆσιν αὐτοῦ· κακῶν γὰρ αὐτὸ ἐπίληθον ἡγοῦνται
τῷ ὄντι· καὶ οὖν καὶ ἐν τοῖς δώροις τοῖς μέγα
τιμίοις τῷ Περσῶν βασιλεῖ ὁ Ἰνδῶν πέμπει καὶ
τοῦτο. ὁ δὲ καὶ τῶν ἄλλων ἁπάντων προτιμᾷ
λαβὼν καὶ ἀποθησαυρίζει κακῶν ἀνιάτων ἀντίπαλόν τε καὶ ἀμυντήριον, εἰ ἀνάγκη καταλάβοι.
οὔκουν οὐδὲ ἔχει τις ἐν Πέρσαις αὐτὸ ἄλλος, ὅτι
μὴ βασιλεύς τε αὐτὸς καὶ μήτηρ ἡ βασιλέως. καὶ
διὰ ταῦτα ἀντικρίνοντες βασανίσωμεν τῶν φαρμάκων τοῦ τε Ἰνδικοῦ καὶ τοῦ Αἰγυπτίου ὁπότερον
ἦν προτιμότερον· ἐπεὶ τὸ μὲν ἐφ᾽ ἡμέραν [1]
ἀνεῖργέ [2] τε καὶ ἀνέστελλε τὰ δάκρυα τὸ Αἰγύπτιον, τὸ δὲ λήθην κακῶν παρεῖχεν αἰώνιον τὸ
Ἰνδικόν· καὶ τὸ μὲν γυναικὸς δῶρον ἦν, τὸ δὲ
ὄρνιθος ἢ ἀπορρήτου φύσεως δεσμῶν τῶν ὄντως
βαρυτάτων ἀπολυούσης δι᾽ ὑπηρέτου τοῦ προειρημένου. καὶ Ἰνδοὺς κτήσασθαι αὐτὸ εὐτυχήσαν-
τας,[3] ὡς τῆς ἐνταυθοῖ φρουρᾶς ἀπολυθῆναι ὅταν
ἐθέλωσιν.

42. Ὁ ὄρνις ὁ ἀτταγᾶς (μέμνηται δὲ καὶ
Ἀριστοφάνης αὐτοῦ ἐν Ὄρνισι τῷ δράματι),
οὗτός τοι τὸ ἴδιον ὄνομα ᾗ σθένει φωνῇ φθέγγεται
καὶ ἀναμέλπει αὐτό. λέγουσι δὲ καὶ τὰς καλουμέ-
νας μελεαγρίδας τὸ αὐτὸ δήπου δρᾶν τοῦτο, καὶ
ὅτι Μελεάγρῳ τῷ Οἰνέως προσήκουσι κατὰ γένος
μαρτυρεῖσθαι καὶ μάλα εὐστόμως. λέγει δὲ ὁ
μῦθος, ὅσαι ἦσαν οἰκεῖαι τῷ Οἰνείδῃ νεανίᾳ,
ταύτας ἐς δάκρυά τε ἄσχετα καὶ πένθος ἄτλητον

---

[1] ἡμέραν αὐτήν.   [2] ἀνεῖχε.
[3] εὐτυχήσαντάς ⟨φασιν⟩ Warmington.

who desire it. The Indians accordingly do their utmost to obtain possession of it, for they regard it as in fact ' causing them to forget their troubles ' [Hom. *Od.* 4. 221]. And so the Indian King includes this also among the costly presents which he sends to the Persian King, who receives it and values it above all the rest and stores it away, to counteract and to remedy ills past curing, should necessity arise. But there is not another soul in Persia save the King and the King's mother who possesses it. So let us compare the Indian and Egyptian drug *a* and see which of the two was to be preferred. On the one hand the Egyptian drug repelled and suppressed sorrow for a day, whereas the Indian drug caused a man to forget his troubles for ever. The former was the gift of a woman, the latter of a bird or else of Nature, which mysteriously releases men from a truly intolerable bondage through the aforesaid agency. And the Indians are fortunate in possessing it so that they can free themselves from this world's prison whenever they wish.

42. The bird called ' Francolin ' (Aristophanes The mentions it in his comedy of the *Birds* [249, *etc.*]) proclaims and sings its own name as loudly as it can. And they say that Guinea-fowls, as they are called, The Guineado the same and testify to their kinship with fowl Meleager the son of Oeneus in the clearest tones. The legend goes that all the women who were related to the son of Oeneus dissolved into unassuageable tears and sorrow past bearing, and mourned for him

a In Hom. *Od.* 4. 219-32 Helen mixes a drug, thought to have been opium in some form, in the wine of Telemachus to make him forget his sorrow for his father.

ἐκπεσεῖν καὶ θρηνεῖν, οὐδέν τι τῆς λύπης ἄκος
προσιεμένας, οἴκτῳ δὲ ἄρα τῶν θεῶν ἐς ταῦτα τὰ
ζῷα ἀμεῖψαι τὸ εἶδος. ταῖς δὲ ἴνδαλμά τε καὶ
σπέρμα τοῦ τότε πένθους ἐντακῆναι, καὶ ἐς νῦν
ἔτι Μελέαγρόν τε ἀναμέλπειν, καὶ ὡς αὐτῷ προσή-
κουσιν ᾄδειν καὶ τοῦτο μέντοι. ὅσοι δὲ ἄρα
αἰδοῦνται τὸ θεῖον,[1] οὐκ ἄν ποτε τῶνδε τῶν
ὀρνίθων ἐπὶ τροφῇ[2] προσάψαιντο. καὶ ἥτις ἡ
αἰτία ἴσασί τε οἱ τὴν νῆσον οἰκοῦντες τὴν Λέρον
καὶ ἔνεστι μαθεῖν ἀλλαχόθεν.

43. Πέπυσμαι δὲ ὑπὲρ τῶν μυρμήκων καὶ
ταῦτα. οὕτως ἄρα αὐτοῖς τὸ ἐθελουργὸν καὶ τὸ
ἐθελόπονον πάρεστιν ἀπροφασίστως καὶ ἄνευ τινὸς
ὑποτιμήσεως ἐθελοκακούσης καὶ σκήψεως, ἐς ἣν
ὑποικουρεῖ τὸ ῥάθυμον, ὡς κἂν[3] ταῖς πανσελήνοις
μηδὲ νύκτωρ βλακεύειν μηδὲ ἐλινύειν, ἀλλ' ἔχεσθαι
τῆς σπουδῆς. ὦ ἄνθρωποι, μυρίας προφάσεις τε
καὶ σκήψεις ἐς τὸ ῥᾳστωνεύειν ἐπινοοῦντες. καὶ
τί δεῖ καταλέγειν τε καὶ ἐπαντλεῖν τὸν τοσοῦτον[4]
ὄχλον; κεκήρυκται γὰρ Διονύσια καὶ Λήναια καὶ
Χύτροι καὶ Γεφυρισμοί, καὶ μετελθόντων ἐς τὴν
Σπάρτην ἄλλα καὶ ἐς Θήβας ἄλλα καὶ κατὰ
πόλιν μυρία ἑκάστην τὰ μὲν βάρβαρον τὰ δὲ
Ἑλλάδα.

---

[1] θεῖον καὶ εἰ μᾶλλον τὴν Ἄρτεμιν.
[2] Schn : τροφήν.
[3] Jac : καί or κἄν.   [4] τοιοῦτον.

---

[a] Leros, off the coast of Caria, contained a shrine of Artemis
Parthenos, and there according to the legend the women were
transformed.

and found no cure for their sorrow. So the gods in pity allowed them to change their shape into these birds; and the semblance and seed of their ancient grief have sunk into them so that to this day they raise a strain to Meleager and even sing of how they are his kin.

So then all who reverence the gods would never lay hands on one of these birds for the sake of food. And the reason of this is known to the inhabitants of the island of Leros^a and can be learned from other sources.

43. Here are more facts that I have learned touch- The Ant ing Ants. So indefatigable, so ready to work are they, without making excuses, without any base plea for release, without alleging reasons that are a cloak for indolence, that not even at night when the moon is full do they idle and take holiday, but stick to their occupation.

Look at you men—devising endless pretexts and Greek excuses for idling! What need is there to detail and festivals pour out the full number of these occasions? Proclaimed as holidays are the Dionysia,^b the Lenaea, the Festival of Pots, Causeway Day: go to Sparta, and there are others: others again at Thebes: and an endless number in every city, some in a foreign, others in a Greek city.

<hr>

^b Greater or City Dionysia held about March 28–April 2; Lesser or Country Dionysia, about December 19–22; Lenaea, at the end of January; Χύτροι, feast in honour of the departed, about March 4; all these at Athens. Γεφυρισμός: those who took part in the Eleusinia, in March, indulged in abusive repartee as they passed along the Sacred Way between Athens and Eleusis.

44. Μαρτύριον δὲ τῆς τῶν ζῴων φύσεως, ὅτι οὐ πάνυ τι [1] δυσμεταχείριστά [2] ἐστιν, ἀλλὰ εὖ παθόντα ἀπομνησθῆναι τῆς εὐεργεσίας ἐστὶν ἀγαθά,[3] ἐν τῇ Αἰγύπτῳ οἵ τε αἴλουροι καὶ οἱ ἰχνεύμονες καὶ οἱ κροκόδιλοι καὶ τὸ τῶν ἱεράκων ἔτι φῦλον. ἁλίσκεται δὲ κολακείᾳ τῇ κατὰ γαστέρα, καὶ ἐντεῦθεν ἡμερωθέντα λοιπὸν πραότατα μένει· καὶ οὐκ ἄν ποτε ἐπίθοιτο τοῖς εὐεργέταις τοῖς ἑαυτῶν, τοῦ θυμοῦ τοῦ συμφυοῦς τε καὶ συγγενοῦς ἅπαξ παραλυθέντα. ἄνθρωπος δὲ καὶ λόγου μετειληχὸς ζῷον καὶ φρονήσεως ἀξιωθὲν καὶ αἰδεῖσθαι λαχὸν καὶ ἐρύθημα πιστευθὲν φίλου γίνεται βαρὺς πολέμιος, καὶ ὅσα ἀπόρρητα ἐπιστεύθη, ταῦτα δι' αἰτίαν βραχυτάτην καὶ τὴν παρατυχοῦσαν ἐς ἐπιβουλὴν ἐξέπτυσε τὴν τοῦ πεπιστευκότος.

45. Θαυμάσαι λόγον ἄξιόν φησιν Εὔδημος, καὶ τῷ γε ἀνδρὶ τῷδε ὁ λόγος οὗτός ἐστι. νεανίας θηρατικός, συμβιοῦν τοῖς τῶν ζῴων ἀγριωτάτοις οἷός τε, ἐκ νέων μέντοι καὶ βρεφῶν πεπωλευμένοις,[4] εἶχε συντρόφους τε καὶ συσσίτους ἑαυτοῖς γεγενημένους κύνα καὶ ἄρκτον καὶ λέοντα. καὶ ταῦτα μὲν χρόνου πρὸς ἄλληλα εἰρήνην ἄγειν καὶ φίλα νοεῖν σφίσι λέγει ὁ Εὔδημος· μιᾶς δὲ τυχεῖν ἡμέρας τὸν κύνα προσπαίζοντα τὴν ἄρκτον καὶ ὑπαικάλλοντα καὶ ἐρεσχελοῦντα, τὴν δὲ οὐκ εἰωθότως ἐκθηριωθῆναι καὶ ἐμπεσεῖν τῷ κυνί, καὶ λαφύξαι τοῖς ὄνυξι τοῦ δειλαίου τὴν γαστέρα καὶ διασπάσασθαι αὐτόν· ἀγανακτῆσαι δὲ τῷ συμβάντι ὁ αὐτός φησι τὸν λέοντα καὶ οἰονεὶ μισῆσαι τὸ ἄσπονδον τῆς ἄρκτου καὶ ἄφιλον, καὶ τὸν

44. In Egypt the Cats, the Ichneumons, the Croco- Animals remember kind actions diles, and moreover the Hawks afford evidence that animal nature is not altogether intractable,[1] but that when well-treated they are good at remembering kindness. They are caught by pandering to their appetites, and when this has rendered them tame they remain thereafter perfectly gentle: they would never set upon their benefactors once they have been freed from their congenital and natural temper. Man however, a creature endowed with reason, credited with understanding, gifted with a sense of honour, supposed capable of blushing, can become the bitter enemy of a friend and for some trifling and casual reason blurt out confidences to betray the very man who trusted him.

45. Eudemus has a story to fill one with amaze- The story of a Lion, a Bear, and a Dog ment, and this is the story he tells. A young hunter who was able to spend his life among the wildest of[3] animals, after they had been trained from the day when they were young cubs, had living with him and sharing each other's food a Dog, a Bear, and a Lion. And for a time, Eudemus says, they lived in peace and mutual amity. But it happened one day that the Dog was playing with the Bear, fawning upon it and teasing it, when the Bear became unwontedly savage, fell upon the Dog, and with its claws ripped the poor creature's belly open and tore him to pieces. The Lion, says the writer, was indignant at what had occurred and seemed to detest the Bear's implaca-

---

[1] οὐ πάντῃ.   [2] δυσμεταχείριστος.
[3] ἀγαθὰ ἀγριώτατα ζῴων.
[4] Jac: πεπωλευμένους.

κύνα οἷα ἑταῖρον ποθῆσαι καὶ ἐς δικαίαν προελθεῖν
ὀργήν, καὶ ἐπιθεῖναι τῇ ἄρκτῳ τὴν δίκην, καὶ τὰ
αὐτὰ δρᾶσαι αὐτήν, ἅπερ οὖν εἰργάσατο τὸν κύνα
ἐκείνη. Ὅμηρος μὲν οὖν φησιν

ὡς ἀγαθὸν καὶ παῖδα καταφθιμένοιο λιπέσθαι·

ἔοικε δὲ ἡ φύσις δεικνύναι ὅτι καὶ φίλον ἑαυτῷ τι-
μωρὸν καταλιπεῖν, ὦ φίλε Ὅμηρε, κέρδος ἐστίν.
οἷόν τι καὶ περὶ Ζήνωνος καὶ Κλεάνθους νοοῦμεν,
εἴ τι ἀκούομεν.

46. Ἐν Ἰνδοῖς γίνεται θηρία τὸ μέγεθος ὅσον
γένοιντο ἂν οἱ κάνθαροι, καὶ ἔστιν ἐρυθρά· κιν-
ναβάρει δὲ εἰκάσειας [1] ἄν, εἰ πρῶτον θεάσαιο
αὐτά. πόδας ⟨δὲ⟩[2] ἔχει ταῦτα μηκίστους, καὶ
προσάψασθαι μαλακά ἐστι. φύεται δὲ ἄρα ἐπὶ
τῶν δένδρων τῶν φερόντων τὸ ἤλεκτρον, καὶ
σιτεῖται τὸν τῶν φυτῶν καρπὸν τῶνδε. θηρῶσι
δὲ αὐτὰ οἱ Ἰνδοὶ καὶ ἀποθλίβουσι, καὶ ἐξ αὐτῶν
βάπτουσι τάς τε φοινικίδας καὶ τοὺς ὑπ᾽ αὐταῖς
χιτῶνας καὶ πᾶν ὅ τι ἂν ἐθέλωσιν ἄλλο ἐς τήνδε
τὴν χρόαν ἐκτρέψαι τε καὶ χρῶσαι. κομίζεται
δὲ ἄρα ἡ τοιάδε ἐσθὴς καὶ τῷ τῶν Περσῶν βασιλεῖ. καὶ τό γε εὐειδὲς τῆς ἐσθῆτος δοκεῖ τοῖς
Πέρσαις θαυμαστόν, ἀντικρινομένη [3] δὲ ταῖς [4]
Περσῶν ἐπιχωρίοις κρατεῖ κατὰ πολὺ καὶ ἐκπλήτ-

---

[1] εἰκάσαις.
[2] ⟨δὲ⟩ add. H.
[3] καὶ ἀντικρινομένη.
[4] τοῖς.

bility and want of affection: it was smitten with grief for the Dog as for a companion, and being filled with righteous anger, punished the Bear by treating it exactly as the Bear had treated the Dog. Now Homer says [*Od.* 3. 196]

‘ So good a thing it is that when a man dies a son should be left.’

And Nature seems to show that there is an advantage, my dear Homer, in leaving a friend behind to avenge one. Something of the same kind, we believe, occurred with Zeno and Cleanthes, if there is some truth in what we hear.[a]

46 (i). In India are born insects [b] about the size of beetles, and they are red. On seeing them for the first time you might compare them to vermilion. They have very long legs and are soft to the touch. They flourish on those trees which produce amber, and feed upon the fruit of the same. And the Indians hunt them and crush them and with their bodies dye their crimson cloaks and their tunics beneath and everything else that they wish to convert and stain to that colour. Garments of this description are even brought to the Persian king, and their beauty excites the admiration of the Persians, and indeed when set against their native garments far surpasses them and amazes people, according to

The Lac insect

[a] Cleanthes succeeded his master Zeno as head of the Stoic school at Athens, 263 B.C.

[b] This is the *Tachardia lacca* of India and S Asia, an insect allied to the cochineal and kermes insects. It exudes a resinous secretion (on to the twigs of certain trees, esp. those of the species *Ficus*) which is lac. The crimson dye is the red fluid in the ovary of the female.

τει, ὥς φησι Κτησίας· ἐπεὶ καὶ τῶν ᾀδομένων
Σαρδιανικῶν[1] ὀξυτέρα τέ ἐστι καὶ τηλαυγεστέρα.

Γίνονται δὲ ἐνταῦθα τῆς Ἰνδικῆς, ἔνθα οἱ κάνθαροι, καὶ οἱ καλούμενοι κυνοκέφαλοι, οἷς τὸ
ὄνομα ἔδωκεν ἡ τοῦ σώματος ὄψις τε καὶ φύσις·
τὰ δὲ ἄλλα ἀνθρώπων ἔχουσι, καὶ ἠμφιεσμένοι
βαδίζουσι δορὰς θηρίων. καὶ εἰσι δίκαιοι, καὶ
ἀνθρώπων λυποῦσιν οὐδένα, καὶ φθέγγονται μὲν
οὐδὲ ἕν, ὠρύονται δέ, τῆς γε μὴν Ἰνδῶν φωνῆς
ἐπαΐουσι. τροφὴ δὲ αὐτοῖς τῶν ζῴων τὰ ἄγρια·
αἱροῦσι δὲ αὐτὰ ῥᾷστα, καὶ γάρ εἰσιν ὤκιστοι, καὶ
ἀποκτείνουσι καταλαβόντες, καὶ ὀπτῶσιν οὐ πυρί,
ἀλλὰ πρὸς τὴν εἵλην τὴν τοῦ ἡλίου ἐς μοίρας
διαξήναντες. τρέφουσι δὲ καὶ αἶγας καὶ οἷς.
καὶ σῖτον μὲν ποιοῦνται τὰ ἄγρια, πίνουσι δὲ τὸ
ἐκ τῶν θρεμμάτων γάλα ὧν τρέφουσι. μνήμην
δὲ αὐτῶν ἐν τοῖς ἀλόγοις ἐποιησάμην, καὶ εἰκότως·
ἔναρθρον γὰρ καὶ εὔσημον καὶ ἀνθρωπίνην φωνὴν
οὐκ ἔχουσιν.

47. Χλωρὶς ὄνομα ὄρνιθος, ἥπερ οὖν οὐκ ἂν
ἀλλαχόθεν ποιήσαιτο τὴν καλιὰν ἢ ἐκ τοῦ λεγομέ-
νου συμφύτου· ἔστι δὲ ῥίζα τὸ σύμφυτον εὑρεθῆναί
τε καὶ ὀρύξαι χαλεπή. στρωμνὴν δὲ ὑποβάλλεται
τρίχας καὶ ἔρια. καὶ ὁ μὲν θῆλυς ὄρνις οὕτω
κέκληται, ὁ δὲ ἄρρην, χλωρίωνα καλοῦσιν αὐτόν,
καὶ ἔστι τὸν βίον μηχανικός, μαθεῖν τε πᾶν ὅ τι

[1] τῶν Σ.

Ctesias, because the colour is even stronger and
more brilliant than the much-vaunted wares of
Sardes.

(ii). And in the same part of India as the beetles, The Dog-
are born the 'Dog-heads,' as they are called—a name heads
which they owe to their physical appearance and
nature. For the rest they are of human shape and
go about clothed in the skins of beasts; and
they are upright and injure no man; and though
they have no speech they howl; yet they understand the Indian language. Wild animals are their
food, and they catch them with the utmost ease, for
they are exceedingly swift of foot; and when they
have caught them they kill and cook them, not over
a fire but by exposing them to the sun's heat after
they have shredded them into pieces. They also
keep goats and sheep, and while their food is the
flesh of wild beasts, their drink is the milk of the
animals they keep. I have mentioned them along
with brute beasts, as is logical, for their speech is
inarticulate, unintelligible, and not that of man.

47. Golden Oriole *a* is the name of a bird which The
declines to build its nest with anything but comfrey, Golden
as it is called. Comfrey is a root which is hard to find Oriole
and hard to dig up. For bedding it lays down hairs
and wool. *Chloris* is the name given to the hen, but
the cock-bird they call *chlorion*, and it is clever at
getting a livelihood; it is quick to learn anything

---

a Ael. has confused the habits of two different birds: it
is the *Greenfinch*, the χλωρίς of Arist. *HA* 615 b 32, that
builds its nest of comfrey, etc. But Ael. uses the word to
signify the *Golden Oriole*, a migratory bird, which the Greenfinch is not.

οὖν ἀγαθὸς καὶ τλήμων ὑπομεῖναι τὴν ἐν τῷ
μανθάνειν βάσανον, ὅταν ἁλῷ. καὶ διὰ μὲν τοῦ
χειμῶνος ἄφετον καὶ ἐλεύθερον οὐκ ἂν ἴδοι τις
αὐτόν, θεριναὶ [1] δὲ ὅταν ὑπάρξωνται [2] τροπαὶ
τοῦ ἔτους, τηνικαῦτ᾽ ἂν [3] ἐπιφαίνοιτο. Ἀρκτοῦ-
ρός τε ἐπέτειλεν,[4] ὁ δὲ ἀναχωρεῖ ἐς τὰ οἰκεῖα,
ὁπόθεν καὶ δεῦρο ἐστάλη.

48. Ὑπὸ θυμοῦ τεθηγμένον ταῦρον καὶ ὑβρί-
ζοντα ἐς κέρας καὶ σὺν ὁρμῇ ἀκατασχέτῳ [5]
φερόμενον οὐχ ὁ βουκόλος ἐπέχει, οὐ φόβος
ἀναστέλλει, οὐκ ἄλλο τοιοῦτον, ἄνθρωπος δὲ
ἵστησιν αὐτὸν καὶ παραλύει τῆς ὁρμῆς τὸ δεξιὸν
αὐτοῦ γόνυ διασφίγξας ταινίᾳ καὶ ἐντυχὼν αὐτῷ.

49. Ἡ πάρδαλις πέντε ἔχει δακτύλους ἐν τοῖς
ποσὶ τοῖς προσθίοις, ἐν δὲ τοῖς κατόπιν τέτταρας.
ἡ δὲ θήλεια εὐρωστοτέρα τοῦ ἄρρενος. ἐὰν δὲ
γεύσηται ἀγνοοῦσα τοῦ καλουμένου παρδαλιάγχου
(πόα δέ ἐστιν), ἀπόπατημα ἀνθρώπου ποθὲν
λιχνεύσασα [6] διασώζεται.

50. Οἱ ἵπποι, τὰς κάτω βλεφαρίδας οὔ φασιν
αὐτοὺς ἔχειν. Ἀπελλῆν οὖν τὸν Ἐφέσιον αἰτίαν
λέγουσιν ἔχειν, ἐπεί τινα ἵππον γράφων οὐ
παρεφύλαξε τὸ ἴδιον τοῦ ζῴου. οἱ δὲ οὐκ Ἀπελλῆν φασι ταύτην τὴν αἰτίαν ἐνέγκασθαι, ἀλλὰ
Μίκωνα,[7] ἀγαθὸν μὲν ἄνδρα γράψαι τὸ ζῷον
τοῦτο, σφαλέντα δ᾽ οὖν ἐς μόνον τὸ εἰρημένον.

---

[1] Schn : ἠριναί MSS, H. [2] ὑπάρχωνται.
[3] τηνικαῦτα. [4] Ἀρκτούρου τε ἐπιτολαί.
[5] καὶ ἀκατασχέτως. [6] Radermacher : ἀνιχνεύσασα MSS, H.
[7] Meursius : Νίκωνα.

whatsoever, and will patiently endure the ordeal of
learning when in captivity. In the winter season you
will not see it abroad and free, but at the occurrence of
the summer solstice, that is when it will appear. As
soon as Arcturus has risen [a] the bird returns to its
native haunts whence it came to us.

48. When once a Bull has been provoked to anger
and is threatening violence with his horns and rushing
on with irresistible speed, the herdsman cannot control him, fear cannot check him, nor anything else;
only a man may bring him to a halt and stay his
onrush if he tie a scarf round his own right knee and
face the Bull.

How to
check an
angry Bull

49. The Leopard has five toes on its fore-paws and
four on its hind-paws. But the female is stronger
than the male. If it unwittingly eats what is called
'leopard's-choke'[b] (this is a herb), it licks some
human excrement and preserves its life.

The
Leopard

50. Horses, they say, have no lower eyelashes, so
that Apelles[c] of Ephesus incurred blame for ignoring
this peculiarity in his picture of a horse. But others
assert that it was not Apelles who was charged with
this fault but Micon, a man of great skill in depicting
this animal, although on this one point he made a
mistake.

The Horse,
its eyelashes

[a] The morning rising of Arcturus in the region of Rome is
on September 20.

[b] Aconite.

[c] Apelles, the most renowned of Grecian painters, contemporary of Alexander the Great.—Micon, fl. middle of 5th
cent. B.C. at Athens, famous as painter and sculptor.

51. Τὸν οἶστρόν φασιν ὅμοιον εἶναι μυίᾳ μεγίστῃ καὶ εἶναι στερεὸν καὶ εὐπαγῆ καὶ ἔχειν κέντρον ἰσχυρὸν ἠρτημένον τοῦ σώματος, προΐεσθαι δὲ καὶ ἦχον βομβώδη. τὸν μὲν οὖν μύωπα ὅμοιον φῦναι[1] τῇ καλουμένῃ κυνομυίᾳ, βομβεῖν δὲ τοῦ οἴστρου μᾶλλον, ἔχειν δὲ ἔλαττον τὸ κέντρον.

52. Ὄνους ἀγρίους οὐκ ἐλάττους ἵππων τὰ μεγέθη ἐν Ἰνδοῖς γίνεσθαι πέπυσμαι. καὶ λευκοὺς μὲν τὸ ἄλλο εἶναι σῶμα, τήν γε μὴν κεφαλὴν ἔχειν πορφύρᾳ παραπλησίαν, τοὺς δὲ ὀφθαλμοὺς ἀποστέλλειν κυανοῦ χρόαν. κέρας δὲ ἔχειν ἐπὶ τῷ μετώπῳ ὅσον πήχεως τὸ μέγεθος καὶ ἡμίσεος προσέτι, καὶ τὸ μὲν κάτω μέρος τοῦ κέρατος εἶναι λευκόν, τὸ δὲ ἄνω φοινικοῦν, τό γε μὴν μέσον μέλαν δεινῶς. ἐκ δὴ τῶνδε τῶν ποικίλων κεράτων πίνειν Ἰνδοὺς ἀκούω, καὶ ταῦτα οὐ πάντας, ἀλλὰ τοὺς τῶν Ἰνδῶν κρατίστους, ἐκ διαστημάτων αὐτοῖς χρυσὸν περιχέαντας,[2] οἱονεὶ ψελίοις[3] τισὶ κοσμήσαντας βραχίονα ὡραῖον ἀγάλματος. καί φασι νόσων ἀφύκτων ἀμαθῆ καὶ ἄπειρον γίνεσθαι[4] τὸν ἀπογευσάμενον ἐκ τοῦδε τοῦ κέρατος· μήτε γὰρ σπασμῷ ληφθῆναι ἂν αὐτὸν μήτε τῇ καλουμένῃ ἱερᾷ νόσῳ, μήτε μὴν διαφθαρῆναι φαρμάκοις. ἐὰν δέ τι καὶ πρότερον ᾖ πεπωκὼς κακόν, ἀνεμεῖν τοῦτο, καὶ ὑγιᾶ γίνεσθαι[5] αὐτόν. πεπίστευται δὲ τοὺς ἄλλους τοὺς ἀνὰ πᾶσαν τὴν γῆν ὄνους καὶ ἡμέρους καὶ ἀγρίους καὶ τὰ ἄλλα[6] μώνυχα θηρία ἀστραγάλους οὐκ ἔχειν, οὐδὲ μὴν ἐπὶ τῷ ἥπατι χολήν, ὄνους δὲ τοὺς

---

[1] Schn : φῆναι or φύεσθαι.
[2] Reiske : περιχέοντας.　　　　　　[3] ψελλίοις.

51. They say that the Gadfly is like a fly of the The Gadfly
largest size; it is robust and compact and has a
strong sting attached to its body and emits a buzzing
sound. The Horsefly on the other hand is like the The Horse-
dog-fly, as it is called, but though its buzz is louder fly
than the Gadfly its sting is smaller.[a]

52. I have learned that in India are born Wild The Wild
Asses as big as horses. All their body is white ex- Ass of India
cept for the head, which approaches purple, while
their eyes give off a dark blue colour. They have a
horn on their forehead as much as a cubit and a half its horn
long; the lower part of the horn is white, the upper
part is crimson, while the middle is jet-black. From
these variegated horns, I am told, the Indians drink,
but not all, only the most eminent Indians, and round
them at intervals they lay rings of gold, as though
they were decorating the beautiful arm of a
statue with bracelets. And they say that a man
who has drunk from this horn knows not, and
is free from, incurable diseases: he will never be
seized with convulsions nor with the sacred sickness,[b] as it is called, nor be destroyed by poisons.
Moreover if he has previously drunk some deadly
stuff, he vomits it up and is restored to health.

It is believed that Asses, both the tame and the
wild kind, all the world over and all other beasts with
uncloven hoofs are without knucklebones and without
gall in the liver; whereas those horned Asses of

---

[a] Cp. 6. 37, and see *Stud. ital. di fil. class.* 12. 441.
[b] Epilepsy.

[4, 5] γενέσθαι.   [6] τὰ ἄλλα τά.

Ἰνδοὺς λέγει Κτησίας τοὺς ἔχοντας τὸ κέρας
ἀστραγάλους φορεῖν, καὶ ἀχόλους μὴ εἶναι· λέγονται δὲ οἱ ἀστράγαλοι μέλανες εἶναι, καὶ εἴ τις
αὐτοὺς συντρίψειεν,[1] εἶναι τοιοῦτοι καὶ τὰ ἔνδον.
εἰσὶ δὲ καὶ ὤκιστοι οἵδε οὐ μόνον τῶν ὄνων,
ἀλλὰ καὶ ἵππων καὶ ἐλάφων· καὶ ὑπάρχονται μὲν
ἡσυχῇ τοῦ δρόμου, κατὰ μικρὰ δὲ ἐπιρρώννυνται,
καὶ διώκειν ἐκείνους τοῦτο δὴ τὸ ποιητικὸν
μεταθεῖν τὰ ἀκίχητά ἐστιν. ὅταν γε μὴν ὁ θῆλυς
τέκῃ, καὶ περιάγηται τὰ ἀρτιγενῆ, σύννομοι
αὐτοῖς οἱ πατέρες αὐτῶν φυλάττουσι[2] τὰ βρέφη.
διατριβαὶ δὲ τοῖς ὄνοις τῶν Ἰνδικῶν πεδίων τὰ
ἐρημότατά ἐστιν. ἰόντων[3] δὲ τῶν Ἰνδῶν ἐπὶ τὴν
ἄγραν αὐτῶν, τὰ μὲν ἁπαλὰ καὶ ἔτι νεαρὰ ἑαυτῶν
νέμεσθαι κατόπιν ἐῶσιν, αὐτοὶ δὲ ὑπερμαχοῦσι,
καὶ ἵασι τοῖς ἱππεῦσιν ὁμόσε, καὶ τοῖς κέρασι
παίουσι. τοσαύτη δὲ ἄρα ἡ ἰσχὺς ἡ τῶνδέ ἐστιν.
οὐδὲν ἀντέχει αὐτοῖς παιόμενον, ἀλλὰ εἴκει καὶ
διακόπτεται καὶ ἐὰν τύχῃ κατατέθλασται[4] καὶ
ἀχρεῖόν ἐστιν. ἤδη δὲ καὶ ἵππων πλευραῖς ἐμπε-
σόντες διέσχισαν καὶ τὰ σπλάγχνα ἐξέχεαν.
ἔνθεν τοι καὶ ὀρρωδοῦσιν αὐτοῖς πλησιάζειν οἱ
ἱππεῖς· τὸ γάρ τοι τίμημα τοῦ γενέσθαι πλησίον
θάνατός ἐστιν οἴκτιστος αὐτοῖς, καὶ ἀπόλλυνται
καὶ αὐτοὶ καὶ οἱ ἵπποι. δεινοὶ δέ εἰσι καὶ λακτίσαι.
δήγματα δὲ ἄρα ἐς τοσοῦτον καθικνεῖται αὐτῶν,
ὡς ἀποσπᾶν τὸ περιληφθὲν πᾶν. ζῶντα μὲν οὖν
τέλειον οὐκ ἂν λάβοις, βάλλονται δὲ ἀκοντίοις καὶ
οἰστοῖς, καὶ τὰ κέρατα[5] ἐξ αὐτῶν Ἰνδοὶ νεκρῶν
σκυλεύσαντες ὡς εἶπον περιέπουσιν. ὄνων δὲ

---

[1] συντρίψει or -τρίψαι.   [2] φυλάττονται.

India, Ctesias says, have knucklebones and are not its knuckle-
bones without gall. Their knucklebones are said to be black, and if ground down are black inside as well. And these animals are far swifter than any ass or even than any horse or any deer. They begin to run, it is true, at a gentle pace, but gradually gather strength until to pursue them is, in the language of poetry, to chase the unattainable.

When the dam gives birth and leads her new-born colts about, the sires herd with, and look after, them. And these Asses frequent the most desolate plains in India. So when the Indians go to hunt them, the hunted by
the Indians Asses allow their colts, still tender and young, to pasture in their rear, while they themselves fight on their behalf and join battle with the horsemen and strike them with their horns. Now the strength of these horns is such that nothing can withstand their blows, but everything gives way and snaps or, it may be, is shattered and rendered useless. They have in the past even struck at the ribs of a horse, ripped it open, and disembowelled it. For that reason the horsemen dread coming to close quarters with them, since the penalty for so doing is a most lamentable death, and both they and their horses are killed. They can kick fearfully too. Moreover their bite goes so deep that they tear away everything that they have grasped. A full-grown Ass one would never capture alive: they are shot with javelins and arrows, and when dead the Indians strip them of their horns, which, as I said, they decorate.

---

[3] ἐστιν. ἰόντων] ἐπιόντων.

[4] κατέθλασται.

[5] κέρατα οὕτω τά.

Ἰνδῶν ἄβρωτόν ἐστι ⟨τὸ⟩[1] κρέας· τὸ δὲ αἴτιον, πέφυκεν εἶναι πικρότατον.

53. Εἶναι δὲ ἄλογα μὲν ζῷα, φυσικὴν δὲ ἔχειν ἀριθμητικὴν μὴ διδαχθέντα Εὔδημός φησι, καὶ ἐπάγει μαρτύριον ἐκεῖνο τῶν ἐν τῇ Λιβύῃ ζῴων. τὸ δὲ ὄνομα οὐ λέγει· ἃ δὲ λέγει, ταῦτά ἐστιν. ὅ τι ἂν θηράσῃ, ποιεῖν μοίρας ἕνδεκα, καὶ τὰς μὲν δέκα σιτεῖσθαι, τὴν δὲ ἑνδεκάτην ἀπολείπειν (ὅτῳ δὲ καὶ ἀντὶ τοῦ καὶ ἐννοίᾳ τίνι σκοπεῖν ἄξιον) ἀπαρχήν γέ τινα ἢ δεκάτην, ὡς ἂν εἴποις. οὐκοῦν ἐκπλαγῆναι δίκαιον τὴν αὐτοδίδακτον σοφίαν ⟨τήνδε⟩[2]· τὴν γάρ τοι[3] μονάδα καὶ δυάδα καὶ τοὺς ἑξῆς ἀριθμοὺς ζῷον οἶδεν ἄλογον· ἀνθρώπῳ δὲ δεῖ πόσων μὲν τῶν μαθημάτων, πόσων δὲ τῶν πληγῶν, ἵνα ἢ μάθῃ ταῦτα εὖ καὶ καλῶς ἢ πολλάκις μὴ μάθῃ;

54. Λέγουσιν Αἰγύπτιοι (καὶ ῥᾳθύμως αὐτῶν οὐκ ἀκούουσιν ἄνδρες φιλόσοφοι) ἔν τινι νομῷ τῶν Αἰγυπτίων, ὅνπερ οὖν ἐξ Ἡρακλέους τοῦ Διὸς ὀνομάζουσι, παῖδα ὡραῖον ὡς ἂν Αἰγύπτιον, χηνῶν ποιμένα, ἐράστριαν ἀσπίδα λαχεῖν, καὶ μέντοι ⟨καὶ⟩[4] παρ' αὐτῇ εἶναι θαυμαστόν. εἶτα φοιτῶσαν τῷ ἐρωμένῳ ὄναρ προλέγειν τὰς ἐπιβουλὰς τὰς ἐς αὐτὸν πανουργουμένας ἐκ θατέρου θηρίου, ὅπερ ἦν αὐτῇ σύννομον, ὡς ἂν εἴποι τις, ζηλοτυπίᾳ τῇ πρὸς τὸν παῖδα ὑπὲρ τῆς νύμφης[5] ταῦτα πειρωμένου δρᾶν τοῦ ἄρρενος· τὸν δὲ

[1] ⟨τό⟩ add. H.
[2] ⟨τήνδε⟩ add. H.
[3] Schn : τὴν δέ γε.
[4] ⟨καί⟩ add. H.

But the flesh of Indian Asses is uneatable, the reason being that it is naturally exceedingly bitter.

53. Eudemus declares that animals though devoid of reason have a natural instinct for numbers, even though untaught, and adduces as evidence this animal from Libya. Its name he does not mention, but what he says is this. Whatever it catches it divides into eleven portions; ten of these it eats, but the eleventh it leaves (it is worth considering for whose benefit, from what cause, and with what intent) as a kind of first-fruits or tithe, so to say. Hence one's amazement at this self-taught skill is justifiable: a brute beast understands 1, 2, and the following numbers; then think of all the instruction, all the whippings a human being needs if he is to learn these things well and truly—or often, if he is not to learn them.

A calculating animal

54. The Egyptians assert (and scholars do not lend an indifferent ear to what they say) that in a certain district of Egypt which they name after Heracles[a] the son of Zeus, a good-looking boy, as Egyptian boys go, who herded geese, was beloved and even admired by a female Asp. It would keep company with its favourite and warn him in a dream as he slept of the plots that another savage creature, its fellow you might say, was hatching against him: the male Asp was attempting his life, being as it were jealous of the boy on account of its wedded bride.[5] And the

Asp in love with a Gooseherd

[a] Nomos Heracleotes in Middle Egypt, of which the capital was Heracleopolis.

[5] τῆς νύμφης τῆς ἀσπίδος.

ὑπακούοντα [1] πείθεσθαι καὶ φυλάττεσθαι. Ὅμηρος μὲν οὖν ἔδωκεν ἵππῳ φωνήν, ἀσπίδι δὲ ἡ φύσις, ᾗ νόμων οὐδὲν μέλει, φησὶν Εὐριπίδης.

55. Καμήλους ἔτη βιοῦν καὶ πεντήκοντα ἀκήκοα, τὰς δὲ ἐκ Βάκτρων πέπυσμαι προϊέναι καὶ ἐς δὶς τοσαῦτα. καὶ οἵ γε ἄρρενες καὶ πολεμικοί, ἐκτέμνουσιν αὐτοὺς οἱ Βάκτριοι, τὴν ὕβριν καὶ τὸ ἀκολασταίνειν ἀφαιροῦντες, τὴν δὲ ῥώμην αὐτοῖς φυλάττοντες. κάονται [2] δὲ αἱ θήλειαι τὰ ἐξάπτοντα ἐς οἶστρον μέρη αὐτάς.

56. Φώκην Εὔδημος λέγει ἐρασθῆναι ἀνδρὸς σπογγιὰς θηρεύειν συνειθισμένου, καὶ προϊοῦσαν τῆς θαλάττης ἔνθα ἦν ὕπαντρος πέτρα ὁμιλεῖν αὐτῷ. τῶν δὲ ὁμοτέχνων ἦν ἄρα οὗτος αἴσχιστος, ἀλλὰ ἐδόκει τῇ φώκῃ ὡραιότατος εἶναι. καὶ θαῦμα ἴσως οὐδέν, ἐπεὶ καὶ ἄνθρωποι πολλάκις τῶν ἧττον καλῶν ἠράσθησαν, ἐς τοὺς ὡραιοτάτους οὐ παθόντες οὐδὲ ἕν, ἀλλ᾽ ἀμελήσαντες αὐτῶν.

57. Ἀριστοτέλης [3] λέγει τὸν ὑπὸ ὕδρου πληγέντα παραχρῆμα ὀσμὴν βαρυτάτην ἀπεργάζεσθαι, ὡς μὴ οἷόν τε εἶναι προσπελάσαι αὐτῷ τινα. λήθην τε καταχεῖσθαι τοῦ πληγέντος [4] ὁ αὐτὸς λέγει καὶ μέντοι καὶ ἀχλὺν κατὰ τῶν ὀμμάτων πολλήν, καὶ λύτταν ἐπιγίνεσθαι καὶ τρόμον εὖ [5] μάλα ἰσχυρόν, καὶ ἀπόλλυσθαι διὰ τρίτης αὐτόν.

---

[1] ἐπακούοντα.   [2] καίονται.
[3] Ἀπολλόδωρος Wellmann.
[4] τῷ πληγέντι.
[5] Reiske : εὐθύς.

boy would listen and obey and be on his guard. Now Homer [*Il.* 19. 404] allowed a horse to speak, and Nature, who according to Euripides ' recks nought of laws ' [*fr.* 920 N], did the same to an Asp.

55. I have heard that Camels live for fifty years, but I have ascertained that those from Bactria live as much as twice that number. The males which are used in battle, the Bactrians castrate, thereby ridding them of their violent and intemperate disposition while preserving their strength. But in the case of the females they cauterize those parts which inflame them to lust. *The Camel of Bactria*

56. Eudemus asserts that a Seal fell in love with a man whose habit was to dive for sponges, and that it would emerge from the sea and consort with him where there was a rocky cavern. Now this man was the ugliest of his fellows, but in the eyes of the Seal the handsomest. Perhaps there is nothing to wonder at, for even human beings have frequently loved the less beautiful of their kind, being quite unaffected by the best-looking and paying no attention to them. *Seal in love with a Diver*

57. Aristotle says[a] that when a man has been bitten by a Water-snake he at once exhales a most foul odour, so much so that nobody can come near him. He says also that forgetfulness descends upon the bitten man and a thick mist upon his eyes, and that madness ensues and a violent trembling, and that after three days he dies. *The Water-snake, its bite*

[a] Not in any extant work. Wellmann (*Hermes* 26. 334) would substitute the name of Apollodorus for that of Aristotle, which he regards as a slip on the part of Ael. Cp. Nic. *Th.* 425.

58. Τὴν οἰνάδα ὄρνεον εἰδέναι χρὴ οὖσαν, οὐ μὴν ὥς τινες ἄμπελον. λέγει δὲ Ἀριστοτέλης μεῖζον μὲν αὐτὸ εἶναι φάττης, περιστερᾶς γε μὴν ἧττον. καλοῦνται δὲ ὡς ἀκούω καὶ ἐν τῇ Σπάρτῃ οἰναδοθῆραί τινες. λέγοιτο δ᾽ ἂν καὶ κίρκη διαλλάττειν κίρκου οὐ μόνον τῷ γένει ἀλλὰ καὶ τῇ φύσει.[1]

59. Κύανος ⟨τὸ⟩[2] ὄνομα, ὄρνις τὴν φύσιν, ἀπάνθρωπος τὸν τρόπον, μισῶν μὲν τὰς ἀστικὰς διατριβὰς καὶ τὰς κατ᾽ οἰκίαν αὐλίσεις, φεύγων δὲ καὶ τὰς ἐν ἀγροῖς διατριβὰς καὶ ὅπου καλύβαι τε καὶ ἀνθρώπων αὔλια, χαίρων δὲ ἐρημίαις καὶ ἡδόμενος ὀρείοις κορυφαῖς καὶ πάγοις ἀποτόμοις. ἀλλ᾽ οὐδὲ ἠπείροις φιληδεῖ οὐδὲ[3] νήσοις ἀγαθαῖς, Σκύρῳ δὲ καὶ εἴ τις τοιαύτη ἑτέρα ἄγαν λυπρὰ καὶ ἄγονος καὶ ἀνθρώπων χηρεύουσα ὡς τὰ πολλά.

60. Σπίνοι δὲ ἄρα σοφώτεροι καὶ ἀνθρώπων τὸ μέλλον προεγνωκέναι. ἴσασι γοῦν καὶ χειμῶνα μέλλοντα, καὶ χιόνα ἐσομένην προμηθέστατα ἐφυλάξαντο. καὶ τοῦ καταληφθῆναι δέει ἀποδιδράσκουσιν ἐς τὰ ἀλσώδη χωρία,[4] καὶ αὐτοῖς τὰ δάση κρησφύγετα ὡς ἂν εἴποις ἐστίν.

---

[1] λέγοιτο . . . φύσει] λέγοιτο δ᾽ ἄν τι καὶ κ. διαλλάττον τοῦ κ. ὄρνεον . . . γένει φασὶν ἀλλὰ καὶ τὴν φύσιν.

[2] ⟨τό⟩ add. H.

[3] οὔτε . . . οὔτε.

[4] χωρία καὶ τὰ δασέα.

58. You must know that the *Oenas* (Rock-dove) is a The Rock-
bird and not, as some maintain, a vine. And dove
Aristotle says [*HA* 544 b 6] that it is larger than a
ring-dove but smaller than a pigeon. In Sparta too,
I hear, there are men called *Oenadotherae* (Rock-
dove-catchers).

The Circe may be said to differ from the falcon not The Circe
only in sex but in its nature too.

59. ' Blue-fowl ' [a] is its name; it is a bird; its ways The
are apart from man; it hates to linger in cities or to Blue-fowl
lodge in a house; it even avoids lingering in fields or
where there are cottages and huts belonging to man;
it likes desolate places and delights in mountain peaks
and precipitous crags. It has no love even for the
mainland or for pleasant islands, but for Scyros and
any equally dreary, barren spot, generally destitute
of human beings.

60. Chaffinches, it seems, are cleverer than man at The
predicting the future. For instance, they can tell Chaffinch
when winter is coming, and they take the most careful precautions against an impending snowfall, and
for fear of being overtaken they flee to the wood-
lands where the thick foliage affords them, as you
might say, an asylum.

[a] Perh. the ' Syrian Nuthatch.'

BOOK V

E

1. Γῆν τὴν Παριανῶν καὶ τὴν γείτονα Κύζικον
ὄρνιθας οἰκεῖν μέλανας ἰδεῖν φασι, τὸ δὲ σχῆμα
εἴποις ἱέρακας αὐτοὺς ἄν. ἄγευστοι δέ εἰσι
σαρκῶν, καὶ σωφρονοῦσι περὶ τὴν γαστέρα, καὶ
αὐτοῖς τὰ σπέρματα εἶναι δεῖπνον ἀπόχρη. ὅταν
δὲ ὑπάρξηται τὸ μετόπωρον, ἐς τὴν Ἰλιάδα γῆν
ἀγέλη τῶνδε τῶν ὀρνίθων (καλοῦσι δὲ αὐτοὺς
μέμνονας) εὐθὺ τοῦ Μεμνονείου τάφου φοιτῶσι.
λέγουσι δὲ οἱ τὴν Τρωάδα ἔτι οἰκοῦντες ἠρίον
εἶναί τι τῷ Ἠοῦς [1] Μέμνονι ἄνετον·[2] καὶ αὐτὸν
μὲν τὸν νεκρὸν ἐς τὰ Σοῦσα τὰ οὕτω Μεμνόνεια
ὑμνούμενα ὑπὸ τῆς μητρὸς κομισθέντα μετέωρον
ἐκ τῶν φόνων τυχεῖν κηδεύσεως τῆς προσηκούσης
αὐτῷ, ἐπονομάζεσθαι [3] δέ οἱ τὴν στήλην τὴν
ἐνταῦθα ἄλλως. οὐκοῦν τοὺς ὄρνιθας τοὺς ἐπωνύ-
μους τοῦ ἥρωος τοῦ προειρημένου ἀφικνεῖσθαι
κατὰ πᾶν ἔτος, καὶ διαιρεῖσθαί τε καὶ διασχίζεσθαι
ἐς ἔχθραν καὶ διαφοράν, καὶ μάχεσθαι μάχην καρ-
τεράν,[4] ἔστ᾽ ἂν οἱ μὲν αὐτῶν ἀποθάνωσιν οἱ
ἡμίσεις, οἱ δὲ ἀπέλθωσιν οἱ κρατήσαντες ἔνθεν
⟨τοι⟩[5] καὶ ἀφίκοντο. ὅπως ⟨μὲν⟩[6] οὖν ταῦτα
δρᾶται καὶ ὁπόθεν, οὔ μοι σχολὴ φιλοσοφεῖν νῦν,

---

[1] τῷ τῆς Ἠοῦς mss, H, τῆς del. De Stefani.
[2] εἰς τιμήν.
[3] Schn : ὀνομάζεσθαι.
[4] καρτερὰν καὶ ἐς τοσοῦτον.

284

BOOK V

1. They say that the country about Parium [a] and The Ruff
its neighbour Cyzicus are inhabited by birds black in
appearance; from their shape you would say that
they were hawks. But they do not touch flesh, are
temperate in their appetite, and for them seeds are a
sufficient meal. And when late autumn sets in, a
flock of these birds (they call them *Memnons*) [b] resort
to the land round Ilium, making straight for the tomb
of Memnon. And the people who still inhabit the
Troad assert that there is a tomb there dedicated to
Memnon the son of Eos (Dawn); and since the actual
dead body was borne through the air by his mother
from the midst of the carnage to Susa (celebrated for
this reason as ' Memnonian '), where it was awarded
a becoming burial, the monument in the Troad is
called after him to no purpose. And so year by year
the birds named after the aforesaid hero arrive and
separate themselves into hostile factions and fight
violently until half their number are killed, when the
victors depart and return whence they came. How
this all comes to pass and for what reason, I have at
the moment no leisure to speculate, nor yet to
track down the mysteries of Nature. This however I

---

[a] Town at the western end of the S coast of the Propontis;
Cyzicus is some 40 mi. further E.
[b] Ruffs.

---

5 ⟨τοι⟩ add. H.   6 ⟨μέν⟩ add. H.

οὐδὲ μὴν τὰ τῆς φύσεως ἀπόρρητα ἀνιχνεύειν·
εἰρήσεται δὲ ἐκεῖνο. ἐπιτάφιον τῷ παιδὶ τῷ τῆς
Ἡοῦς καὶ Τιθωνοῦ τοῦτον ὅσα ἔτη τὸν ἀγῶνα
ἀθλοῦσιν οἱ προειρημένοι ὄρνιθες· Πελίαν δὲ
ἅπαξ ἐτίμησαν Ἕλληνες ἀγῶνι καὶ Ἀμαρυγκέα
καὶ μέντοι καὶ Πάτροκλον καὶ τὸν ἀντίπαλον
Μέμνονος τὸν Ἀχιλλέα.

2. Ἐν τῇ Κρήτῃ γλαῦκα μὴ γίνεσθαί φασι τὸ
παράπαν, ἀλλὰ καὶ ἐσκομισθεῖσαν ἔξωθεν ἀπο-
θνήσκειν. ἔοικε δὲ ὁ Εὐριπίδης ἀβασανίστως
πεποιηκέναι τὸν Πολύειδον ὁρῶντα τήνδε τὴν
ὄρνιν καὶ ἐξ αὐτῆς τεκμηράμενον ὅτι εὑρήσει τὸν
τεθνεῶτα τῷ Μίνωι υἱόν.[1] πυνθάνομαι δὲ ἔγωγε
λόγους Κρῆτας ᾄδειν καὶ διδάσκειν ἐκεῖνα πρὸς
τοῖς ἤδη διηνυσμένοις. δῶρον λαβεῖν τὴν γῆν
τὴν Κρητικὴν ἐκ Διός, οἷα δήπου τροφὸν καὶ τὴν
κρύψιν τὴν ὑμνουμένην ἀποκρύψασαν αὐτόν, ἐλευ-
θέραν εἶναι θηρίου πονηροῦ καὶ ἐπὶ λύμῃ γεγεννη-
μένου[2] παντός, καὶ μήτε αὐτὴν τίκτειν μήτε
ἔξωθεν κομισθὲν τρέφειν. καὶ τὴν μὲν ἀποδείκνυσθαι τοῦ δώρου τὴν ἰσχύν· τῶν γάρ τοι
προειρημένων ἄγονον εἶναι· εἰ δὲ ἐπὶ πείρᾳ τις
ἢ ἐλέγχῳ τῆς ἐκ Διὸς χάριτος τῶν ὀθνείων τι
ἐσαγάγοι, τὸ δὲ ἐπιψαῦσαν μόνον τῆς γῆς ἀπόλ-

---

[1] εὑρήσει καὶ τὸν Γλαῦκον τὸν τεθνεῶτα τοῦ Μίνω (τῷ Μίνωι
V) τὸν υἱόν.
[2] γεγενημένου.

will mention. The aforesaid birds engage in this contest around the tomb of the son of Eos and Tithonus year after year, whereas the Greeks held but one contest in honour of Pelias,[a] of Amarynceus, and even of Patroclus, and of Achilles the adversary of Memnon.

2. They say that the Owl is not found at all in Crete, and moreover that if it is introduced from abroad it dies. So it seems that Euripides uncritically represented Polyeidus[b] as seeing this bird and thereby conjecturing that he would discover the dead son of Minos. And I myself have ascertained that the Cretan histories, beside the facts already told, relate in verse and prose how Crete received from Zeus a boon—seeing that the island had nursed him and effected that famous concealment of him—, namely that it should be free of all noxious creatures born to do harm, that it should neither produce them nor support them if introduced from abroad. And the island proves how potent this boon was, for it produces none of the aforesaid creatures. But if a man by way of trying and testing the extent of Zeus's favour imports one of these alien creatures, it has but to touch

---

[a] King of Iolcus; his son Acastus paid him the honour of funeral games.—Amarynceus, acc. to a later legend, sent help to the Greeks against Troy; see Hom. *Il.* 23. 630.—For the funeral games of Patroclus see Hom. *Il.* 23.—The death of Achilles is referred to but not described in Hom. *Od.* 24. 37.

[b] Polyeidus (*i.e.* the much-knowing), son of Coeranus and descendant of Melampus, famous as seer and wonder-worker, divined through the presence of an owl that the body of Glaucus, the son of Minos, lay dead in a cask of honey and restored him to life. See Nauck *TGF*[2], p. 558.

λύσθαι. οὐκοῦν τοὺς θηρῶντας τοὺς ὄφεις ἐν τῇ πλησίον Λιβύῃ τοιαῦτα παλαμᾶσθαι. ἡμερώσαντες ἄγουσιν ἐς θαῦμα οἵδε οἱ γόητες τῶν δακετῶν θηρίων [1] πολλά, καὶ σὺν αὐτοῖς ἐπάγονται φόρτον γῆς τῆς Λιβύσσης σφίσι τὸ ἀρκοῦν ἐς τὴν χρείαν. προμηθείᾳ δὲ τῶν ὄφεων τοῦτο δρῶσιν, ἵνα μὴ ἀπόλωνται· καὶ διὰ ταῦτα ἐς τὴν νῆσον τὴν προειρημένην ὅταν ἀφίκωνται, οὐ πρότερον κατατίθενται τὰ ζῷα, πρὶν ἢ ὑποσπεῖραι τὴν ξένην γῆν ἣν ἐπάγονται. καὶ ἐπὶ τούτοις ἀθροίζουσι τὰ πλήθη, καὶ μέντοι καὶ τοὺς ἀνοήτους τε καὶ πολλοὺς ἐκπλήττουσιν. ἕως μὲν οὖν ἕκαστον αὐτῶν κατὰ χώραν μένει συνεσπειραμένον τε καὶ ἱδρυμένον, καὶ ἐπανίσταται μέν, οὐ μὴν ὑπερβάλλει τὴν οἰκείαν κόνιν καὶ σύντροφον, ἐς τοσοῦτον ζῇ· ἐὰν δὲ ἐκφοιτήσῃ ἐς τὴν ὀθνείαν καὶ ἑαυτῷ ξένην γῆν τὴν ἐχθραίνουσαν αὐτῷ, ἀποθνήσκει, καὶ εἰκότως. εἰ γὰρ τὸ ἐκ τοῦ Διὸς νεῦμα ἀτελὲς οὔτε πρὸς τὴν Θέτιν ἐγένετο οὔτε πρὸς ἄλλον τινὰ γένοιτο ἄν, σχολῇ δήπου πρὸς τὴν αὐτοῦ τροφὸν ἐκεῖνο φανεῖται ἄκυρον.

3. Ὁ ποταμὸς ὁ Ἰνδὸς ἄθηρός ἐστι, μόνος δὲ ἐν αὐτῷ τίκτεται σκώληξ φασί. καὶ τὸ μὲν εἶδος αὐτῷ ὁποῖον δήπου καὶ τοῖς ἐκ τῶν ξύλων γεννωμένοις τε καὶ τρεφομένοις, ἑπτὰ δὲ πήχεων [2] τὸ μῆκος προήκουσιν οἱ ἐκεῖθι, εὑρεθεῖεν δ' ἂν καὶ μείζους ἔτι καὶ ἐλάττους· τὸ πάχος δὲ αὐτῶν δεκαετὴς παῖς γεγονὼς μόλις ταῖς χερσὶ περιβάλλειν ἀρκέσει.[3] τούτοις δὴ ἄνω μὲν εἷς ὀδοὺς προσπέφυκε, κάτω δὲ ἄλλος, τετράγωνοι δὲ ἄμφω, πυγόνος δὲ τὸ μῆκος. τοσοῦτον δὲ ἄρα τῶν

the soil and it dies. Accordingly snake-hunters from the neighbouring Libya use devices of this kind. These charmers of venomous reptiles tame a great number and bring them for people to wonder at, and with them they import a load of soil from Libya sufficient for their need. This they do by way of precaution, to prevent the snakes from meeting their death. With this object, when they arrive at the aforesaid island they do not put down their snakes until they have laid a bed of the imported soil. This done, they collect crowds and fill the unintelligent majority with amazement. Now as long as each snake remains coiled up and settled in its place, or rises up without however crossing the limit of its own native dust, so long it lives. If however it strays on to the alien soil which is strange and hostile to it, it dies, and naturally so. For if the will of Zeus did not fail of effect in the case of Thetis, and would not fail in the case of any other person, far less, I think, will it prove ineffectual when his own nurse is concerned.

3. The river Indus is devoid of savage creatures; the only thing that is born in it is a worm, so they say, in appearance like those that are engendered in, and feed upon, timber. But these creatures attain to a length of as much as seven cubits, though one might find specimens both larger and smaller. Their bulk is such that a ten-year-old boy could hardly encircle it with his arms. A single tooth is attached to the upper jaw, another to the lower, and both are square and about eighteen inches long; and such is

---

¹ θηρία. ² πηχῶν MSS *always*.

³ ἀρκέσειε most MSS, ἰσχύσει V, ⟨ἂν⟩ ἀρκέσειε *Jac.*

289

ὀδόντων αὐτοῖς τὸ κράτος ἐστί· πᾶν ὅ τι ἂν ὑπ'
αὐτοῖς λάβωσι συντρίβουσι ῥᾷστα, ἐάν τε λίθος ᾖ
ἐάν τε ἥμερον ζῷον ἢ ἄγριον. καὶ μεθ' ἡμέραν
μὲν κάτω καὶ ἐν ⟨τῷ⟩[1] βυθῷ τοῦ ποταμοῦ δια-
τρίβουσι, τῷ πηλῷ καὶ τῇ ἰλύι φιληδοῦντες, καὶ
ἐντεῦθεν οὐκ εἰσὶν ἔκδηλοι· νύκτωρ δὲ προΐασιν
ἐς τὴν γῆν, καὶ ὅτῳ ἂν περιτύχωσιν, ἢ ἵππῳ ἢ
βοῒ ἢ ὄνῳ, συντρίβουσιν αὐτόν, εἶτα σύρουσιν ἐς
τὰ ἑαυτῶν ἤθη, καὶ ἐσθίουσιν ἐν τῷ ποταμῷ, καὶ
πάντα βρύκουσι[2] τὰ μέλη πλὴν τῆς τοῦ ζῴου
κοιλίας. εἰ δὲ αὐτοὺς καὶ ἐν ἡμέρᾳ πιέζοι λιμός,
εἴτε κάμηλος πίνοι ἐπὶ τῆς ὄχθης εἴτε βοῦς,
ὑπανερπύσαντες καὶ λαβόμενοι ἄκρων τῶν χειλέων
μάλα εὐλαβῶς, ὁρμῇ βιαιοτάτῃ καὶ ἕλξει ἐγκρατεῖ
ἐς τὸ ὕδωρ ἄγουσι, καὶ δεῖπνον ἴσχουσι. δορὰ δὲ
ἕκαστον περιαμπέχει τὸ πάχος καὶ δύο δακτύλων.
ἄγρα δὲ αὐτῶν[3] καὶ θήρα τὸν τρόπον τόνδε
τετέχνασται. ἄγκιστρον παχὺ καὶ ἰσχυρὸν ἁλύ-
σει σιδηρᾷ προσηρτημένον καθιᾶσι, προσδήσαντες
αὐτῷ λευκολίνου ταλαντιαῖον[4] ὅπλον, ἐρίῳ κατει-
λήσαντες καὶ τὸ[5] καὶ τό, ἵνα μὴ διατράγῃ ὁ
σκώληξ αὐτά, ἀναπήξαντες δὲ ἐς τὸ ἄγκιστρον
ἄρνα ἢ ἔριφον, εἶτα μέντοι ἐς τὸ τοῦ ποταμοῦ
ὕδωρ μεθιᾶσιν. ἔχονται δὲ ἄνθρωποι τοῦ ὅπλου
καὶ τριάκοντα, καὶ ἕκαστος ἀκόντιόν τε ἐνηγκύλη-
ται καὶ μάχαιραν παρήρτηται. καὶ παράκειται
ξύλα εὐτρεπῆ, παίειν εἰ δέοι· κρανείας δέ ἐστι
ταῦτα, ἰσχυρὰ ἄγαν. εἶτα περισχεθέντα τῷ ἀγκίστρῳ καὶ τὸ δέλεαρ καταπιόντα τὸν σκώληκα
ἀνέλκουσι,[6] θηραθέντα δὲ ἀποκτείνουσι, καὶ πρὸς
τὴν εἴλην κρεμῶσι τριάκοντα ἡμερῶν. λείβεται

[1] ⟨τῷ⟩ add. H.   [2] Schn: βρυκῶσι.

the strength of their teeth that they can crush with the greatest ease anything that they get between them, be it stone, be it animal, tame or wild. During the daytime they live at the bottom of the river, wallowing in the mud and slime; for that reason they are not to be seen. But at night they emerge on to the land, and whatever they encounter, whether horse or ox or ass, they crush and then drag down to their haunts and eat it in the river, devouring every member of the animal excepting its paunch. If however they are assailed by hunger during the day as well, and should a camel or an ox be drinking on the bank, they slide furtively up and seizing firmly upon its lips, haul it along with the utmost force and drag it by sheer strength into the water, where they feast upon it. Each one is covered with a hide two fingers thick. The following means have been devised for hunting and capturing them. Men let _{its capture} down a stout, strong hook attached to an iron chain, and to this they fasten a rope of white flax weighing a talent, and they wrap wool round both chain and rope to prevent the worm biting through them. On the hook they fix a lamb or a kid, and then let them sink in the river. As many as thirty men hold on to the rope and each of them has a javelin ready to hurl and a sword at his side. Wooden clubs are placed handy, should they need to deal blows, and these are of cornel-wood and very hard. Then when the worm is secured on the hook and has swallowed the bait, the men haul, and having captured it and killed it, hang it up in the sun for thirty days. From the body

---

[3] κατ' αὐτῶν. [4] πλατέος.
[5] Jac : καὶ τὸ ἄγκιστρον. [6] ἕλκουσι.

δὲ ἐξ αὐτοῦ ἔλαιον παχὺ ἐς ἀγγεῖα κεράμου·
ἀφίησι δὲ ἕκαστον ζῷον ἐς κοτύλας δέκα. τοῦτο
δὴ τὸ ἔλαιον ⟨τῷ⟩[1] βασιλεῖ τῶν Ἰνδῶν κομίζουσι,
σημεῖα ἐπιβαλόντες·[2] ἔχειν γὰρ αὐτοῦ ἄλλον
οὐδὲ ὅσον ῥανίδα ἐφεῖται. ἀχρεῖον δέ ἐστι τὸ
λοιπὸν τοῦ ζῴου σκῆνος. ἔχει δὲ ἄρα τὸ ἔλαιον
ἰσχὺν ἐκείνην. ὅντινα ἂν ξύλων σωρὸν καταπρῆσαί
τε καὶ ἐς ἀνθρακιὰν στορέσαι θελήσῃς, κοτύλην
ἐπιχέας τοῦδε ἐξάψεις, μὴ πρότερον ὑποχέας
πυρὸς σπέρμα· εἰ δὲ καταπρῆσαι ἄνθρωπον ἢ
ζῷον, σὺ μὲν ἐπιχεῖς, τὸ δὲ παραχρῆμα ἐνεπρήσθη.
τούτῳ τοί φασι τὸν τῶν Ἰνδῶν βασιλέα καὶ τὰς
πόλεις αἱρεῖν τὰς ἐς ἔχθραν προελθούσας οἱ, καὶ
μήτε κριοὺς μήτε χελώνας μήτε τὰς ἄλλας ἐλεπό-
λεις ἀναμένειν, ἐπεὶ καταπιμπρὰς ᾕρηκεν· ἀγγεῖα
γὰρ κεραμεᾶ ὅσον κοτύλην ἕκαστον χωροῦντα
ἐμπλήσας αὐτοῦ καὶ ἀποφράξας ἄνωθεν ἐς τὰς
πύλας σφενδονᾷ. ὅταν δὲ[3] τύχῃ ⟨τῶν⟩[4] θυρίδων,
τὰ μὲν ἀγγεῖα προσαράττεται καὶ ἀπερράγη, καὶ
τὸ ἔλαιον κατώλισθε, καὶ τῶν θυρῶν πῦρ κατεχύθη,
καὶ ἄσβεστόν ἐστι. καὶ ὅπλα δὲ κάει καὶ ἀνθρώπους μαχομένους, καὶ ἄπλετόν[5] ἐστι τὴν ἰσχύν.
κοιμίζεται δὲ καὶ ἀφανίζεται πολλοῦ φορυτοῦ
καταχυθέντος.[6] λέγει ὁ Κνίδιος Κτησίας ταῦτα.

4. Ἡ φώκαινα[7] ὅμοιον δελφῖνι ζῷόν ἐστιν,
ἔχει δὲ γάλα καὶ αὐτή. χρόαν δὲ οὐκ ἔστι μέλαινα,
κυανῷ δὲ εἴκασται τῷ βαθυτάτῳ, ἀναπνεῖ δὲ οὐ
βραγχίοις, ἀλλὰ φυσητῆρι· τοῦτο γὰρ καὶ καλοῦ-

---

[1] ⟨τῷ⟩ add. H.   [2] ἐπιβάλλοντες.
[3] τε.   [4] ⟨τῶν⟩ add. H.

there drips a thick oil into earthenware vessels; and the oil from its body
each worm yields up to ten *cotylae*.[a] This oil they seal
and bring to the Indian King; no one else is permitted
to have so much as a drop. The rest of the carcase is
of no use. Now the oil has this power: should you
wish to burn a pile of wood and to scatter the embers,
pour on a *cotyle* and you will set it alight without previously applying a spark. And if you want to burn a
man or an animal, pour some oil over him and at once
he is set on fire. With this, they say, the Indian King
even takes cities that have risen against him; he
does not wait for battering-rams or penthouses or
any other siege-engines, for he burns them down and
captures them. He fills earthen vessels, each holding
one *cotyle*, with oil, seals them, and slings them from
above against the gates. When the vessels touch
the embrasures they are dashed into fragments;
the oil oozes down; fire pours over the doors, and
nothing can quench it. And it burns weapons and
fighting men, so tremendous is its force. It is however allayed and put out if piles of rubbish are
poured over it.

Such is the account given by Ctesias of Cnidus.

4. The Porpoise is a creature like the dolphin, and The Porpoise
it too has milk. Its colour is not black but resembles
very deep blue. It breathes not through gills but
through a blow-hole, for that is the name they give

---

[a] 1 κοτύλη = about ½ pint.

---

[5] *Triller*: ἄπληστον.
[6] πολλῷ φορυτῷ καταχυθέντι.
[7] *Schn*: φάλαινα.

σίν οἱ τοῦ πνεύματος τὴν ὁδόν. διατριβὴ δὲ ὁ
Πόντος αὐτῇ καὶ ἡ ἐκεῖ θάλαττα· πλανᾶται δὲ
⟨τῶν⟩[1] ἠθῶν ἐκείνων ἐξωτέρω ἡ φώκαινα[2]
ἥκιστα.

5. Τὸν ἄρρενα ἡ θήλεια νικήσασα ὄρνις[3] ἐν τῇ
μάχῃ, ἁβρύνεταί τε ὑφ' ἡδονῆς καὶ καθίησι
κάλλαια,[4] οὐκ ἐς τοσοῦτον μὲν ἐς ὅσον καὶ οἱ
ἀλεκτρυόνες, καθίησι δ' οὖν, καὶ φρονήματος
ὑποπίμπλαται, καὶ βαίνει μακρότερα.

6. Φιλοίκειον ὁ δελφὶς ζῷον πεπίστευται. καὶ
τὸ[5] μαρτύριον, Αἶνός ἐστι πόλις Θρῇσσα. ἔτυχεν
οὖν ἁλῶναι δελφῖνα καὶ τρωθῆναι μέν, οὐ μὴν ἐς
θάνατον, ἀλλ' ⟨ὡς⟩[6] ἔτι βιώσιμα εἶναι τῷ
ἑαλωκότι. οὐκοῦν ἐρρύη μὲν αἷμα, ἤσθοντο δὲ
οἱ ἀθήρατοι, καὶ ἀφίκοντο ἐς τὸν λιμένα ἀγέλη,
καὶ κατεσκίρτων, καὶ ⟨δῆλοι⟩[7] ἦσάν τι δρασείον-
τες οὐκ ἀγαθόν.[8] οἱ τοίνυν Αἴνιοι ἔδεισαν καὶ
ἀφῆκαν τὸν ἑαλωκότα. καὶ ἐκεῖνοι κομισάμενοι
ὡς ἕνα τῶν κηδεστῶν[9] ᾤχοντο ἀπιόντες. σπανίως[10] δὲ ἄνθρωπος ἢ οἰκείῳ δυστυχήσαντι ἢ
οἰκείᾳ κοινωνὸς σπουδῆς καὶ φροντίδος.

7. Ἐν Αἰγύπτῳ πίθηκος, ὥς φησιν Εὔδημος,
ἐδιώκετο, αἴλουροι δὲ ἦσαν οἱ διώκοντες. ἀνὰ
κράτος οὖν ἀποδιδράσκων ὥρμησεν εὐθὺ δένδρου
τινός, οἱ δὲ καὶ αὐτοὶ ἀνέθορον[11] ὤκιστα· ἔχονται

---

[1] ⟨τῶν⟩ add. H.
[2] Schn : φάλαινα.
[3] ὄρνιν.
[4] κάλλη.
[5] τούτου τό.
[6] ⟨ὡς⟩ add. H.
[7] ⟨δῆλοι⟩ add. Cobet.

to its air-passage. The Porpoise frequents Pontus and the sea round about, and rarely strays beyond its familiar haunts.

5. When a Hen has defeated a cock-bird in battle it gives itself airs from sheer delight and lets down its wattles, not however to the same extent as cocks, although it does so and is filled with pride and struts more grandly.

The victorious Hen

6. The Dolphin is believed to love its own kin, and here is the evidence. Aenus is a city in Thrace. Now it happened that a Dolphin was captured and wounded, not indeed fatally, but the captive was still able to live. So when its blood flowed the dolphins which had not been caught saw this and came thronging into the harbour and leaping about and were plainly bent on some mischief. At this the people of Aenus took fright and let their captive go, and the dolphins, escorting as it might be some kinsman, departed.

A captured Dolphin

But a human being will hardly attend or give a thought to a relative, be it man or woman, in misfortune.

7. In Egypt, says Eudemus, a Monkey was being pursued and Cats were the pursuers. So the Monkey fled as fast as he could and made straight for a tree. But the Cats also ran up very swiftly, for they cling to

Monkey and Cats

---

[8] ἀγαθόν· ἐν ἔθει δὲ ἦν, ὡς τὸ εἰκός, καὶ αὐτοὺς νήχεσθαι καὶ παῖδας αὐτῶν.

[9] ὡς . . . κηδεστῶν] ὡς ἑταίρων ἕνα τῶν κ. ἢ γένει προσηκόντων.

[10] σπάνιον.　　　　[11] συνέθορον.

γὰρ τῶν φλοιῶν, καὶ ἔστι καὶ τούτοις ἐς δένδρα
ἐπιβατά.[1] ὁ δὲ ὡς ἡλίσκετο εἷς ὤν, καὶ ταῦτα
ὑπὸ πολλῶν, ἐκπηδᾷ τοῦ πρέμνου, καὶ κλάδου
τινὸς ἐπηρτημένου[2] καὶ μετεώρου λαμβάνεται
ἄκρου ταῖς χερσί, καὶ ἐγκρατῶς εἴχετο οὐκ ἐπ᾽
ὀλίγον· οἱ δὲ αἴλουροι, ὡς οὐκ ἦν ἐφικτὰ αὐτοῖς
ἔτι, ἐπ᾽ ἄλλην θήραν κατέδραμον. ὁ δὲ κατὰ
πολλὴν τὴν σπουδὴν διεσώζετο, ἑαυτῷ ὀφείλων
ὡς τὸ εἰκὸς ζωάγρια.

8. Ἀριστοτέλης ὄφεσιν ἔχθραν εἶναι τὴν Ἀστυ-
παλαιέων γῆν λέγει, καθάπερ καὶ τὴν Ῥήνειαν
ταῖς γαλαῖς ὁ αὐτὸς ὁμολογεῖ ἡμῖν. κορώνη δὲ
ἐς τὴν Ἀθηναίων ἀκρόπολιν οὐκ ⟨ἔστιν⟩[3] ἐπιβατά.[4] ἡμιόνων δὲ Ἦλιν μητέρα οὐκ ἐρεῖς, ἢ τὸ
λεχθὲν ψεῦδός ἐστιν.

9. Ῥηγίνοις καὶ Λοκροῖς ἐς τὴν γῆν τὴν
ἀλλήλων παριέναι καὶ γεωργεῖν ἔνσπονδόν ἐστιν.
οὐ μὴν ὁμολογοῦσι τούτοις οὐδὲ ἐς μίαν νοοῦσι
καὶ τὴν αὐτὴν οἱ τέττιγες οἱ τῶνδε καὶ τῶνδε,
ἐπεὶ τὸν μὲν Λοκρὸν ἐν Ῥηγίῳ σιγηλότατον
ἕξεις, τὸν δὲ Ῥηγῖνον ἐν τοῖς Λοκροῖς ἀφωνότατον.
καὶ τίς ἡ αἰτία τῆς τοιαύτης ἀντιδόσεως[5] ἐγὼ
μὲν οὐκ οἶδα οὐδὲ ἄλλος, εἰ μὴ μάτην θρασύνοιτο·
οἶδε δέ, ὦ Ῥηγῖνοι καὶ Λοκροί, μόνη ἡ φύσις.
ποταμὸς γοῦν τῆς τε Ῥηγίνων καὶ τῆς Λοκρίδος

---

[1] ἐπιβατόν.
[2] ὑπηρτημένου.
[3] ⟨ἔστιν⟩ add. H.
[4] ἐπιβατόν.
[5] τοιαύτης ἀμοιβηδὸν εἰς τὴν σιωπὴν ἀντιδόσεως.

the bark and can also climb trees. But as he was going to be caught, being one against many, he leapt from the trunk and with his paws seized the end of an overhanging branch high up and clung to it for a long while. And since the Cats could no longer get at him, they descended to go after other prey. So the Monkey was saved by his own considerable exertions, and it was to himself, as was proper, that he owed the reward for his rescue.

8. Aristotle says[a] that the soil of Astypalaea[b] is unfriendly to snakes; just as, according to the same writer, Rhenea is to martens. No crow can go up on to the Acropolis at Athens. Say that Elis is the mother of mules,[c] and you say what is false.

Places hostile to certain animals

9. There is an agreement between the people of Rhegium and of Locris[d] that they shall have access to, and shall cultivate, one another's lands. But the Cicadas of the two territories do not agree to this and are not of one and the same mind, for you will find the Locrian Cicada is completely silent in Rhegium, and the Cicada from Rhegium is absolutely voiceless among the Locrians. What the cause of such an exchange may be neither I nor anyone else, save an idle boaster, can say. Only to Nature, you men of Rhegium and of Locris, is it known. At any rate there is a river[e] separating the territories of Rhegium

The Cicadas of Locris and Rhegium

---

[a] The passage is not in his extant works; *fr.* 315 (Rose *Arist. pseudepigraphus*, p. 331).

[b] Astypalaea and Rhenea are islands of the Cyclades.

[c] Cp. Hdt. 4. 30.

[d] The two towns lay some 35 mi. apart in the ' toe ' of Italy.

[e] The Caecinus acc. to Paus. 6. 6. 4, the Halex acc. to Strabo 6. 260 and others.

ἐστὶ μέσος, καὶ εἴργονταί γε οὐδὲ πλεθριαίῳ
διαστήματι[1] αἱ ὄχθαι, καὶ ὅμως οὐδέτεροι[2]
διαπέτονται αὐτόν. καὶ ἐν Κεφαλληνίᾳ[3] ποταμός
ἐστιν, ὅσπερ οὖν τῆς τε εὐγονίας τῶν τεττίγων
καὶ τῆς ἀγονίας αἴτιος.

10. Τὸν βασιλέα αὐτῶν αἱ μέλιτται πρᾶον ὄντα
καὶ ἥμερον καὶ ὁμοῦ τι καὶ ἄκεντρον ὅταν αὐτὰς
ἀπολίπῃ μεταθέουσί τε καὶ διώκουσι φυγάδα τῆς
ἀρχῆς ὄντα. ῥινηλατοῦσι δὲ αὐτὸν ἀπορρήτως,
καὶ ἐκ τῆς ὀσμῆς τῆς περὶ αὐτὸν αἱροῦσι, καὶ ἐς
τὴν βασιλείαν ἐπανάγουσιν ἑκοῦσαί τε καὶ βουλό-
μεναι καὶ τοῦ τρόπου ἀγάμεναι. Πεισίστρατον
δὲ ἐξήλασαν Ἀθηναῖοι καὶ Συρακόσιοι[4] Διονύ-
σιον καὶ ἄλλοι ἄλλους, τυράννους τε καὶ παρανό-
μους ὄντας καὶ τέχνην βασιλικὴν ἀποδείξασθαι
μὴ δυναμένους, ἥπερ οὖν φιλανθρωπία τε καὶ τῶν
ὑπηκόων ἐστὶ προστασία.

11. Μέλει τῷ βασιλεῖ τῶν μελιττῶν κεκοσμῆ-
σθαι τὸ σμῆνος τὸν τρόπον τοῦτον. τὰς μὲν
προστάττει ὑδροφορεῖν, τὰς δὲ ἔνδον κηρία δια-
πλάττειν, τήν γε μὴν τρίτην μοῖραν ἐπὶ τὴν
νομὴν προϊέναι· εἶτα μέντοι ἀμείβουσι τὰ ἔργα
ἐκ περιόδου κάλλιστά πως[5] ἀποκριθείσης.[6] αὐτὸς
δὲ ὁ βασιλεύς, ἀπόχρη οἱ τούτων πεφροντικέναι
καὶ νομοθετεῖν ὅσα προεῖπον κατὰ τοὺς μεγάλους
ἄρχοντας, οὓς οἱ φιλόσοφοι φιλοῦσιν ὀνομάζειν

---

[1] διαστήματι μέσῳ.   [2] οὐθέτεροι.
[3] Κεφαληνίᾳ.
[4] Συρακού- MSS always.
[5] δέ πως.

and Locris, and the banks are not so much as a hundred feet apart; for all that the Cicadas of neither side fly across it. And in Cephallenia there is a river which occasions both fertility and barrenness among Cicadas.

10. Bees when forsaken by their King, who is at once gentle and inoffensive and also stingless, give chase and pursue after the deserter from the post of rule. They track him down in some mysterious way and detect him by means of the smell he diffuses and bring him back to his kingdom of their own free will, indeed eagerly, for they admire his disposition. But the Athenians drove out Pisistratus,[a] and the Syracusans Dionysius,[b] and other states their rulers, since they were tyrants and broke the laws and could not exhibit the art of kingship which consists in loving one's fellow-men and protecting one's subjects.

Bees and their King

11. It is the concern of the King Bee that his hive should be regulated in the following manner. To some bees he assigns the bringing of water, to others the fashioning of honeycombs within the hive, while a third lot must go abroad to gather food. But after a time they exchange duties in a precisely determined rotation. As to the King himself, it is enough for him to take thought and to legislate for the matters that I mentioned above after the manner of great rulers to whom philosophers like to ascribe simul-

The King Bee and his state

[a] Tyrant of Athens 560 B.C., twice expelled but regained power and held it till his death, 527 B.C.

[b] See below, ch. 15 n.

---

[6] Gow: ἀποκριθεῖσαι (so H) φιλοῦσιν οἰκουρεῖν αἱ πρεσβύταται MSS, φιλοῦσιν . . . πρεσβύταται del. H.

πολιτικούς τε καὶ βασιλικοὺς τοὺς αὐτούς· τὰ δὲ
ἄλλα ἡσυχάζει καὶ τοῦ αὐτουργεῖν ἀφεῖται. ἐὰν δὲ
ᾖ λῷον ταῖς μελίτταις μεταστῆναι, τηνικαῦτα καὶ
ὁ ἄρχων ἀπαλλάττεται. καὶ ἐὰν μὲν ἔτι νέος ᾖ,
ἡγεῖται, αἱ δὲ λοιπαὶ ἕπονται·[1] ἐὰν δὲ πρεσβύτερος, φοράδην ἔρχεται, κομιζουσῶν αὐτὸν μελιττῶν
ἄλλων. αἱ μέλιτται δὲ ὑπὸ συνθήματι ἐς ὕπνον
τρέπονται. ὅταν δὲ δοκῇ καιρὸς εἶναι καθεύδειν,
ὁ[2] βασιλεὺς μιᾷ προστάττει ὑποσημῆναι κατα-
δαρθάνειν. καὶ ἡ μὲν πεισθεῖσα τοῦτο ἐκήρυξεν,
αἱ δὲ ἐς κοῖτον τρέπονται ἐντεῦθεν, τέως βομβοῦ-
σαι. ἕως ⟨μὲν⟩[3] οὖν περίεστιν ὁ βασιλεύς,
εὐθενεῖται[4] τὸ σμῆνος, καὶ ἀταξία πᾶσα ἠφάνισθη,
καὶ οἱ μὲν κηφῆνες ἀγαπητῶς ἐν τοῖς ἑαυτῶν
κυττάροις ἡσυχάζουσιν, αἱ δὲ[5] πρεσβύτεραι διαιτῶνται ἰδίᾳ, καὶ αἱ νέαι ἰδίᾳ, καὶ καθ' ἑαυτὸν ὁ
βασιλεύς, καὶ αἱ σχαδόνες ἐφ' ἑαυτῶν εἰσι, καὶ ἡ
τροφὴ καὶ αἱ ἄφοδοι χωρίς· ἐπειδὰν δὲ ὁ βασιλεὺς
ἀπόληται, ἀταξίας τε καὶ ἀναρχίας μεστὰ πάντα·
οἵ τε γὰρ κηφῆνες τοῖς τῶν μελιττῶν κυττάροις
ἐντίκτουσι, τά τε λοιπὰ ἐν ἀλλήλοις φυρόμενα
εὐθενεῖσθαι τῷ σμήνει τὸ λοιπὸν οὐκ ἐπιτρέπει·
διαφθείρονται δὲ τελευτῶσαι ἐρημίᾳ ἄρχοντος.
βίον δὲ καθαρὸν ζῇ μέλιττα, καὶ ζῴου οὐκ ἂν
οὐδενὸς πάσαιτο ποτε· καὶ οὐ δεῖται Πυθαγόρου
συμβούλου οὐδὲ ἕν, ἀπόχρη δὲ ἄρα σῖτον αὐτῇ
εἶναι τὰ ἄνθη. ἔστι δὲ καὶ σωφροσύνην ἀκροτάτη.
χλιδὴν γοῦν καὶ θρύψιν μεμίσηκε. καὶ τὸ μαρτύριον, τὸν χρισάμενον μύρῳ διώκει τε καὶ ἐλαύνει
ὡς πολέμιον ἀνήκεστα δράσαντα. οἶδε δὲ καὶ τὸν

---

[1] ἄγονται.   [2] ὁ μέν.
[3] ⟨μέν⟩ add. H.

taneously the qualities of a citizen and of a king. For the rest he lives at ease and abstains from physical labour. If however it is expedient for the bees to change their dwelling, then the ruler departs, and if he happens to be still young, he leads the way and the rest follow; if however he is elderly, he is carried on his way and conveyed by other bees.

At a signal bees retire to slumber. When it seems to be time to go to sleep the King commands one bee to give the signal for going to rest. And the bee obeys and gives the word, whereupon the bees that have been buzzing till then retire to bed. Now so long as the King survives, the swarm flourishes and all disorder is suppressed. The drones gladly remain at rest in their cells, the older bees dwell in their quarters apart, the young in theirs, the King by himself, and the larvae in their own place. Their food and their excrement are in separate places. But when the King dies, disorder and anarchy fill the place; the drones produce offspring in the cells of the bees; the general confusion no longer permits the swarm to thrive, and finally the bees perish for want of a ruler.

The Bee leads a blameless life and would never touch animal food. It has no need of Pythagoras for counsellor, but flowers afford it food enough. It is in the highest degree temperate; at any rate it abhors luxury and delicate living; witness the fact that it pursues and drives away a man who has perfumed himself, as if he were some enemy who has perpetrated actions past all remedy. It recognises too a

The Bee, its temperate life

ἐλθόντα [1] ἐξ ἀκολάστου ὁμιλίας,[2] καὶ διώκει καὶ
ἐκεῖνον οἷα δήπου ἔχθιστον. καὶ ἀνδρείας δὲ εὖ
ἥκουσι καὶ ἄτρεπτοί εἰσιν. οὐδὲ ἓν γοῦν ζῷον
ἀποδιδράσκουσιν, οὐδὲ μὴν κάκῃ εἴκουσι, χωροῦσι
δὲ ὁμόσε. καὶ πρὸς μὲν τοὺς μὴ ἐνοχλοῦντας
μηδὲ ἄρχοντας ἀδίκων μηδὲ τῷ σμήνει προσιόντας
κακούργως καὶ σὺν ἐπιβουλῇ εἰρηναῖα αὐταῖς καὶ
ἔνσπονδά ἐστι, πόλεμος δὲ ἀκήρυκτος τὸ ἀδόμενον
τοῦτο ἐπὶ τοὺς λυποῦντας ἐξάπτεται, καὶ ὅστις
ἥκει κεράσων τὸ μέλι αὐταῖς, ἐς τοὺς ἐχθροὺς
ἠρίθμηται οὗτος. παίουσι δὲ καὶ τοὺς σφῆκας
κακῶς. λέγει δὲ Ἀριστοτέλης ὅτι καὶ ἱππεῖ [3]
ποτε ἐντυχοῦσαι πρὸς τῷ σμήνει ἀπέκτειναν αὐτὸν
ἐπιθέμεναι κατὰ τὸ καρτερὸν αἱ μέλιτται αὐτῷ
ἵππῳ. ἤδη μέντοι καὶ πρὸς ἀλλήλας διαφέρονται,
καὶ αἱ δυνατώτεραι κρατοῦσι τῶν ἡττόνων. κρατοῦσι δὲ ὡς ἀκούω αὐτῶν οἵ τε φρῦνοι καὶ οἱ ἐκ
τῶν τελμάτων βάτραχοι οἵ τε μέροπες καὶ αἱ
χελιδόνες, πολλάκις γε μὴν καὶ οἱ σφῆκες. ὅστις
δὲ τούτων ἐκράτησε, Καδμείαν ὥς γε εἰπεῖν τὴν
νίκην ἐνίκησε· παιόμενοι γὰρ καὶ κεντούμενοι
κακῶς ἀπαλλάττουσιν· εἰσὶ γὰρ οὐ μεῖον τῷ
θυμῷ ἢ τοῖς κέντροις ὡπλισμέναι. οὐκ ἀμοιροῦσι
δὲ οὐδὲ τῆς ἐς τὸ προμηθὲς σοφίας, καὶ Ἀριστοτέλης τεκμηριοῖ ὃ λέγω. ἔστι δὲ τοιοῦτον.
ἐλθοῦσαι μέλιτται [4] ἐπί τι σμῆνος οὐκ οἰκεῖον

---

[1] προσελθόντα.
[2] ἀκολασίας τε καὶ ὁμιλίας τῆς πρός τινα.
[3] Reiske : ἵππῳ.   [4] αἱ μ.

---

[a] The ' horseman ' is an addition of Aelian's.
[b] Two explanations are given : (i) Cadmus slew a dragon
set by Ares to guard a well. From its teeth sprang armed

man who comes from an unchaste bed, and him also
it pursues, as though he were its bitterest foe. And its courage
Bees are well-endowed with courage and are un-
daunted. For instance, there is not a single animal
from which they flee; they are not mastered by
cowardice but go to the attack. Towards those who
do not trouble them or start to injure them or who
do not approach the hive bent on mischief and with
evil intent they show themselves peaceful and
friendly; but against those who would injure them
the fires of a truceless war, as the phrase goes, are
kindled; and anyone who comes to plunder their
honey is reckoned among their enemies. And they
sting even wasps severely. And Aristotle records its sting
[*HA* 626 a 21] how Bees once finding a horseman *a*
near the hive attacked him violently and slew both
him and his horse. And further, they fight with one
another, and the stronger party defeats the weaker.
But I learn that toads and frogs from pools, bee- its enemies
eaters, and swallows defeat them, and frequently
wasps do so too. Yet the victor achieves what you
might call a Cadmean victory,*b* for he comes off badly
from their blows and stings, since the Bees are armed
with courage no less than with stings. But Bees are
not without a share of the wisdom of foresight, and
Aristotle vouches for my statement [*HA* 626 b 12]
thus. Some Bees came to a hive that was not theirs
but a different one and proceeded to plunder the

men who would have fallen upon C. had he not prevailed upon
them to kill one another. (ii) Eteocles the defender, and
Polynices the assailant of Thebes, the city founded by Cadmus,
slew each other in battle. The Thebans were victorious but
were later driven out by the descendants of the 'Septem
contra Thebas.'

ἀλλὰ ἕτερον, εἶτα τὸ μηδέν σφισι προσῆκον
ἐκεράιζον μέλι. αἱ δὲ καίτοι συλώμεναι τὸν σφέτερον πόνον, ὅμως ἐνεκαρτέρουν ἡσυχῇ ἀτρεμοῦσαι, εἶτα μέντοι τὸ μέλλον ἐγκρατῶς ἐκαραδόκουν.
ἐπεὶ δὲ ὁ μελιττουργὸς τὰς πολλὰς τῶν ἐχθρῶν
ἀπέκτεινεν, αἱ ἔνδον καταγνοῦσαι ὅτι ἄρα δύνανται
ἀξιόπιστοι εἶναι πρὸς τὴν μάχην τὴν ἰσοπαλῆ,
προελθοῦσαι κατ' ἠμύναντο, καὶ δίκας ἀπῄτησαν
ὑπὲρ ὧν ἐσυλήθησαν οὐδαμῶς μεμπτάς.

12. Καὶ τοῦτο δὲ φιλεργίας [1] τῆς τῶν [2] μελιττῶν μαρτύριον.[3] ἐν γοῦν τοῖς χειμεριωτάτοις τῶν
χωρίων μετὰ Πλειάδων δυσμὰς ἐς ἰσημερίαν
ἠρινὴν διατελοῦσιν οἰκουροῦσαί τε καὶ ἔνδον
ἀτρεμοῦσαι ἀλέας πόθῳ καὶ φυγῇ ῥίγους αἱ
μέλιτται· τὸν δὲ ἄλλον χρόνον τοῦ ἔτους πάντα
ἀργίαν τε [4] καὶ ἡσυχίαν μισοῦσι, καὶ καμεῖν εἰσιν
ἀγαθαί. καὶ οὐκ ἄν ποτε ἴδοις βλακεύουσαν
μέλιτταν τῆς ὥρας ἐκείνης ἔξω ἐν ᾗ μαλκίει [5] τὰ
μέλη.

13. Γεωμετρίαν δὲ καὶ κάλλη σχημάτων καὶ
ὡραίας πλάσεις αὐτῶν ἄνευ τέχνης τε καὶ κανόνων
καὶ τοῦ καλουμένου ὑπὸ τῶν σοφῶν διαβήτου [6]
ἀποδείκνυνται αἱ μέλιτται. ὅταν δὲ ἐπιγονὴ ᾖ
καὶ εὐθενῇ ταῖς μελίτταις τὸ σμῆνος, ἐκπέμπουσιν [7] ὥσπερ οὖν αἱ μέγισταί τε καὶ πολυανδρού-
μεναι τῶν πόλεων. οἶδε δὲ ἄρα ἡ μέλιττα καὶ

---

[1] τῆς φιλεργίας.
[2] Jac : τῆς μ. L, τῶν μ. other MSS.
[3] τὸ μαρτύριον.
[4] μέν.

honey which did not belong to them. But the Bees which were being despoiled of their labours nevertheless remained quiet and waited patiently to see what would happen. Then, when the beekeeper had killed the greater number of the enemy, the Bees in the hive realised that they were in fact sufficient to sustain an equal combat and emerged to strike back, and the penalty which they exacted for the robbery left nothing to cavil at.

12. Here is further evidence of the industry of Bees. In the coldest countries from the time when the Pleiads have set[a] until the vernal equinox they continue at home and stay quiet in the hive, longing for the warmth and shunning the cold. But for the rest of the year they abhor indolence and repose and are good at hard labour. And you would never see a Bee idling unless it were during the season when their limbs are numb with cold.

The Bee, its industry

13. Bees practise geometry and produce their graceful figures and beautiful conformations without any theory or rules of art, without what the learned call a 'compass.' And when their numbers increase and the swarm thrives they send out colonies just as the largest and most populous cities do. Now the Bee knows when there is rain that threatens to persist, and when there will be a gale. But if surprised

The Bee, its skill

its colonies

as weather-prophet

[a] About the beginning of November.

---

[5] *Schn*: μαλακιεῖ.
[6] διαβήτου τὸ κάλλιστον σχημάτων ἐξάγωνόν τε καὶ ἐξάπλευρον καὶ ἰσογώνιον.
[7] καὶ εἰς ἀποικίαν ἐκπέμπουσιν.

ὑετοῦ ἀπειλοῦντος ἐπιδημίαν καὶ σκληρὸν πνεῦμα
ἐσόμενον. εἰ δὲ αὐτῇ παρὰ δόξαν γένοιτο τὸ τοῦ
πνεύματος, ὄψει φέρουσαν λίθον ἑκάστην ἄκροις
τοῖς ποσὶν ἔρμα εἶναι.[1] ὅπερ δὲ ὁ θεῖος Πλάτων
περὶ τῶν τεττίγων λέγει καὶ τῆς ἐκείνων φιλῳδίας
τε καὶ φιλομουσίας, τοῦτ' ἂν καὶ περὶ τοῦ τῶν
μελιττῶν χοροῦ εἴποι τις. ὅταν γοῦν σκιρτήσω-
σιν ἢ πλανηθῶσιν, ἐνταῦθα οἱ σμηνουργοὶ κροτοῦσι
κρότον τινὰ ἐμμελῆ τε καὶ συμμελῆ· αἱ δὲ ὡς
ὑπὸ Σειρῆνος ἕλκονται, καὶ μέντοι καὶ ὑποστρέ-
φουσιν ἐς ἤθη τὰ οἰκεῖα αὖθις.

14. Ἐν τῇ Γυάρῳ[2] τῇ νήσῳ Ἀριστοτέλης
λέγει μῦς εἶναι καὶ μέντοι καὶ τὴν γῆν σιτεῖσθαι
τὴν σιδηρῖτιν. Ἀμύντας δὲ καὶ τοὺς ἐν Τερηδόνι
(γῆς[3] δέ ἐστιν αὕτη τῆς Βαβυλωνίας) τὴν αὐτὴν
προσφέρεσθαι λέγει.

Ἐν Λάτμῳ δὲ τῆς Καρίας ἀκούω σκορπίους
εἶναι, οἵπερ οὖν τοὺς μὲν πολίτας σφίσι παίουσιν
ἐς θάνατον, τοὺς δὲ ξένους ἡσυχῇ καὶ ὅσον παρασχεῖν ὀδαξησμόν, ἐμοὶ δοκεῖν[4] τοῦ Ξενίου Διὸς
τοῖς ἀφικνουμένοις τὸ δῶρον τοῦτο ἀποκρίναντος.

15. Βασιλεύονται δὲ ἄρα καὶ σφῆκες, ἀλλ' οὐ
τυραννοῦνται ὡς ἄνθρωποι. καὶ τὸ μαρτύριον,
ἄκεντροι καὶ οἵδε εἰσί. καὶ οἱ μὲν ὑπήκοοι τὰ
ἔργα πλάττειν αὐτοῖς νόμον ἔχουσιν, οἱ δὲ ἄρχοντές
εἰσι διπλάσιοι μὲν τὸ μέγεθος, πρᾶοι δὲ καὶ οἷοι
μήτε ἑκόντες λυπεῖν ἔχειν μήτε ἄκοντες. τίς οὖν
οὐκ ἂν μισήσειε[5] Διονυσίους τοὺς ἐν Σικελίᾳ καὶ

---

[1] εἶναι καὶ μὴ ἀνατρέπεσθαι.
[2] *Holstein* : Πάρῳ.

by a wind, you will see every Bee carrying a pebble between the tips of its feet by way of ballast. What the divine Plato says [*Phaedr.* 230 c, 259 b] of cicadas and their love of song and music one might equally say of the choir of Bees. For instance, when they frolic and roam abroad, then the bee-keepers make a clashing sound, melodious and rhythmical, and the Bees are attracted as by a Siren and come back again to their own haunts.

its love of song

14 (i). In the island of Gyarus *a* Aristotle says [*Mir.* 832 a 22] that there are Rats and that they actually eat iron ore. And Amyntas says that the Rats of Teredon (this is in Babylonia) adopt the same food.*b*

Rats in Gyarus and Teredon

(ii). I am told that on Latmus in Caria there are Scorpions which inflict a fatal sting on their fellow-countrymen; strangers however they sting lightly and just enough to produce an itching sensation. This in my opinion is a boon bestowed upon visitors by Zeus, Protector of the Stranger.

Scorpions on mt Latmus

15. Wasps also are subject to a King, but not, as men are, to a despot. Witness the fact that their Kings also are stingless. And their subjects have a law that they shall construct their combs for them. But although the rulers are twice the size of a subject, yet they are gentle and of a nature incapable of doing an injury either willingly or unwillingly. Who then would not detest the Dionysii of

The King Wasp

a One of the Cyclades, some 40 mi. SEE of Attica.
b Cp. 17. 17.

---

[3] *Holstein* : γῆ.   [4] *Schn* : δοκεῖ.
[5] μισήσῃ *or* -αι.

Κλέαρχον τὸν ἐν Ἡρακλείᾳ καὶ Ἀπολλόδωρον
τὸν Κασανδρέων λευστῆρα καὶ τὸν Λακεδαιμονίων
λυμεῶνα τὸν Νάβιν, εἴγε οἱ μὲν ἐθάρρουν τῷ
ξίφει, τῷ δὲ ἀκέντρῳ καὶ τῇ πραότητι οἱ τῶν
σφηκῶν βασιλεῖς;

16. Λέγονται δὲ οἱ τῶν σφηκῶν κεκεντρωμένοι
καὶ ἐκεῖνο δρᾶν. ὅταν θεάσωνται νεκρὰν ἔχιδναν,
οἱ δὲ ἐμπίπτουσι καὶ φαρμάττουσι τὸ κέντρον.
ὅθεν μοι δοκοῦσι μαθεῖν καὶ οἱ ἄνθρωποι μάθημα,
καὶ τοῦτο οὐκ ἀγαθόν. καὶ μέντοι καὶ μαρτυρεῖ
ἐν Ὀδυσσείᾳ Ὅμηρος λέγων

φάρμακον ἀνδροφόνον διζήμενος, ὄφρα οἱ εἴη
ἰοὺς χρίεσθαι χαλκήρεας,

ἢ καὶ νὴ Δία εἴ τι δεῖ τῷ περὶ Ἡρακλέους λόγῳ
προσέχειν, ⟨ὡς⟩[1] ἐκεῖνος ἔβαψε τῷ τῆς Ὕδρας ἰῷ
τοὺς ὀιστούς, οὕτω τοι καὶ ἐκεῖνοι τῇ βαφῇ τὰ
κέντρα ὑποθήγουσιν.[2]

17. Ἔστω δὲ[3] καὶ τῇ μυίᾳ παρ' ἡμῶν γέρας
μὴ ἀμοιρῆσαι[4] τῆς μνήμης τῆς ἐνταῦθα· φύσεως
γάρ τοι καὶ ἐκείνη πλάσμα ἐστίν. αἱ μυῖαι αἱ
Πισάτιδες κατὰ τὴν τῶν Ὀλυμπίων ἑορτὴν ὡς
ἂν εἴποις σπένδονται καὶ τοῖς ἀφικνουμένοις καὶ

---

[1] ⟨ὡς⟩ add. Jac. [2] ἐπιθήγουσιν.
[3] δέ τι.
[4] γέρας καὶ εἰκότως εἰ μὴ ἀμοιρήσει.

---

[a] Dionysius the elder, c. 430–367 B.C., elected general and
ruler of Syracuse, extended his power over Sicily and parts of
Magna Graecia; represented as a tyrant of the worst kind.—
Dionysius the younger succeeded his father, 367 B.C. Ejected

Sicily,[a] Clearchus of Heraclea, Apollodorus the oppressor of Cassandrea, Nabis the scourge of Sparta, if they trusted in the sword, when the King Wasps trust to their lack of sting and to their gentle nature?

16. This is what Wasps that are armed with a sting are said to do. When they observe a dead viper they swoop upon it and draw poison into their sting. It is from this source, I fancy, that men have acquired that knowledge, and no good knowledge either. And Homer is witness to the fact when he says in the *Odyssey* [1. 261]

'Seeking a deadly drug, that he might have wherewithal to smear his bronze-tipped arrows.'

Or again, to be sure (if one can trust the story), just as Heracles dipped his arrows in the venom of the Hydra, so do Wasps dip and sharpen their sting.

17. Let not the Fly lack the honour of a mention in this record of mine, for it too is Nature's handiwork.

The Flies of Pisa at the season of the Olympic festival make peace, so to speak, both with visitors

The Wasp and its poison

The Fly

from Sicily, he made himself Tyrant of Locris—and deserved the title. Recovered Syracuse by treachery but was again expelled in 345 B.C., by Timoleon.—Clearchus by championing the cause of the people against the nobles of Heraclea obtained the tyranny. After a reign of 12 years marked by signal cruelty he was murdered, 353 B.C.—Apollodorus, tyrant of Cassandrea, 3rd cent. B.C., became a byword for cruelty; conquered and executed by Antigonus Gonatas.—Nabis usurped the kingship of Sparta, which he exercised with the utmost savagery; defeated by Philopoemen and Flamininus in his efforts to regain lost territory; finally murdered, 192 B.C.

τοῖς ἐπιχωρίοις. ἱερείων γοῦν καταθυομένων τοσούτων καὶ αἵματος ἐκχεομένου καὶ κρεμαμένων
κρεῶν αἱ δὲ ἀφανίζονται ἑκοῦσαι, καὶ τοῦ γε
Ἀλφειοῦ περαιοῦνται ἐς τὴν ἀντιπέρας ὄχθην.
καὶ ἐοίκασι τῶν γυναικῶν τῶν ἐπιχωρίων διαλλάττειν οὐδὲ ὀλίγον, εἰ μὴ ἄρα τι ἐγκρατέστεραι
αἱ μυῖαι ἐκεῖναι τῶν γυναικῶν ὁμολογοῦνται τοῖς
ἔργοις· τὰς μὲν γὰρ ὁ τῆς ἀγωνίας καὶ τῆς κατ'
αὐτὴν σωφροσύνης νόμος ἐλαύνει τὰς γυναῖκας,[1]
αἱ μυῖαι δὲ ἑκοῦσαι τοῖς ἱεροῖς ἀφίστανται, καὶ ἐν
μὲν ταῖς ἱερουργίαις καὶ παρὰ τὸν τῶν ἄθλων
χρόνον τὸν νενομισμένον ἀπαλλάττονται. λῦτο δ'
ἀγών, αἱ δὲ ἐπιδημοῦσιν, ὥσπερ οὖν καθόδου
τυχοῦσαι ψηφίσματι φυγάδες, εἶτα ἐπιρρέουσιν ἐς
τὴν Ἦλιν αἱ μυῖαι αὖθις.[2]

18. Ὁ ὀρφὼς[3] θαλάττιον ζῷόν ἐστι, καὶ εἰ
ἕλοις καὶ ἀνατέμοις, οὐκ ἂν ἴδοις τεθνεῶτα παραχρῆμα αὐτόν, ἀλλὰ ἐπιλαμβάνει τῆς κινήσεως καὶ
οὐκ ἐπ' ὀλίγον. διὰ χειμῶνος δὲ ἐν τοῖς φωλεοῖς
οἰκουρῶν χαίρει· διατριβαὶ δὲ ἄρα αἱ πρὸς τῇ γῇ
μᾶλλον φίλαι αὐτῷ.

19. Λύκος ὁμόσε ταύρῳ χωρεῖν καὶ ἰέναι οἱ
κατὰ πρόσωπον ἥκιστός ἐστι,[4] δέδοικε δὲ τὰ
κέρατα καὶ τὰς ἀκμὰς αὐτῶν ἐκνεύει. καὶ ὡς ἐξ
εὐθείας οἱ μαχούμενος ἀπειλεῖ· οὐ μὴν δρᾷ τοῦτο,
ἀλλὰ ὥσπερ οὖν ἐπιθησόμενος ὑποφαίνει, εἶτα
μέντοι προσπεσόντος ὁ δὲ ἑαυτὸν ἐξελίξας ἐς τὰ
νῶτα ἀνέθορε, καὶ ἐγκρατῶς ἔχεται τοῦ θηρὸς ὁ

---

[1] τὰς γυναῖκας del. Cobet.
[2] αὖθις ὡς αἱ γυναῖκες.

and with the local inhabitants. At any rate, despite
the multitude of sacrifices, the quantity of blood shed
and of flesh hung out, the Flies disappear of their avoids the
own free will and cross to the opposite bank of the Olympic
Alpheus. And they appear to differ not a whit from Games
the women there, except that their behaviour shows
them to be more self-restrained than the women.
For while women are excluded by the rules of training and of continence at that season, the Flies of their
own free will abstain from the sacrifices and absent
themselves while the ceremonies are in progress and
during the recognised period of the Games. ' Then
was the assembly ended ' [Hom. *Il*. 24. 1] and the
Flies come home, just like exiles whom a decree has
allowed to return, and once again they stream into
Elis.

18. The Great Sea Perch is a marine creature, and The Great
if you were to catch and cut it up, you would not then Sea Perch
and there see it dead, but it retains the power of
movement, and for a considerable time. All through
the winter it likes to remain at home in its caverns,
and its favourite resorts are near the land.

19. The Wolf does not dare to close with a Bull and Wolf and
to meet it face to face; he is afraid of its horns and Bull
avoids their points. So he makes a feint of attacking
the Bull frontally; he does not however attack but
gives the appearance of being about to try; and
then when the Bull makes a rush at him, the Wolf
slips aside and leaps on its back and clings with might
and main, beast wrestling with beast. And the Wolf

---

³ ὀρφός.   ⁴ ἐστι καὶ εἰκότως.

θὴρ ὁ ἀντίπαλος, καὶ κατισχύει αὐτοῦ σοφίᾳ
φυσικῇ τὸ ἐνδέον ἀνακούμενος ὁ λύκος.

20. Ὄνος ὁ θαλάττιος ἐν τῇ γαστρὶ τὴν καρδίαν
ἔλαχεν ἔχειν, ὡς οἱ δεινοὶ τὰ τοιαῦτα [1] ὁμολογοῦσιν ἡμῖν καὶ διδάσκουσιν.

21. Ὁ ταῶς οἶδεν ὀρνίθων ὡραιότατος ὤν, καὶ
ἔνθα οἱ τὸ κάλλος κάθηται, καὶ τοῦτο οἶδε, καὶ
ἐπ᾽ αὐτῷ κομᾷ καὶ σοβαρός ἐστι, καὶ θαρρεῖ τοῖς
πτεροῖς, ἅπερ οὖν αὐτῷ καὶ κόσμον περιτίθησι,
καὶ πρὸς τοὺς ἔξωθεν φόβον ἀποστέλλει, καὶ ἐν
ὥρᾳ θερείῳ σκέπην οἴκοθεν καὶ οὐκ ᾐτημένην
οὐδὲ ὀθνείαν παρέχεται. ἐὰν γοῦν θελήσῃ φοβῆσαί
τινα, ἐγείρας τὰ οὐραῖα εἶτα διεσείσατο καὶ
ἀπέστειλεν ἦχον, καὶ ἔδεισαν οἱ παρεστῶτες, ὡς
ὁπλίτου τὸν ἐκ τῶν ὅπλων πεφοβημένοι δοῦπον.
ἀνατείνει δὲ τὴν κεφαλὴν καὶ ἐπινεύει σοβαρώτατα,
ὥσπερ οὖν ἐπισείων τριλοφίαν. δεηθείς γε μὴν
ψυχάσαι, τὰ πτερὰ ἐγείρει, καὶ ἐς τοὔμπροσθεν
ἐπικλίνας συμφυᾶ σκιὰν ἀποδείκνυται τοῦ ἰδίου
σώματος τὴν ἀκμὴν τὴν ἐκ τῆς ἀκτῖνος ἀποστέγων.
εἰ δὲ εἴη καὶ ἄνεμος κατόπιν, ἡσυχῇ διίστησι τὰ
πτερά· καὶ τὸ πνεῦμα ⟨τὸ⟩ [2] διαρρέον αὔρας οἱ
μαλακὰς καὶ ἡδίστας ἐπιπνέον ἀναψύχειν τὸν ὄρνιν δίδωσιν. ἐπαινεθεὶς δὲ αἰσθάνεται, καὶ ὥσπερ
οὖν ἢ παῖς καλὸς ἢ γυνὴ ὡραία τὸ μάλιστα πλεονεκτοῦν [3] τοῦ σώματος ἐπιδείκνυσιν, οὕτω τοι καὶ
ἐκεῖνος τὰ πτερὰ ἐν κόσμῳ καὶ κατὰ στοῖχον ὀρ-
θοῖ, καὶ ἔοικεν ἀνθηρῷ λειμῶνι ἢ γραφῇ πεποικιλ-

----

[1] Jac: δεινότατοι αὐτά.   [2] ⟨τὸ⟩ add. H.

overpowers it and by native cunning makes good his lack of strength.

20. The Hake has its heart in its belly, as ex- The Hake perts in these matters agree and inform us.

21. The Peacock knows that it is the most beautiful The Peacock of birds; it knows too wherein its beauty resides; it prides itself on this and is haughty, and gathers confidence from the plumes which are its ornament and which inspire strangers with terror. In summertime they afford it a covering of its own, unsought, not adventitious. If, for instance, it wants to scare somebody it raises its tail-feathers and shakes them and emits a scream, and the bystanders are terrified, as though scared by the clang of a hoplite's armour. And it raises its head and nods most pompously, as though it were shaking a triple plume at one. When however it needs to cool itself it raises its feathers, inclines them in a forward direction and displays a natural shade from its own body, and wards off the fierceness of the sun's rays. But if there is a wind behind it, it gradually expands its feathers, and the breeze which streams through them, blowing gently and agreeably, enables the bird to cool itself. It knows when it has been praised, and as some handsome boy or lovely woman displays that feature which excels the rest, so does the Peacock raise its feathers in orderly succession; and it resembles a flowery meadow or a picture made beautiful by the many hues of the paint, and painters must be prepared to sweat in order to represent its special

---

[3] πλεονεκτοῦν εἰς ὥραν.

μένῃ πολυχροίᾳ τῇ τῶν φαρμάκων, καὶ ἱδρὼς πρόκειται ζωγράφοις εἰκάσαι τῆς φύσεως τὸ ἴδιον.
καὶ ὅπως ἔχει τῆς ἐς τὴν ἐπίδειξιν ἀφθονίας
παρίστησιν· ἐᾷ γὰρ ἐμπλησθῆναι τῆς θέας τοὺς
παρεστῶτας, καὶ ἑαυτὸν περιάγει δεικνὺς φιλοπόνως τὸ τῆς πτερώσεως πολύμορφον, ὑπὲρ τὴν τῶν
Μήδων ἐσθῆτα καὶ τὰ ⟨τῶν⟩[1] Περσῶν ποικίλματα
τὴν ἑαυτοῦ στολὴν ἀποδεικνύμενος ἐκεῖνός γε
σοβαρώτατα. λέγεται δὲ ἐκ βαρβάρων ἐς Ἕλληνας
κομισθῆναι. καὶ χρόνου πολλοῦ σπάνιος ὢν εἶτα
ἐδείκνυτο τῶν ἀνθρώπων τοῖς φιλοκάλοις μισθοῦ,
καὶ Ἀθήνησί γε ταῖς νουμηνίαις ἐδέχοντο καὶ
ἄνδρας καὶ γυναῖκας ἐπὶ τὴν ἱστορίαν αὐτῶν, καὶ
τὴν θέαν πρόσοδον εἶχον. ἐτιμῶντο δὲ τὸν
ἄρρενα καὶ τὸν θῆλυν δραχμῶν μυρίων, ὡς Ἀν-
τιφῶν ἐν τῷ πρὸς Ἐρασίστρατον λόγῳ φησί. δεῖ
δὲ καὶ διπλῆς οἰκίας τῇ τροφῇ αὐτῶν, καὶ φρουρῶν
τε καὶ μελεδωνῶν. Ὁρτήσιος δὲ ὁ Ῥωμαῖος
καταθύσας ἐπὶ δείπνῳ ταῶν πρῶτος ἐκρίθη.
Ἀλέξανδρος δὲ ὁ Μακεδὼν ἐν Ἰνδοῖς ἰδὼν τούσδε
τοὺς ὄρνιθας ἐξεπλάγη, καὶ τοῦ κάλλους θαυμάσας
ἠπείλησε τῷ καταθύσαντι ταῶν ἀπειλὰς βαρυτάτας.

22. Ἐς τοὺς ψυκτῆρας ὅταν οἱ μύες ἐμπέσωσιν,
ἀνανεῦσαι καὶ ἀνελθεῖν οὐ δυνάμενοι, τὰς ἀλλήλων
οὐρὰς ἐνδακόντες εἶτα ἐφέλκουσι τὸν δεύτερον ὁ
πρῶτος καὶ ὁ δεύτερος τὸν τρίτον. οὕτω μὲν δὴ
καὶ τούτους ἀλλήλοις συμμαχεῖν καὶ ἐπικουρεῖν ἡ
σοφωτάτη φύσις ἐξεπαίδευσεν.

characteristics. And it proves how ungrudgingly it exhibits itself by permitting bystanders to take their fill of gazing, as it turns itself about and industriously shows off the diversity of its plumage, displaying with the utmost pride an array surpassing the garments of the Medes and the embroideries of the Persians. It is said to have been brought to Greece from foreign lands. And since for a long while it was a rarity, it used to be exhibited to men of taste for a fee, and at Athens the owners used on the first day of each month to admit men and women to study them, and they made a profit by the spectacle. They used to value the cock and the hen at ten thousand drachmas,[a] as Antiphon says in his speech against Erasistratus.[b] For their maintenance a double establishment and custodians and keepers are needed. Hortensius the Roman was judged to have been the first man to slaughter a Peacock for a banquet. But Alexander of Macedon was struck with amazement at the sight of these birds in India, and in his admiration of their beauty threatened the severest penalties for any man who slew one.

22. When Mice fall into cooling-vessels, since they cannot get out by swimming, they fasten their teeth into one another's tails, and then the first pulls the second and the second the third. In this way has Nature in her supreme wisdom taught them to combine and help one another.

Mouse saved from drowning

[a] About £375.
[b] The speech is lost, but see Athen. 9. 397 c, d.

---

[1] ⟨τῶν⟩ *add. H.*

AELIAN

23. Ἐλλοχῶσιν οἱ κροκόδιλοι τοὺς ὑδρευομένους ἐκ τοῦ Νείλου τὸν τρόπον τοῦτον. φρύγανα ἑαυτοῖς ἐπιβαλόντες [1] καὶ δι' αὐτῶν ἐμβλέποντες εἶτα ὑπονέουσι τοῖς φρυγάνοις. οἱ δὲ ἀφικνοῦνται κεράμια ἢ κάλπεις ἢ πρόχους κομίζοντες. εἶτα ἀρυτομένους [2] αὐτοὺς ὑπεκδύντες τῶν φρυγάνων καὶ τῇ ὄχθῃ προσαναπηδήσαντες ἁρπαγῇ βιαιοτάτῃ συλλαβόντες ἔχουσι δεῖπνον. κακίας δὴ καὶ πανουργίας κροκοδίλων συμφυοῦς εἴρηταί μοι τὰ νῦν ταῦτα.

24. Λαγὼς δέδοικε κύνας καὶ μέντοι καὶ ἀλώπηξ. καί που ⟨καὶ⟩ [3] σῦν ἐγείρουσιν [4] ἐκ τῆς λόχμης αἱ αὐταὶ τῇ ὑλακῇ, καὶ λέοντα ἐπιστρέφουσι, καὶ ἔλαφον διώκουσιν· ὀρνίθων δὲ οὐδὲ εἷς ὥραν ποιεῖται κυνός, ἀλλ' αὐτοῖς πρὸς αὐτοὺς [5] ἔνσπονδά ἐστι. μόνη δὲ ἡ ὠτὶς πέφρικε κύνας. τὸ δὲ αἴτιον, βαρεῖαί τέ εἰσι καὶ σαρκῶν ὄγκον περιφέρουσιν. οὔκουν αὐτὰς αἴρει τε καὶ ἐλαφρίζει τὰ πτερὰ ῥᾳδίως, καὶ διὰ τοῦτο ταπειναὶ πέτονται καὶ κάτω περὶ γῆν, βρίθοντος τοῦ ὄγκου αὐτάς. αἱροῦνται δὲ ὑπὸ τῶν κυνῶν πολλάκις. ὅπερ ἑαυταῖς συνειδυῖαι, ὅταν ἀκούσωσιν ὑλακῆς, ἐς τοὺς θάμνους καὶ τὰ ἕλη καταθέουσι, προβαλλόμεναι ἑαυτῶν ταῦτα, καὶ ῥυόμεναι σφᾶς ἐκ τῶν παρόντων καὶ μάλα εὐπόρως.

25. Ὀψὲ τοὺς γειναμένους ἄνθρωπος γνωρίζειν ἄρχεται, διδασκόμενος καὶ οἱονεὶ καταναγκαζόμενος [6] ἐς πατέρα ὁρᾶν καὶ μητέρα ἀσπάζεσθαι καὶ οἰκείοις προσμειδιᾶν· οἱ δὲ ἄρνες περὶ τὰς μητέρας

---

[1] ἐπιβάλλοντες.   [2] ἀρυομένους.

316

23. This is the way in which Crocodiles lie in wait The Crocodile
for those who draw water from the Nile: they cover
themselves with driftwood and, spying through it,
swim up beneath it. And the people come bringing
earthen vessels or pitchers or jugs. Then, as men
draw water, the creatures emerge from the driftwood, leap against the bank, and seizing them with
overpowering force make a meal of them. So much
for the innate wickedness and villainy of Crocodiles.

24. The Hare dreads Hounds, and so too does the The Bustard and Hounds
Fox. And Hounds, I fancy, with their barking will
rouse a boar from the brake, and will bring a lion to
bay, and pursue a stag. Yet there is not a single
bird that cares anything for a Hound, but there is
peace between them. The Bustard alone is afraid
of Hounds, the reason being that these birds are
heavy and carry a burden of flesh about with them.
Their wings do not easily lift them and carry them
through the air, so they fly low along the ground,
weighed down by their bulk. Hence they are frequently captured by Hounds. And since they are
aware of this, whenever they hear the bark of
Hounds, they run away into thickets and swamps,
using these as a protection and escaping instant
danger without difficulty.

25. The human child is slow to recognise its The Lamb
parents: it is taught and, one might say, compelled
to look at its father, to greet its mother, and to smile
upon its relatives. Whereas Lambs from the day of

---

³ ⟨καί⟩ add. H.   ⁴ Jac : συνεγείρουσιν.
⁵ παρ' αὐτῶν.   ⁶ Reiske : καταδόμενος.

πηδῶσιν ἀπὸ γενεᾶς, καὶ ἴσασι τό τε ὀθνεῖον καὶ
τὸ οἰκεῖον, καὶ παρὰ τῶν νομέων μαθεῖν δέονται
οὐδὲ ἕν.

26. Μιμηλότατόν ἐστιν ὁ πίθηκος ζῷον, καὶ
πᾶν ὅ τι ἂν ἐκδιδάξῃς τῶν διὰ τοῦ σώματος
πραττομένων ὁ δὲ εἴσεται ἀκριβῶς, ἵνα ἐπιδεί-
ξηται [1] αὐτό. ὀρχεῖται [2] γοῦν, ἐὰν μάθῃ, καὶ
αὐλεῖ, ἐὰν ἐκδιδάξῃς. ἐγὼ δὲ καὶ ἡνίας κατέχοντα
εἶδον καὶ ἐπιβάλλοντα τὴν μάστιγα καὶ ἐλαύνοντα.
καὶ ἄλλο δ᾿ ἄν τι μαθὼν καὶ ἄλλο οὐ διαψεύσαιτο
τὸν διδάξαντα· οὕτως ἄρα ἡ φύσις ποικίλον τε καὶ
εὐτράπελόν ἐστιν.

27. Ἴδιαι δὲ καὶ διάφοροι τῶν ζῴων καὶ αἵδε [3]
αἱ φύσεις. τοὺς ἐν τοῖς Βισάλταις λαγὼς διπλᾶ
ἥπατα ἔχειν Θεόπομπος λέγει. τὰς δ᾿ ἐν Λέρῳ
μελεαγρίδας ὑπὸ μηδενὸς ἀδικεῖσθαι τῶν γαμψω-
νύχων ὀρνέων λέγει Ἴστρος. τοὺς δὲ ἐν Νευροῖς
βοῦς Ἀριστοτέλης φησὶν ἐπὶ τῶν ὤμων ἔχειν τὰ
κέρατα, Ἀγαθαρχίδης δὲ τὰς ἐν Αἰθιοπίᾳ ὗς
κέρατα ἔχειν. Σώστρατος δὲ τοὺς ἐν τῇ Κυλλήνῃ
κοσσύφους πάντας λέγει λευκούς. Ἀλέξανδρος
δὲ ὁ Μύνδιος ⟨τὰ⟩ [4] ἐν τῷ Πόντῳ πρόβατα
πιαίνεσθαι ὑπὸ τοῦ πικροτάτου φησὶν ἀψινθίου.
τὰς δὲ ἐν τῷ Μίμαντι γινομένας αἶγας ἐξ μηνῶν
μὴ πίνειν, ὁρᾶν δὲ ἐς τὴν θάλατταν μόνον καὶ
κεχηνέναι καὶ τὰς αὔρας τὰς ἐκεῖθεν δέχεσθαι ὁ
αὐτὸς λέγει. αἶγας δὲ Ἰλλυρίδας ὁπλὴν ἀκούω

---

[1] να μαθὼν καὶ ἀποδείξηται.
[2] καὶ ὀρχεῖται.
[3] Perh. καὶ τῶνδε H.   [4] ⟨τά⟩ add. Jac.

their birth gambol about their dams and know what
is strange and what is akin to them. They have no
need to learn anything from their shepherds.

26. The Monkey is a most imitative creature, and The Monkey
any bodily action that you teach it it acquires exactly,
so as to be able to display its accomplishment. For
instance, it will dance, once it has learnt, and if you
teach it, will play the pipe. And I myself have even
seen it holding the reins, laying on the whip, and
driving a chariot. And once it has learnt whatever
it may be, it would never disappoint its teacher. So
versatile and so adaptable a thing is Nature.

27. Here are further examples of the peculiar and Peculiarities
diverse natures of animals. Theopompus reports of certain
animals
that in the country of the Bisaltae [a] the Hares have
a double liver. According to Istcr the Guinea-fowls
of Leros are never injured by any bird of prey.
Aristotle says [b] that among the Neuri [c] the Oxen
have their horns on their shoulders, and Agatharcides
says that in Ethiopia the Swine have horns. Sostratus asserts that all Blackbirds on Cyllene [d] are white.
Alexander of Myndus says that in Pontus the Flocks
grow fat upon the bitterest wormwood. He states
also that Goats born on Mimas [e] do not drink for
six months; all they do is to look towards the sea
with their mouths open and to drink in the breezes
from that quarter. I learn that the Goats of Illyria

[a] Macedonian tribe living on W coast of the gulf of the
Strymon.
[b] Not in any surviving work; *fr.* 313 (Rose p. 331).
[c] Tribe living between the rivers Boug and Dnieper.
[d] Mountain in N Arcadia.
[e] Mountain on coast of Ionia, W of Smyrna.

ἔχειν, ἀλλ᾽ οὐ χηλήν. Θεόφραστος δὲ δαιμονιώ-
τατα λέγει ἐν τῇ Βαβυλωνίᾳ γῇ τοὺς ἰχθῦς
ἀνιόντας ἐκ τοῦ ποταμοῦ εἶτα μέντοι ἐν τῷ ξηρῷ
τὰς νομὰς ποιεῖσθαι πολλάκις.

28. Ἴδιον δὲ ἄρα ⟨ὁ⟩[1] πορφυρίων πρὸς τῷ
ζηλοτυπώτατος εἶναι καὶ ἐκεῖνο[2] δήπου κέκτηται.
φιλοίκειον αὐτὸν εἶναί φασιν καὶ τὴν συντροφίαν
τῶν συννόμων ἀγαπᾶν. ἐν οἰκίᾳ γοῦν τρέφεσθαι
πορφυρίωνα καὶ ἀλεκτρυόνα ἤκουσα, καὶ σιτεῖσθαι
μὲν τὰ αὐτά, βαδίζειν δὲ τὰς ἴσας βαδίσεις καὶ
κοινῇ κονίεσθαι. οὐκοῦν ἐκ τούτων φιλίαν τινὰ
θαυμαστὴν αὐτοῖς ἐγγενέσθαι. καί ποτε ἑορτῆς
ἐπιστάσης ὁ δεσπότης ἀμφοῖν τὸν ἀλεκτρυόνα
καταθύσας εἱστιάθη σὺν τοῖς οἰκείοις· ὁ δὲ
πορφυρίων τὸν σύννομον οὐκ ἔχων καὶ τὴν ἐρημίαν μὴ φέρων ἑαυτὸν ἀτροφίᾳ διέφθειρεν.

29. Ἐν Αἰγίῳ τῆς Ἀχαίας ὡραίου παιδός,
Ὠλενίου τὸ γένος, ὄνομα Ἀμφιλόχου, ἤρα χήν.
Θεόφραστος λέγει τοῦτο. σὺν τοῖς Ὠλενίων δὲ
φυγάσιν ἐφρουρεῖτο ἐν Αἰγίῳ ὁ παῖς. οὐκοῦν ὁ
χὴν αὐτῷ δῶρα ἔφερε. καὶ ἐν Χίῳ Γλαύκης τῆς
κιθαρῳδοῦ ὡραιοτάτης οὔσης εἰ μὲν ἤρων ἄνθρωποι, μέγα οὐδέπω· ἠράσθησαν δὲ καὶ κριὸς καὶ
χήν, ὡς ἀκούω, τῆς αὐτῆς.

---

[1] ⟨ὁ⟩ add. H.   [2] ὁ ὄρνις καὶ ἐκεῖνο.

---

[a] Aegium, one of the principal cities of Achaia, stood on the
coast near the W end of the Corinthian gulf. It was the
regular meeting-place of the Achaean League.

have a solid, not a cloven hoof. And Theophrastus [*fr.* 171. 2] has the most amazing statement that in Babylonia the fish frequently come out of the river and pasture on dry land.

28. Now the Purple Coot, in addition to being extremely jealous, has, I believe, this peculiarity: they say that it is devoted to its own kin and loves the company of its mates. At any rate I have heard that a Purple Coot and a Cock were reared in the same house, that they fed together, that they walked step for step, and that they dusted in the same spot. From these causes there sprang up a remarkable friendship between them. And one day on the occasion of a festival their master sacrificed the Cock and made a feast with his household. But the Purple Coot, deprived of its companion and unable to endure the loneliness, starved itself to death.

The Purple Coot

29. In Aegium,[a] a city of Achaia, a good-looking boy, an Olenian[b] by birth, of the name of Amphilochus, was loved by a Goose. Theophrastus relates this [*fr.* 109]. The boy was kept under guard with exiles from Olenus in Aegium, and so the Goose used to bring him presents. In Chios Glauce, the harp-player, being a woman of extraordinary beauty, was adored by men, not that there is anything wonderful in that, but I am told that a Ram and a Goose also fell in love with her.

Geese in love with human beings

[b] Olenus was a small town on the NW coast of Achaia, near the mouth of the Pirus. The reference to 'exiles from O.' is obscure; it may signify an effort on the part of the Achaean League to ensure peace among the 12 cities of Achaia. As the League was broken up by Alexander, the event must have occurred earlier.

321

AELIAN

Οἱ δὲ χῆνες διαμείβοντες τὸν Ταῦρον τὸ ὄρος δεδοίκασι τοὺς ἀετούς, καὶ ἕκαστός γε αὐτῶν λίθον
ἐνδακόντες, ἵνα μὴ κλάζωσιν, ὥσπερ οὖν ἐμβαλόν-
τες σφίσι στόμιον, διαπέτονται σιωπῶντες, καὶ
τοὺς ἀετοὺς τὰ πολλὰ ταύτῃ διαλανθάνουσι.
θερμότατος δὲ ἄρα ὢν καὶ διαπυρώτατος τὴν
φύσιν ὁ χὴν φιλόλουτρός ἐστι καὶ νήξεσι χαίρει
καὶ τροφαῖς μάλιστα ταῖς ὑγροτάταις καὶ πόαις
καὶ θριδακίναις καὶ τοῖς λοιποῖς, ὅσα αὐτοῖς
ἔνδοθεν ψύχος ἐργάζεται· εἰ δὲ καὶ ἐξαυαίνοιτο
ὑπὸ ⟨τοῦ⟩[1] λιμοῦ, δάφνης φύλλον οὐκ ἂν φάγοι,
οὐδ᾽ ἂν πάσαιτο ῥοδοδάφνης οὔτε ἑκὼν οὔτε ἄκων·
οἶδε γὰρ ὅτι τεθνήξεται τούτων τινὸς ἐμφαγών.
ἄνθρωποι δὲ ὑπ᾽ ἀσωτίας[2] ἐπιβουλεύονται καὶ ἐς
τροφὴν καὶ ἐς ποτόν.[3] μυρίοι γοῦν καὶ πίνοντές
τι κακὸν κατέπιον, ὡς Ἀλέξανδρος, καὶ ἐσθίοντες,
ὡς Κλαύδιος ὁ Ῥωμαῖος καὶ Βρεττανικὸς ὁ τούτου
παῖς· καὶ κατακοιμηθέντες οὐκ ἐξανέστησαν χρή-
σει φαρμάκου, οἱ μὲν ἑκόντες τοῦτο σπάσαντες, οἱ
δὲ ἐπιβουλευθέντες.

30. Ὁ δὲ χηναλώπηξ, πέπλεκταί οἱ τὸ ὄνομα[4]
ἐκ τῶν ⟨ἑκατέρου⟩[5] τοῦ ζῴου ἰδίων τε καὶ
συμφυῶν. ἔχει μὲν γὰρ τὸ εἶδος τὸ τοῦ χηνός,
πανουργίαν[6] δὲ δικαιότατα ἀντικρίνοιτο ἂν τῇ
ἀλώπεκι. καὶ ἔστι μὲν χηνὸς βραχύτερος, ἀνδρειότερος δέ, καὶ χωρεῖν ὁμόσε δεινός. ἀμύνεται γοῦν
καὶ ἀετὸν καὶ αἴλουρον καὶ τὰ λοιπά, ὅσα αὐτοῦ
ἀντίπαλά ἐστιν.

---

[1] ⟨τοῦ⟩ add. H.
[3] Ges : ὕπνον.
[5] ⟨ἑκατέρου⟩ add. H.
[2] Pauw : ὑπὸ σοφίας.
[4] ὄνομα καὶ εἰκότως.
[6] πανουργίᾳ.

When Geese cross the Taurus range they go in fear Geese and
Eagles of the eagles; so each of them bites on a pebble to prevent it from uttering its cry, just as though they had gagged themselves, and so they cross in silence and by these means generally slip past the eagles. The Goose being of a very hot and fiery habits and
food nature is fond of bathing and delights in swimming, and prefers very moist fare, grass, lettuce, and all other things that generate coolness in its body. But even if it is exhausted with hunger it will not eat a bay-leaf or touch a rose-laurel either willingly or against its will, for it knows that if it eats either of them it will die.

Yet men through their unbridled appetites are the Human
victims of
food and
drink victims of plots against their food and drink. At any rate countless numbers have swallowed some bane while drinking, like Alexander,[a] or in food, like Claudius the Roman,[b] and Britannicus, his son.[c] And having fallen asleep from a dose of poison, they never rose again, some having drunk it deliberately, others because they were the victims of a plot.

30. The Egyptian Goose owes its composite name The
Egyptian
Goose (goose-fox) to the innate peculiarities of the two creatures. It has the appearance of a goose, but for its mischievousness it might most justly be compared to the fox. It is smaller than a goose but more courageous, and is a fierce fighter. For instance, it defends itself against an eagle, a cat, and all other animals that come against it.

[a] Alexander died (323 B.C.) of a fever aggravated by excessive drinking.

[b] Roman Emperor, A.D. 41–54, poisoned by his wife Agrippina.

[c] Poisoned by order of Nero, A.D. 55.

31. Ἴδια δὲ ὄφεως καὶ ἐκεῖνά ἐστι. τὴν καρδίαν κεκλήρωται ἐπὶ τῇ φάρυγγι, τὴν δὲ χολὴν
ἐν τοῖς ἐντέροις, πρὸς δὲ τῇ οὐρᾷ τοὺς ὄρχεις
ἔχει, τὰ δὲ ᾠὰ τίκτει μακρὰ καὶ μαλακά, τὸν δὲ
ἰὸν ἐν τοῖς ὀδοῦσι φέρει.

32. Ταῷ δὲ τῷ ὄρνιθι τῷ προειρημένῳ καὶ
ἐκεῖνα συμφυᾶ καὶ ἴδια, ἅπερ ἐστὶ μαθεῖν ἄξια.
τρία ἔτη γενόμενος κυήσεως ἄρχεται καὶ ὠδῖνα
ἀπολύει καὶ τῆς τῶν πτερῶν πολυχροίας τε καὶ
ὥρας τότε ἄρχεται. ἐπῳάζει δὲ οὐ κατὰ τὸ ἑξῆς,
ἀλλὰ παραλιπὼν δύο ἡμέρας. ἤδη δ᾽ ἂν τέκοι
καὶ ὑπηνέμια ὁ ταώς, ὡς καὶ ὄρνιθες ἕτεροι.

33. Ἡ νῆττα ὅταν τέκῃ, τίκτει μὲν [1] ἐν ξηρῷ,
πλησίον δὲ ἢ τῆς λίμνης ἢ τοῦ τενάγους ἢ ἄλλου
τινὸς ὑδρηλοῦ χώρου καὶ ἐνδρόσου. τὸ δὲ νήτ-
τιον [2] φύσει τινὶ ἰδίᾳ καὶ ἀπορρήτῳ οἶδεν ὅτι μήτε
τῆς [3] μετεώρου φορᾶς οἱ μέτεστι μήτε μὴν τῆς
ἐν τῇ χέρσῳ διατριβῆς. καὶ ἐκ τούτων ἐς τὸ
ὕδωρ πηδᾷ, καὶ ἐξ ὠδίνων ἐστὶ νηκτική, καὶ
μαθεῖν οὐ δεῖται, ἀλλὰ καταδύεται καὶ ἀναδύεται
πάνυ σοφῶς καὶ ὡς ἤδη χρόνου πεπαιδευμένη
τοῦτο. ἀετὸς δέ, ὃν καλοῦσι νηττοφόνον, ἐπιπηδᾷ
τῇ νηχομένῃ ὡς ἁρπασόμενος· ἡ δὲ καταδῦσα
ἑαυτὴν ἠφάνισεν, εἶτα ὑπονηξαμένη ἀλλαχόθι
ἐκκύπτει. ὁ δὲ καὶ ἐκεῖ πάρεστι, καὶ αὖθις
κατέδυ ἐκείνη, καὶ πάλιν ταῦτα καὶ πάλιν. καὶ
δυοῖν θάτερον· ἢ γὰρ καταδῦσα [4] ἀπεπνίγη, ἢ ὁ

---

[1] μέντοι.
[2] νεοττίον.
[3] τῆς ἐν ἀέρι.
[4] Ραυω : καταδύς.

other prey; whereupon the Duck, with nothing to fear, swims once more upon the surface.

34. The Swan has this advantage over men in matters of the greatest moment, for it knows when the end of its life is at hand, and, what is more, in bearing its approach with cheerfulness, it has received from Nature the noblest of gifts. For it is confident that in death there is neither pain nor sorrow. But men are afraid of what they know not, and regard death as the greatest of all ills. Now the Swan has so contented a spirit that at the very close of its life it sings and breaks out into a dirge, as it were, for itself. Even so does Euripides [*fr.* 311 N] sing of Bellerophon, prepared like a hero of high soul for death. For example, he has portrayed him addressing his soul thus: ^{The Swan and death}

'Reverent wast thou ever in life towards the gods; strangers didst thou succour; nor didst thou ever grow weary towards thy friends '—

and so on. So then the Swan too intones its own funeral chant, and either by hymns to the gods or by the rehearsal of its own praises it makes provision for its departure. Socrates also testifies [Pl. *Phaedo* 84 E] to the fact that it sings not from sorrow but rather from cheerfulness, for (he says) a man whose heart is vexed and sore has no leisure for song and melody.

Now death is not the only thing that the Swan faces with courage: it is not afraid of a fight. But though it will not be the first to do an injury, any

---

³ τοῦτο τό.   ⁴ αὐτῷ.

νος ἀνήρ, τῷ δὲ ἄρξαντι [1] οὔτε ἀφίσταται οὔτε
εἴκει. οἱ μὲν οὖν ὄρνιθες οἱ λοιποί, εἰρηναῖα
αὐτοῖς πρὸς αὐτοὺς καὶ ἔνσπονδά ἐστιν,[2] ὁ δὲ
ἀετὸς καὶ ἐπὶ τοῦτον ὥρμησε πολλάκις, ὡς
Ἀριστοτέλης φησί, καὶ οὐδεπώποτε ἐκράτησεν,
ἡττήθη δὲ ἀεὶ μὴ μόνον σὺν τῇ ῥώμῃ τοῦ κύκνου
μαχομένου, ἀλλὰ καὶ σὺν τῇ δίκῃ ἀμυνομένου.

35. Ὁ ἐρωδιὸς τὰ ὄστρεα ἐσθίειν δεινός ἐστι,
καὶ μεμυκότα αὐτὰ καταπίνει, ὥσπερ οὖν οἱ
πελεκᾶνες τὰς κόγχας. καὶ ἐν τῷ καλουμένῳ
πρηγορεῶνι ὑποθερμαίνων ὁ ἐρωδιὸς φυλάττει τὰ
ὄστρεα· τὰ δὲ ὑπὸ τῆς ἀλέας διίσταται, καὶ
ἐκεῖνος αἰσθανόμενος τὰ μὲν ὄστρακα ἀνεμεῖ,
φυλάττει δὲ τὴν σάρκα, καὶ ἔχει τροφὴν ἀναλίσκων
τῇ τῆς πέψεως δυνάμει τὸ ἔσω παρελθὸν ὁλόκληρον.

36. Ὄνομά ἐστιν ὄρνιθος ἀστερίας, καὶ τιθασεύε-
ταί γε ἐν τῇ Αἰγύπτῳ, καὶ ἀνθρώπου φωνῆς
ἐπαΐει. εἰ δέ τις αὐτὸν ὀνειδίζων δοῦλον εἴποι, ὁ
δὲ ὀργίζεται· καὶ εἴ τις ὄκνον καλέσειεν αὐτόν, ὁ
δὲ βρενθύεται καὶ ἀγανακτεῖ, ὡς καὶ ἐς τὸ ἀγεννὲς
σκωπτόμενος καὶ ἐς ἀργίαν εὐθυνόμενος.

37. Εἰ κατέχοι τις ὀπὸν Κυρηναῖον καὶ λάβοιτο
τῆς νάρκης, ἐνταῦθα δήπου τὸ ἐξ αὐτῆς πάθος
ἐκπέφευγε. δράκοντα δὲ θαλάττιον εἰ ἀνασπάσαι

---

[1] ἄρξαντι καὶ ἐπιβουλεύοντι.   [2] Schn : εἰσιν.

[a] 'This is no Heron but some other bird' (Thompson, *Gk. birds*, s.v.).

more than a sober, educated man would be, yet it will not retire and give way before an aggressor. While all other birds are on terms of peace with the Swan, the Eagle has frequently attacked it, as Aristotle says [*HA* 610 a 1, 615 b 1], though it has never yet overcome it, but has always been defeated not only through the strength of the Swan in battle but also because in defending itself the Swan has justice on its side.

35. The Heron is a great eater of oysters and swallows them when closed,[a] as pelicans swallow mussels. And the Heron warms the oysters a little in what is called its ' crop ' and retains them there. Under the influence of the heat the oysters open, and the Heron becoming aware of this, disgorges the shells but retains the flesh ; and it lives by consuming entire, thanks to a strong digestion, all that passes down into it. **The Heron and oysters**

36. There is a bird called *Asterias* (starling ?),[b] and in Egypt, if tamed, it understands human speech. And if anyone by way of insult calls it ' slave,' it gets angry ; and if anyone calls it ' skulker,' it takes umbrage and is annoyed, as though it was being jeered at for its low birth and rebuked for its indolence. **The ' Asterias '**

37. If a man with the juice of silphium on his hands seizes the Torpedo, he avoids the pain which it inflicts. And should you attempt to draw the Great **The Torpedo** **The Great Weever**

[b] Thompson (*Gk. birds*, s.v. ἀστερίας) records *Bittern* as a common but unsatisfactory interpretation, but offers no other.

τῇ δεξιᾷ ἐθέλοις, ὁ δὲ οὐχ ἕψεται, ἀλλὰ μαχεῖ-
ται ¹ κατὰ κράτος· εἰ δὲ τῇ ἀριστερᾷ ἀνάγοις,²
εἴκει καὶ ἑάλωκεν.

38. Χάρμιδος ἀκούω τοῦ Μασσαλιώτου λέγοντος
φιλόμουσον μὲν εἶναι τὴν ἀηδόνα, ἤδη δὲ καὶ
φιλόδοξον. ἐν γοῦν ταῖς ἐρημίαις ὅταν ᾄδῃ πρὸς
ἑαυτήν, ἁπλοῦν τὸ μέλος καὶ ἄνευ κατασκευῆς τὴν
ὄρνιν ᾄδειν· ὅταν δὲ ἁλῷ καὶ τῶν ἀκουόντων μὴ
διαμαρτάνῃ, ποικίλα τε ἀναμέλπειν καὶ τακερῶς
ἑλίττειν τὸ μέλος. καὶ Ὅμηρος δὲ τοῦτό μοι
δοκεῖ ὑπαινίττεσθαι λέγων

ὡς δ' ὅτε Πανδαρέου κούρη χλωρηὶς ἀηδών
καλὸν ἀείδῃσιν ἔαρος νέον ἱσταμένοιο,
δενδρέων ἐν πετάλοισι καθεζομένη πυκινοῖσιν,
ἥ τε θαμὰ τρωπῶσα χέει πολυηχέα φωνήν.

ἤδη μέντοι τινὲς καὶ πολυδευκέα φωνὴν γρά-
φουσι τὴν ποικίλως μεμιμημένην, ὡς τὴν ἀδευκέα
τὴν μηδ' ὅλως ἐς μίμησιν παρατραπεῖσαν.

39. Λέγει Δημόκριτος τῶν ζῴων μόνον τὸν
λέοντα ἐκπεπταμένοις τίκτεσθαι τοῖς ὀφθαλμοῖς,
ἤδη τρόπον τινὰ τεθυμωμένον καὶ ἐξ ὠδίνων
δρασείοντά τι γεννικόν. ἐφύλαξαν δὲ ἄλλοι καὶ
καθεύδων ὅτι κινεῖ τὴν οὐράν, ἐνδεικνύμενος ὡς
τὸ εἰκὸς ὅτι μὴ πάντῃ ἀτρεμεῖ, μηδὲ μὴν κυκλω-
σάμενος αὐτὸν καὶ περιελθὼν ὁ ὕπνος καθεῖλεν,
ὥσπερ οὖν καὶ τῶν ζῴων τὰ λοιπά. τοιοῦτόν τι
φυλάξαντας Αἰγυπτίους ὑπὲρ αὐτοῦ κομπάζειν
φασὶ λέγοντας ὅτι κρείττων ὕπνου λέων ἐστὶν

---

¹ μάχεται.   ² ἄγοις.

Weever from the sea with your right hand, it will not come but will fight vigorously. But if you haul it up with your left hand, it yields and is captured.

38. From a statement of Charmis of Massilia I learn that the Nightingale is fond of music, and even fond of fame. At any rate when it is singing to itself in lonely places, he says, its melody is simple and spontaneous. But in captivity when it has no lack of hearers it lifts up its voice, warbling and trilling its melting music. And Homer seems to me to hint as much when he says [*Od.* 19. 518] *The Nightingale*

' And as when the daughter of Pandareus, the greenwood Nightingale, sings sweet at the first oncoming of spring, as she rests amid the thick leafage of the trees, and ever varying her note pours forth her full-throated music.'

But there are those who write πολυδευκέα φωνήν, that is, ' variously imitating music,' just as ἀδευκέα signifies ' unadapted for imitating.'

39. Democritus asserts that the Lion alone among animals is born with its eyes open [a] and from the hour of birth is already to some extent angry and ready to perform some spirited action. And others have observed that even when asleep the Lion moves his tail, showing, as you might expect, that he is not altogether quiescent, and that, although sleep has enveloped and enfolded him, it has not subdued him as it does all other animals. The Egyptians, they say, claim to have observed in him something of this kind, asserting that the Lion is superior to sleep *The Lion*

[a] See 4. 34.

ἀγρυπνῶν ἀεί. ταύτῃ τοι καὶ ἡλίῳ ἀποκρίνειν
αὐτὸν αὐτοὺς πέπυσμαι· καὶ γάρ τοι καὶ τὸν
ἥλιον θεῶν ὄντα φιλοπονώτατον ἢ ἄνω [1] τῆς γῆς
ὁρᾶσθαι ἢ τὴν κάτω πορείαν ἰέναι μὴ ἡσυχάζοντα.
Ὅμηρόν τε μάρτυρα Αἰγύπτιοι ἐπάγονται λέγοντα
ἠέλιόν τ' ἀκάμαντα. ἔστι δὲ πρὸς τῇ ῥώμῃ καὶ
συνετὸς ὁ λέων. ταῖς γοῦν βουσὶν ἐπιβουλεύει
νύκτωρ φοιτῶν ἐς τὰ αὔλια. Ὅμηρος δὲ ἄρα
ᾔδει καὶ τοῦτο λέγων

βόες ὥς [2]

ἅς τε [3] λέων ἐφόβησε μολὼν ἐν νυκτὸς ἀμολγῷ.

καὶ ἐκπλήττει μὲν ὑπὸ τῆς ἀλκῆς πάσας,[4] μίαν δὲ
ἐξαρπάσας ἔδει.[5] ὅταν δὲ ἐς κόρον ἐμπλησθῇ,
βούλεται μὲν ταμιεύσασθαι καὶ ἐς αὖθις, αἰδὼς δὲ
ἴσχει αὐτὸν φρουρεῖν παραμένοντα, ὡς τροφῆς
χρῄζει λιμὸν δεδιότα. οὐκοῦν περιχανὼν ἐμπνεῖ
μὲν τοῦ καθ' ἑαυτὸν ἄσθματος, καὶ τούτῳ τὴν
φυλακὴν ἐπιτρέπει, ἀπαλλάττεταί γε μὴν αὐτός·
τὰ δὲ ἄλλα ζῷα ἥκοντα καὶ αἰσθανόμενα ὅτου
λείψανόν ἐστι τὸ κείμενον, οὐ τολμᾷ προσάψασθαι,
ἀλλὰ ἀπαλλάττεται δεδιότα δοκεῖν συλᾶν καὶ
περικόπτειν τι τοῦ σφετέρου βασιλέως. τῷ δὲ
ἄρα εἰ μὲν εὐθηρία [6] γένοιτο καὶ εὐερμία, λήθην
τοῦ πρώτου λαμβάνει καὶ ὡς ἕωλον ἀτιμάσας
ἀπαλλάττεται· εἰ δὲ μή, ὡς ἐπ' οἰκεῖον θη-
σαύρισμα παραγίνεται. ὅταν δὲ ὑπερπλησθῇ, κενοῖ
ἑαυτὸν ἡσυχίᾳ καὶ ἀσιτίᾳ, ἢ αὖ πάλιν πιθήκῳ

---

[1] κατὰ τὸ ἢ ἄνω. [2] βόες ὥς MSS omit.
[3] ὡς δ' ὅτε.
[4] ἁπάσας.
[5] ἔδει· ὁ αὐτὸς λέγει ποιητὴς ταῦτα.

and for ever awake. And I have ascertained that
it is for this reason that they assign him to the sun,
for, as you know, the sun is the most hard-working
of the gods, being visible above the earth or pursuing
his course beneath it without pause. And the
Egyptians cite Homer as a witness when he speaks
of the ' untiring sun ' [*Il.* 18. 239]. And in addition
to his strength the Lion shows intelligence. For instance, he has designs upon cattle and goes to their
folds by night. Now Homer was aware of this when
he said [*Il.* 11. 172]:

> ' Like cattle which a lion has scared, coming in
> the dead of night.'

And he strikes terror into them all by his strength,
but seizes only one and devours it. And when he and his prey
has gorged himself, he wishes to preserve the remains for another occasion, yet he is ashamed to
stay and watch over them, as though he were afraid
of starving from want of food. Accordingly with
jaws agape he breathes upon them and trusts to his
breath to guard them while he himself goes on his
way. But when the other beasts arrive and realise
to whom the remains upon the ground belong, they
do not venture to touch them but go their way for
fear of seeming to rob and diminish anything that
belongs to their king. Now if the Lion chances to
be lucky and has good hunting, he forgets his former
prize, disregards it as being stale, and goes away.
Otherwise he returns to it as to a private store. And
when he has eaten more than enough, he empties
himself by lying quiet and abstaining from food, or
alternatively he catches a monkey and eats some of

---

6 εὐθηρία ἑτέρου.

περιτυχὼν καὶ τούτου φαγὼν κενοῦται τὴν γαστέρα
ταῖς ἐκείνου λαπάξας σαρξίν. ἦν δὲ ἄρα δίκαιος
ὁ λέων καὶ οἷος

ἄνδρ᾽ ἐπαμύνασθαι, ὅτε τις πρότερος [1] χαλεπήνῃ.

τῷ γοῦν ἐπιόντι ἀνθίσταται ⟨καὶ⟩[2] τὴν ἀλκαίαν
ἐπισείων καὶ ἑλίττων κατὰ τῶν πλευρῶν εἶτα
ἐγείρει ἑαυτὸν ὥσπερ οὖν ὑποθήγων μύωπι. τόν
γε μὴν βαλόντα μέν, οὐ τυχόντα δὲ τῇ ἴσῃ ἀμυνού-
μενος [3] φοβεῖ μέν, λυπεῖ δὲ οὐδὲ ἕν. ἡμερωθείς
γε μὴν ἐξέτι νεαροῦ πρᾱότατός ἐστι καὶ ἐντυχεῖν
ἡδύς, καὶ ἔστι φιλοπαίστης, καὶ πᾶν ὅ τι οὖν
ὑπομένει πρᾴνως τῷ τροφεῖ χαριζόμενος. Ἄννων
γοῦν λέοντα εἶχε σκευαγωγόν, καὶ Βερενίκη λέων
πρᾶος συνῆν, τῶν κομμωτῶν [4] διαφέρων οὐδὲ ἕν.
ἐφαίδρυνε γοῦν τῇ γλώττῃ [5] τὸ πρόσωπον αὐτῆς,
καὶ τὰς ῥυτίδας ἐλέαινε, καὶ ἦν ὁμοτράπεζος,
πρᾴως τε καὶ εὐτάκτως ἐσθίων καὶ ἀνθρωπικῶς.
⟨καὶ⟩[6] Ὀνόμαρχος δὲ ὁ Κατάνης τύραννος καὶ ὁ
Κλεομένους υἱὸς συσσίτους εἶχον λέοντας.

40. Εὐωδίας τινὸς θαυμαστῆς τὴν πάρδαλιν
μετειληχέναι φασίν, ἡμῖν μὲν ἀπορρήτου, αὐτὴ δὲ
οἶδε τὸ πλεονέκτημα τὸ οἰκεῖον, καὶ μέντοι καὶ
τὰ ἄλλα ζῷα συνεπίσταται τοῦτο ἐκείνῃ, καὶ

---

[1] πρότερον. [2] ⟨καὶ⟩ add. Schn.
[3] ἀμυνόμενος. [4] Pierson : κομμώντων.
[5] γλώττῃ ἡσυχῇ. [6] ⟨καὶ⟩ add. H.

---

[a] Hanno, Carthaginian general, 3rd cent. B.C. Cp. Plut.
Mor. 799 E.

it, voiding and emptying his belly by means of its
flesh.

The Lion is after all upright and one to

' defend himself against the man who should assail
him first ' [Hom. *Il.* 24. 369; *Od.* 16. 72].

Thus, he faces his attacker and by lashing with his
tail and winding it about his flanks rouses himself as
though he were stimulating himself with a spur.
And if a man shoot at him but miss him, he will
defend himself by a fair return: he will scare the
man but do him no harm. If he has been domesticated since the time when he was a cub, he is ex- The Lion
tremely gentle and agreeable to meet, and is fond of tamed
play, and will submit with good temper to any treat-
ment to please his keeper. For instance, Hanno [a]
kept a Lion to carry his baggage; a tame Lion was
the companion of Berenice [b] and was no different
from her tiring-slaves: for example, it would softly
wash her face with its tongue and smooth away her
wrinkles; it would share her table and eat in a
sober, orderly fashion just like a man. And Ono-
marchus, the Tyrant of Catana, and the son of
Cleomenes [c] both had Lions with them as table-
companions.

40. They say that the Leopard has a marvellous The Leopard
fragrance about it. To us it is imperceptible, though
the Leopard is aware of the advantage it possesses, and
other animals besides share with it this knowledge.

[b] Which of the various queens named Berenice is here
referred to, is uncertain; if the queen of Ptolemy III, she
lived *c.* 273–226 B.C.

[c] Nothing more is known of these persons.

ἀλίσκεταί οἱ [1] τὸν τρόπον τοῦτον. ἡ πάρδαλις
τροφῆς δεομένη ἑαυτὴν ὑποκρύπτει ἢ λόχμῃ πολλῇ
ἢ φυλλάδι βαθείᾳ, καὶ ἐντυχεῖν ἐστιν ἀφανής,
μόνον δὲ ἀναπνεῖ. οὐκοῦν οἱ νεβροὶ καὶ ⟨αἱ⟩ [2]
δορκάδες καὶ οἱ αἶγες οἱ ἄγριοι [3] καὶ τὰ τοιαῦτα
τῶν ζῴων ὡς ὑπό τινος ἴυγγος τῆς εὐωδίας
ἕλκεται, καὶ γίνεται πλησίον· ἡ δὲ ἐκπηδᾷ καὶ
ἔχει τὸ θήραμα.

41. Πυνθάνομαι τῶν ζῴων τὰ μηρυκάζοντα
τρεῖς ἔχειν κοιλίας, καὶ ὀνόματα αὐτῶν ἀκούω
κεκρύφαλον ἐχῖνον ἤνυστρον. σηπίαι δὲ καὶ τευθί-
δες δύο νέμονται προβοσκίσιν· οὐ γάρ τοι [4]
χεῖρον οὕτως ὀνομάσαι καὶ ἐκ τῆς χρείας καὶ ἐκ
τοῦ σχήματος ἐπαρθέντα. καὶ ὅταν ᾖ χειμέρια
καὶ κλύδων τεταραγμένος, αἱ δὲ τῶν πετρῶν
λαμβάνονται ταῖς αὐταῖς προβολαῖς, καὶ ἔχονται
ὡς ἀγκύραις πάνυ ἐγκρατῶς, καὶ ἄσειστοί τε καὶ
ἄκλυστοι μένουσιν· εἶτα εἰ γένοιτο ὑπεύδια,
ἀπολύουσί τε ἑαυτὰς καὶ ἐλευθεροῦσι, καὶ νέουσι
πάλιν, εἰδυῖαι μάθημα οὐκ εὐκαταφρόνητον, χειμῶνος φυγὴν καὶ ἐκ τῶν κινδύνων σωτηρίαν.

42. Εἴ σοι βουλομένῳ μαθεῖν ἐστι μελιττῶν
ὀνόματα, οὐκ ἂν βασκήναιμι εἰπεῖν ὅσα πέπυσμαι.
ἡγεμόνες καλοῦνταί τινες καὶ ἄλλαι σειρῆνες καὶ
ἐργοφόροι [5] τινὲς καὶ ἕτεραι πλάστιδες. Νίκανδρος
δὲ † εὐφορεῖν † [6] τοὺς κηφῆνάς φησι. περὶ δὲ τὴν

---

[1] ἐκείνη . . . οἱ] τῇ παρδάλει καὶ ἀλίσκεται ἐκείνη.
[2] ⟨αἱ⟩ add. H.   [3] αἱ αἶγες αἱ ἄγριαι.
[4] τι.   [5] ὑδροφόροι H.
[6] ἀφορεῖν Post, ὑδροφορεῖν Reiske, H, εὐπορεῖν OSchn.

and the Leopard catches them in the following manner. When the Leopard needs food it conceals itself in a dense thicket or in deep foliage and is invisible; it only breathes. And so fawns and gazelles and wild goats and suchlike animals are drawn by the spell, as it were, of its fragrance and come close up. Whereat the Leopard springs out and seizes its prey.

41. I learn that ruminants have three [a] stomachs, and their names, I gather, are κεκρύφαλον (the second stomach, *reticulum*), ἐχῖνος (the third stomach, manyplies), and ἤνυστρον (the fourth stomach, *abomasum*). **Ruminants and their stomachs**

Cuttle-fish and Squids feed themselves with two 'probosces.' (There is no harm in so styling them: their use and their form induce one to do so.) And in stormy weather when there is broken surf, these creatures grip the rocks with their tentacles and cling fast as with anchors, and there they stay, safe from shock and sheltered from the waves. Later, when it grows calm, they let themselves go and are free again to swim about, having learnt what is by no means to be despised, viz., how to avoid a storm and to escape from danger. **Cuttle-fish and their tentacles**

42. If it is your wish to learn the names of Bees, I would not grudge you the knowledge that I have acquired. Some are called 'captains,' others 'sirens,' [b] some again 'workers,' and others 'moulders.' And Nicander says [*fr.* 93] that the Drones **Bees, their various names**

---

[a] Cp. Arist. *HA* 507 b 1; Ael. has omitted to mention the κοιλία μεγάλη, big stomach or paunch.

[b] Thompson on Arist. *HA* 623 b 11 takes 'siren' to be 'some species of the solitary wasp, e.g. *Eumenes, Synagris*, etc.'

τῶν Καππαδόκων γῆν ἄνευ κηρίων τὸ μέλι τὰς
μελίττας ἐργάζεσθαί φασι, παχὺ δὲ εἶναι τοῦτο
κατὰ τὸ ἔλαιον λόγος ἔχει. ἐν Τραπεζοῦντι δὲ τῇ
Ποντικῇ ἐκ τῆς πύξου γίνεσθαι μέλι πέπυσμαι,
βαρὺ δὲ τὴν ὀσμὴν τοῦτο εἶναι, καὶ ποιεῖν μὲν
τοὺς ὑγιαίνοντας ἔκφρονας, τοὺς δὲ ἐπιλήπτους ἐς
ὑγίειαν ἐπανάγειν αὖθις. ἐν Μηδίᾳ δὲ ἀποστάζειν
τῶν δένδρων ἀκούω μέλι, ὡς Εὐριπίδης [1] ἐν τῷ
Κιθαιρῶνί φησιν ἐκ τῶν κλάδων γλυκείας σταγόνας
ἀπορρεῖν. γίνεσθαι δὲ καὶ ἐν Θράκῃ μέλι ἐκ τῶν
φυτῶν ἤκουσα. ἐν δὲ Μυκόνῳ μέλιττα οὐ γίνεται,
ἀλλὰ καὶ ⟨ἔξωθεν⟩ [2] κομισθεῖσα ἀποθνήσκει.

43. Περὶ τὸν Ὕπανιν ποταμὸν γίνεσθαι τὸ ζῷον
τὸ μονήμερον οὕτω καλούμενον Ἀριστοτέλης
φησί, τικτόμενον μὲν ἅμα τῷ κνέφει, [3] ἀποθνῆσκον
δὲ ἐπὶ δυσμὰς ἡλίου τρεπομένου.

44. Ἔχει δὲ ⟨τὸ⟩ [4] δῆγμα ἡ σηπία ἰῶδες καὶ
τοὺς ὀδόντας ἰσχυρῶς ὑπολανθάνοντας. ἦν δὲ
ἄρα δηκτικὸν καὶ ⟨ὁ⟩ [5] ὀσμύλος καὶ ὁ πολύπους·
καὶ δάκοι μὲν ἂν οὗτος σηπίας βιαιότερον, τοῦ δὲ
ἰοῦ μεθίησιν ἧττον.

ὁ Εὐ. ταῖς Βάκχαις.   [2] ⟨ἔξωθεν⟩ add. H.
κνέφᾳ.
⟨τό⟩ add. H.

. . . And they say that all over Cappadocia the Bees produce honey without combs, and the story goes that it is thick like oil. I am informed that at Trapezus in Pontus honey is obtained from box-trees, but that it has a heavy scent and drives healthy people out of their senses, but restores the frenzied to health. I learn that in Media [a] honey drips from the trees, just as Euripides [*Bacc.* 714] says that on Cithaeron sweet drops flow from the boughs. In Thrace too I have heard that honey is produced from plants. On Myconus [b] there are no bees, and moreover if imported from outside they die.

Honey of various kinds

43. Aristotle says [*HA* 552 b 20] that on the banks of the river Hypanis [c] there occurs a creature that goes by the name of ' day-fly,' [d] because it is born in the morning twilight and dies when the sun begins to set.

The ' Day-fly'

44. The Cuttle-fish has a poisonous bite and teeth that are concealed very deep within. It seems also that the Osmylus [e] and the Octopus are given to biting. And the Octopus has a more powerful bite than the Cuttle-fish, although it emits less poison.

The Cuttle-fish

---

[a] Ael. is copying [Arist.] *Mir.* 831 b 26 where the MSS read Λυδίᾳ.

[b] One of the Cyclades.

[c] Mod. Boug.

[d] ' A May-fly, probably . . . the large *Ephemera longicauda* Oliv.' (Thompson on Arist. *loc. cit.*). .

[e] ' A kind of octopus with an unpleasant musky smell : *Eledone moschata* ' (Thompson, *Gk. fishes*).

---

[5] ⟨ὁ⟩ add. H.

AELIAN

45. Τὸν σῦν τὸν ἄγριόν φασι μὴ πρότερον ἐπί
τινα φέρεσθαι πρὶν ἢ τοὺς χαυλιόδοντας ὑποθῆξαι·
μαρτυρεῖ δὲ ἄρα καὶ Ὅμηρος τοῦτο λέγων

θήξας λευκὸν ὀδόντα μετὰ γναμπτῇσι γένυσσιν.

παχύνεσθαι δὲ τὸν σῦν ἀκούω μάλιστα μὴ λούμε-
νον,[1] ἀλλὰ ἐν τῷ βορβόρῳ διατρίβοντά τε καὶ
στρεφόμενον καὶ πίνοντα ὕδωρ τεθολωμένον, καὶ
ἡσυχίᾳ καὶ στέγῃ σκοτωδεστέρᾳ χαίροντα καὶ
τροφαῖς ὅσαι φυσωδέστεραί τέ εἰσι καὶ ὑποπλῆσαι
δύνανται. καὶ Ὅμηρος δὲ ἔοικε ὑποδηλοῦν ταῦτα.
περὶ μὲν οὖν τοῦ καλινδεῖσθαι αὐτοὺς[2] καὶ
φιληδεῖν τοῖς ῥυπαρωτέροις τέλμασι . . .[3] λέγων
σύες χαμαιευνάδες· ὅτι δὲ τῷ τεθολωμένῳ ὕδατι
πιαίνονται . . .[4] φησὶ

μέλαν ὕδωρ
πίνουσαι, τά θ᾽ ὕεσσι τρέφει τεθαλυῖαν ἀλοιφήν.

ὅτι δὲ χαίρουσι τῷ σκότῳ διὰ τούτων ἐλέγχει

πέτρῃ ὕπο γλαφυρῇ εὗδον Βορέω ὑπ᾽ ἰωγῇ.

τὸ δὲ φυσῶδες αἰνίττεται τῆς τροφῆς ὅταν λέγῃ
βάλανον μενοεικέα ἐσθίειν αὐτάς. εἰδὼς δὲ ἄρα
Ὅμηρος ὡς καὶ ἰσχναίνεται καὶ ἐπιτρίβει τὰ κρέα
ὗς ὁρῶν τὸν θῆλυν, πεποίηκε τοὺς ἄρρενας ἰδίᾳ
καθεύδοντας καὶ τὰς θηλείας ἰδίᾳ. ἐν Σαλαμῖνι
δὲ χλωροῦ σίτου καὶ ληίου κομῶντος ἐὰν σῦς

---

[1] λουόμενον.   [2] αὐτόν.
[3] Lacuna.   [4] Lacuna.

---

[a] The chief city in Cyprus. Eustathius on Hom. *Od.* 18. 29
says that there was a law in Cyprus permitting landowners to
remove the teeth of any pig that they found foraging among

45. They say that the Wild Boar does not attack The Wild
a man until he has whetted his tusks. And Homer Boar
testifies to this when he says [*Il.* 11. 416]

'Having whetted the white tusk between his
curved jaws.'

And I learn that the Boar fattens himself chiefly
by not washing but spending his time wallowing in
the mud, drinking the turbid water, and revelling in
the quiet and the darkness of his lair and in all the
more inflating foods that can fill him up. And Homer
appears to imply as much, for touching their wallowing and their fondness for the more muddy ponds . . .
when he says [*Od.* 10. 243] ' hogs that make their
bed upon the ground.' And that they fatten themselves upon turbid water . . . he says [*Od.* 13. 409]

' drinking black water, which fosters the rich fat on
swine.'

And that they delight in darkness he proves in the
following words [*Od.* 14. 533] :

' They slumbered beneath a hollow rock under
shelter from Boreas.'

And he hints at the inflating quality of their food
when he says [*Od.* 13. 409] that they eat ' the satisfy-
ing acorn.' Now Homer knowing that the Boar
grows thin and that his flesh wastes if he looks at the
Sow, has described [*Od.* 14. 13] the Boars as sleeping
in one place and the Sows in another. In Salamis ^a
if a Sow breaks in and grazes the corn when green or

their crops. So Irus threatens to knock out the teeth of
Odysseus, disguised and unknown, whom he regards as an
interloper in the palace in Ithaca.

ἐμπεσοῦσα [1] ἀποκείρῃ, νόμος ἐστὶ Σαλαμινίων
τοὺς ὀδόντας ἐκτρίβειν αὐτῆς. καὶ τοῦτο εἶναι τὸ
παρ' Ὁμήρῳ συὸς ληιβοτείρης φασίν. οἱ δὲ
ἑτέρως νοοῦσι, καὶ λέγουσι χλωροῦ σίτου τὴν ὗν
γευσαμένην ἀσθενεῖς ἔχειν τοὺς ὀδόντας.

46. Ἔδωκε δὲ ἄρα ἡ φύσις ταῖς κυσὶ τραυμά-
των ἀντίπαλον πόαν. εἰ δὲ ἕλμινθες αὐτὰς λυ-
ποῖεν,[2] τοῦ σίτου τὸ καλούμενον λήιον ἐσθίουσαι
ἐκκρίνουσιν αὐτάς. λέγονται δὲ καὶ ὅταν δέωνται
τὴν γαστέρα ἑκατέραν κενῶσαι πόαν τινὰ ἐσθίειν,
καὶ τὸ μέν τι τῆς τροφῆς τὸ ἐπιπολάζον ἀνεμεῖν,
τὰ δὲ περιττὰ κάτωθεν ἐκκρίνεσθαι αὐταῖς φασιν.
ἐντεῦθεν καὶ τὸ συρμαΐζειν Αἰγύπτιοι λέγονται
μαθεῖν. πέρδικες δὲ [3] καὶ πελαργοὶ τρωθέντες
καὶ φάτται τὴν ὀρίγανον, ὡς λόγος, διατρώγουσιν,
εἶτα τοῖς τραύμασιν ἐντιθέντες ἀκοῦνται τὸ σῶμα
καὶ μέντοι ⟨καὶ⟩[4] τῆς ἀνθρώπων ἰατρικῆς δέονται
οὐδὲ ἕν.

47. Οὐ δεήσομαι ἐνταῦθα μάρτυρος πρεσβυτέ-
ρου, ἃ δὲ αὐτὸς ἔγνων ἐρῶ.[5] σαῦρον τῶν χλωρῶν
μὲν ὑπεράγαν, ἁδροτέρων δὲ τὴν ἕξιν συλλαβὼν
ἀνὴρ καὶ κέντρῳ πεποιημένῳ χαλκοῦ πείρας [6]
εἶτα τυφλώσας τὸν σαῦρον καὶ χύτραν κεραμέαν
τῶν νεωστὶ εἰργασμένων διατρήσας πάνυ λεπταῖς
ὀπαῖς, ὡς μὴ εἴργειν μὲν τὸ πνεῦμα, οὐ μὴν
ἐκείνῳ παρασχεῖν ἔκδυσιν, καὶ γῆν ἐγχέας καὶ

---

[1] *Barnes* : πεσοῦσα.   [2] λυποῦσι.
[3] τε.
[4] ⟨καὶ⟩ add. H.

a field of waving corn, there is a law of the Salaminians that her teeth must be destroyed. And they say that the passage in Homer [*Od.* 18. 29] about ' a sow that consumes the crops ' refers to this. Others take a different view and assert that when a Sow has tasted green corn its teeth are weakened.

46. It would appear that Nature has provided grass as a remedy for the wounds of Dogs. And if they are troubled with worms they get rid of them by eating ' standing ' corn, as it is called. And when they need to empty both stomachs *a* they are said to eat some grass, and as much of their food as remains undigested they vomit up, while the remainder is excreted. It is from this source that the Egyptians are said to have learnt the practice of taking purges. But Partridges, Storks, and Ring-doves, when wounded are said to chew marjoram and then to spread it on their wounds and cure their body; and they have no need at all of man's healing art. *Nature's medicines for animals*

47. In this matter I shall have no need of any witness from antiquity but shall narrate what I myself have seen and know.

A man captured a Lizard of the excessively green and unusually large species, and with a point made of bronze he pierced and blinded the Lizard. And after boring some very fine holes in a newly fashioned earthenware vessel so as to admit the air, but small enough to prevent the creature from escaping, he *A Lizard, blinded, regains its sight*

a The expression is used loosely to denote the stomach proper and the intestines, for the dog has but one stomach.

---

5 λέγω.   6 διείρας.

μάλα ἔνδροσον, καὶ τὸ θηρίον ἐμβαλὼν καὶ πόαν [1]
τινὰ ἧς οὐκ εἶπε τὸ ὄνομα καὶ δακτύλιον σιδήρου [2]
πεποιημένον καὶ ἔχοντα λίθον Γαγάτην, ᾧπερ οὖν
ἐνείργαστο γλύμμα σαῦρος, τὴν μὲν χύτραν
ἐπηλύγασεν, ἐννέα ἐμπλάσας σημεῖα, ὧν ἀφῄρει
σφραγῖδα [3] ἐφ᾽ ἡμέρας ἐννέα. καὶ τὴν ἐπὶ
πάσαις [4] ἀφανίσας ἀνοίγει τὸ σκεῦος, καὶ ἔγωγε
εἶδον τὸν σαῦρον ἐμβλέποντα, καὶ εὐωποτάτους [5]
τοὺς ὀφθαλμοὺς τοὺς τέως πεπηρωμένους εἶχε.
καὶ τὸν [6] μέν, ἔνθεν ᾑρέθη, ἐνταῦθα ἀπελύσαμεν,
δακτύλιον δὲ ἐκεῖνον ὁ ἀνὴρ ὁ ταῦτα δράσας
ὀφθαλμοῖς ἀγαθὸν ἔφασκεν εἶναι.

48. Ἐμοὶ δὲ αἴσχιστον δοκεῖ, ὦ ἄνθρωποι,
φιλίαν μὲν τοῖς ζῴοις πρὸς ἄλληλα εἶναι, μὴ
μόνοις τοῖς συννόμοις αὐτῶν μηδὲ μὴν τοῖς ὁμογε-
νέσιν, ἤδη δὲ καὶ τοῖς μηδὲν προσήκουσί σφισι
κατὰ τὸ κοινὸν γένος. ταῖς γοῦν αἰξὶν αἱ οἶς
φίλιαι, περιστερᾷ δὲ πρὸς τρυγόνα φιλία,[7] φίλα
δὲ ἀλλήλοις [8] νοοῦσι φάτται τε καὶ πέρδικες,
ἀλκυόνα δὲ καὶ κηρύλον ποθοῦντε ἀλλήλω πάλαι
ἴσμεν, κορώνην τε ἐρωδιῷ φίλα νοεῖν καὶ λάρον
τῷ καλουμένῳ κολοιῷ καὶ ἰκτίνῳ ἄρπην. πολε-
μοῦσι δὲ αἰώνιον πόλεμον καὶ ἄσπονδον ὡς εἰπεῖν
κορῶναί τε καὶ γλαῦκες· πολέμιοι δὲ ἄρα εἰσὶν
ἰκτῖνός τε καὶ κόραξ, καὶ πυραλλὶς πρὸς τρυγόνα,
καὶ βρένθος καὶ λάρος,[9] πάλιν τε ὁ χλωρεὺς πρὸς

---

[1] ἔκδυσιν . . . πόαν] ἔκδυσιν, τὸ θηρίον ἐμβαλὼν καὶ γῆν
ὑποχέας καὶ μ. ἐ. καὶ πόαν.

[2] Ges : σιδηροῦν.

[3] ⟨μίαν⟩ σφρ.? H.

[4] πάσαις τὴν ἐννάτην.

[5] εὐωποτέρους.

[6] τό.

[7] περιστερὰ . . . φίλη.

heaped some very moist earth into it and put the
Lizard inside together with a certain herb, of which
he did not divulge the name, and an iron ring with
a bezel of lignite engraved with the figure of a lizard.
After stamping nine seals upon the vessel he then
covered it up, removing one seal daily for nine days.
And when he had destroyed the last seal of all he
opened the vessel, and I myself saw the Lizard
having its sight and its eyes, which till then had been
blinded, seeing perfectly well. And we released the
Lizard on the spot where it had been captured, and
the man who had done these things asserted that that
ring of his was good for the eyes.

48. It fills me with shame, you human beings, to
think of the friendly relations that subsist between
animals, not only those that feed together nor even
those of the same species, but even between those
that have no connexion through a common origin.
For instance, Sheep are friends with Goats; there is
friendship between Pigeon and Turtle-dove; Ring-
doves and Partridges entertain friendly feelings towards one another; we have long known that the
Halcyon and the Ceryl desire each other; that the
Crow is friendly disposed towards the Heron, and the
Sea-mew towards the Little Cormorant, as it is called,
and the Shearwater towards the Kite. But there is
war everlasting and without truce, so to say, between
Crows and Owls. Enemies too are the Kite and the
Raven, the Pyrallis and the Turtle-dove, the Bren-
thus [a] and the Sea-mew, and again the Greenfinch(?)

[a] Unknown water-bird. Perh. the ' Avocet,' Gossen § 187.

---

[8] εἰς ἀλλήλους.　　　　[9] *Ges.*: πάγρος.

τρυγόνα, καὶ αἰγυπιοὶ καὶ ἀετοί, καὶ κύκνοι καὶ
δράκοντες, καὶ πρὸς βουβαλίδας καὶ ταύρους [1]
λέοντες. ἔχθιστα [2] δὲ ἄρα ἐλέφας καὶ δράκων
ἦν, καὶ πρὸς ἀσπίδα ὁ ἰχνεύμων, ὁ δὲ αἴγιθος τῷ
ὄνῳ· ὁ μὲν γὰρ ὠγκήσατο, ῥήγνυται δὲ τῷ
αἰγίθῳ τὰ ᾠά, καὶ οἱ νεοττοὶ ἐκπίπτουσιν ἀτελεῖς·
ὁ δὲ τιμωρῶν τοῖς τέκνοις ἐπιπηδᾷ τῶν ὄνων τοῖς
ἕλκεσι, καὶ ἐσθίει αὐτά. μισεῖ δὲ ἀλώπηξ κίρκον
καὶ ταῦρος κόρακα, καὶ ὁ ἄνθος [3] τὸν ἵππον.
χρὴ δὲ εἰδέναι τὸν πεπαιδευμένον καὶ μηδὲν
μάτην ἀκούοντα ὅτι καὶ δελφὶς φαλλαίνῃ διάφορος,
λάβρακές γε μὴν κεστρεῦσι, μύραιναι δὲ γόγγροις,
καὶ ἄλλα ἄλλοις.

49. Αἱ ἄρκτοι τῶν θηρατῶν τοὺς ἐς [4] στόμα
πεσόντας καὶ τὸ πνεῦμα ἐς ἑαυτοὺς ὤσαντας
ὀσφρησάμεναι ὡς νεκροὺς παραλιμπάνουσι, καὶ
δοκεῖ τοῦτο τὸ ζῷον νεκρὸν βδελύττεσθαι. μισοῦσι
δὲ καὶ οἱ μύες τοὺς ἐν ταῖς ἑαυτῶν διαίταις καὶ
καταδρομαῖς ἀποθανόντας, καὶ μέντοι ⟨καὶ⟩[5]
χελιδὼν ἐκβάλλει χελιδόνα νεκράν.[6] μύρμηκες δέ,
καὶ ἐκείνοις ἐκφορᾶς νεκρῶν μέλειν καὶ καθαίρειν
τοὺς σφετέρους χηραμοὺς ἡ σοφωτάτη φύσις
ἔδωκεν, ἐπεὶ καὶ τοῦτο ἴδιον τῶν ἀλόγων, τὰ
ὁμογενῆ τε καὶ ὁμοφυᾶ τεθνεῶτα τῶν ὀφθαλμῶν
ἀποφέρειν θᾶττον. λέγουσι δὲ Αἰθιόπων λόγοι
αἱμυλίας τε καὶ κόμπου Ἑλληνικοῦ ἄγευστοι ὅτι
ἄρα ἐλέφαντα θεασάμενος ἐλέφας νεκρὸν οὐκ ἂν
παρέλθοι μὴ τῇ προβοσκίδι γῆν ἀρυσάμενος καὶ

---

[1] καὶ ταύρους del. H (1876). [2] ἔχθιστον.
[3] Ges: ἄνθιος. [4] ἐπί Schn.

and the Turtle-dove, the Aegypius and the Eagle,
Swans and Water-snakes(?),[a] and Lions are the
enemies of Antelopes and Bulls. The bitterest hate
exists between the Elephant and the Python,[b] between the Asp and the Ichneumon, between the Blue
Tit and the Ass, for directly the Ass brays the Blue
Tit's eggs are smashed and the young ones are spilt,
still imperfect. And so to avenge its offspring the
Blue Tit leaps upon the Ass's sore places and feeds
on them. The Fox detests a Falcon and the Bull a
Raven, and the Buff-backed Heron the Horse. And
an educated man who attends to what he hears
should know that the Dolphin is at feud with the
Whale, the Basse too with the Mullet, and the Moray
with the Conger Eel, and so on.

49. When Bears have sniffed at hunters who have Animals'
fallen on their face and knocked the breath out of dislike of
dead bodies
themselves, they leave them for dead, and it seems
that these creatures are disgusted by a dead body.
Mice also hate those that die in their holes and lurking-
places; and a Swallow too ejects a dead Swallow
from its nest. Ants also, thanks to the supreme
wisdom of Nature, are careful to carry away dead
bodies and to cleanse their nests, for it is characteristic of brute beasts that, when one of their own
species and kind has died, they speedily remove it
out of sight. And Ethiopian histories, which are
untainted by the pretentious plausibility of the
Greeks, tell us that if one Elephant sees another The
lying dead, it will not pass by without drawing up Elephant
and its dead

---

[a] See Arist. *HA* 602 b 25.   [b] Lit. ' dragon.'

[5] ⟨καί⟩ *add. H.*   [6] νεκρὰν καὶ μέλιτται.

ἐπιβαλών, ὡς ὁσίαν τινὰ ἀπόρρητον ὑπὲρ τῆς
φύσεως τῆς κοινῆς ἐκτελῶν·[1] εἶναι γὰρ τὸ μὴ
δρᾶσαι τοῦτο ἐναγές. ἀπόχρη δέ οἱ καὶ κλάδον
ἐπιβαλεῖν, καὶ ἄπεισι τὸ κοινὸν ἁπάντων τέλος
μὴ ἀτιμάσας. ἀφῖκται δὲ λόγος ἐς ἡμᾶς καὶ
ἐκεῖνος. ὅταν ἐλέφαντες ἀποθνήσκωσιν ἐκ τραυ-
μάτων ἢ βληθέντες ἐν πολέμῳ ἢ ἐν θήρᾳ παθόντες
τοῦτο, τῆς πόας τῆς παρατυχούσης ἢ τῆς κόνεως
τῆς ἐν ποσὶν ἀνελόμενοι, ἐς τὸν οὐρανὸν ἀναβλέ-
πουσι καὶ βάλλουσί τι τῶν προειρημένων, καὶ
φωνῇ τῇ σφετέρᾳ κινύρονταί τε καὶ ποτνιῶνται,
ὥσπερ οὖν τοὺς θεοὺς μαρτυρόμενοι ἐφ' οἷς
ἐκδίκως τε καὶ ἐκνόμως ὑπομένουσιν.

50. Ἴδια δὲ ἄρα τῶν ζῴων καὶ ταύτῃ[2] δήπου
καταγνῶναι πάρεστι. τοὺς γοῦν ὄρνεις τοὺς
ἠθάδας καὶ τοὺς ἐν ποσὶ τρεφομένους τε καὶ
ἐξεταζομένους ὁρῶμεν ἵππους καὶ ὄνους καὶ βοῦς
καὶ καμήλους θαρροῦντας· εἰ δὲ καὶ ἐλέφαντί που
πράῳ καὶ ἡμέρῳ συντρέφοιντο, οἱ δὲ οὐκ ὀρρωδοῦσιν, ἀλλὰ καὶ δι' αὐτῶν ἐκείνων ἔρχονται. ἤδη
δὲ ἀλεκτρυόνες καὶ ἐπὶ τὰ νῶτα αὐτῶν ἀναπέτον-
ται· τοσοῦτον αὐτοῖς τοῦ θάρσους περίεστι καὶ
τοῦ ἀδεοῦς. πτοίαν δὲ αὐτοῖς ἐντίθησι καὶ δέος
ἰσχυρὸν γαλῆ παραδραμοῦσα. καὶ μυκήσεων μὲν
καὶ ὀγκήσεων οὐ ποιοῦνται ὥραν, κρίξασαν δὲ
ἄρα μόνον πεφρίκασι τὴν προειρημένην. χηνῶν
δὲ καὶ[3] κύκνων[4] καὶ στρουθῶν τῶν μεγάλων ἤ
τι ἢ οὐδὲν φροντίζουσιν, ἱέρακα δὲ βραχύτατον[5]
ὄντα ὀρρωδοῦσιν. ὁ δὲ ἀλεκτρυὼν ᾄσας φοβεῖ

---

[1] ἐκτελῶν καὶ φεύγων ἄγος.

some earth with its trunk and casting it upon the corpse, as though it were performing some sacred and mysterious rite on behalf of their common nature; and that to fail in this duty is to incur a curse. It is enough for it even to cast a branch upon the body; and with due respect paid to the common end of all things the Elephant goes on its way.

And there has reached us also the following story. A dying When Elephants are dying of wounds, stricken either Elephant in battle or in hunting, they pick up any grass they may find or some of the dust at their feet, and looking upwards to the heaven, cast some of these objects in that direction and wail and cry aloud in indignation in their own language, as though they were calling the gods to witness how unjustly and how wrongfully they are suffering.

50 (i). By the following cases also, I think, one may Confidence recognize traits peculiar to animals. For instance, and fear in Animals we see domestic fowls that are reared at the feet, and have experience, of horses, asses, cows, or camels, showing no fear of them. And if they are fed along with, say, a tame and gentle elephant, they are not afraid but even move about among those creatures. And cockerels even fly up on to their backs, such are their resulting courage and freedom from fear. But they are fluttered and terrified if a marten runs by. To the lowing of cattle or the braying of an ass they pay no attention; but a marten has but to chatter and they tremble. For geese, swans, and ostriches they care little or nothing, but are in terror of a hawk although it is very small. With its crowing a cock

---

² ταῦτα.   ³ τε καί.
⁴ Reiske : κυνῶν.   ⁵ βραχύτερον.

μὲν λέοντα, ἀναιρεῖ δὲ βασιλίσκον· οὐ μὴν φέρει [1]
οὔτε αἰλούρους οὔτε ἰκτίνους. αἱ δὲ περιστεραὶ
ἀετῶν μὲν κλαγγὴν καὶ γυπῶν θαρροῦσι, κίρκων
δὲ καὶ ἁλιαέτων οὐκέτι.

Ἡ δὲ ποίμνη [2] καὶ ὁ ἔριφος καὶ πῶλιον πᾶν ἐπὶ
τὰς μητρῴας θηλὰς ἔρχεται γεννηθέντα παραχρῆμα,
καὶ μέντοι καὶ τῶν οὐθάτων σπῶντα ἐμπίπλαται·
πολυπραγμονεῖ δὲ τὸ τεκὸν οὐδὲ ἕν, ἀλλὰ ἕστηκεν.
ὕπτια δὲ παραβάλλει τὰς θηλὰς τοῖς βρέφεσι τὰ
σχιζόποδα πάντα, λύκοι καὶ κύνες καὶ λέαιναι
καὶ παρδάλεις.

51. Πολυφωνότατα δὲ τὰ ζῷα καὶ πολύφθογγα
ὡς ἂν εἴποις ἡ φύσις ἀπέφηνεν,[3] ὥσπερ οὖν καὶ
τοὺς ἀνθρώπους. ὁ γοῦν Σκύθης ἄλλως φθέγγεται
καὶ ὁ Ἰνδὸς ἄλλως, καὶ ὁ Αἰθίοψ ἔχει φωνὴν
συμφυᾶ [4] καὶ οἱ Σάκαι· φωνὴ δὲ Ἑλλὰς ἄλλη, καὶ
Ῥωμαία ἄλλη. οὕτω τοι καὶ τὰ ζῷα ἄλλο ἄλλως
προΐεται τὸν συγγενῆ τῆς γλώττης ἦχόν τε καὶ
ψόφον· τὸ μὲν γὰρ βρυχᾶται, μυκᾶται δὲ ἄλλο,
καὶ χρεμέτισμα ἄλλου καὶ ὄγκησις ⟨ἄλλου⟩,[5]
ἄλλου βληχηθμός τε καὶ μηκασμός,[6] καί τισι μὲν
ὠρυγμός, τισὶ δὲ ὑλαγμὸς φίλον, καὶ ἄλλῳ [7]
ἀρράζειν· κλαγγαὶ δὲ [8] καὶ ῥοῖζοι καὶ κριγμοὶ
καὶ ᾠδαὶ καὶ μελῳδίαι καὶ τραυλισμοὶ καὶ μυρία
ἕτερα δῶρα τῆς φύσεως ἴδια τῶν ζῴων ἄλλα
ἄλλων.

52. Ἀνὰ τὴν χώραν τὴν Αἰγυπτίαν ἀσπίδες
φωλεύουσι τοῦ Νείλου πλησίον ἐπὶ τῆς ὄχθης
ἑκατέρας. καὶ τὸν μὲν ἄλλον χρόνον φιλοχωροῦσι

---

[1] οὐ φέρει μήν.   [2] *Abresch* : λίμνη.

scares a lion and is fatal to a basilisk, and yet it cannot endure cats or kites. And pigeons are not afraid at the cry of eagles and vultures, but they are at the cry of falcons and of sea-eagles.

(ii). The lamb, the kid, and every foal directly it is born goes for its dam's teats and sucks the dugs until it is full. And the parent shows no concern but stands still. Whereas all animals with parted toes, wolves, hounds, lions, leopards, lie down to give their young suck. Animals
suckling
their young

51. Nature has made animals with an immense variety of voice and of speech, as it were, even as she has men. For instance, the Scythian speaks one language, the Indian another; the Ethiopian has a natural language, so too have the Sacae; the language of Greece and that of Rome are different. And so it is with animals: each has a different way of producing the tone and the sound natural to its tongue. Thus, one roars, another lows, a third whinnies, ⟨another⟩ brays, yet another baas and bleats; while to some howling is customary, to others barking, and to another snarling. Screaming, whistling, hooting, singing, warbling, twittering, and countless other gifts of Nature are peculiar to different animals. The various
sounds made
by animals

52. In the Egyptian countryside Asps have their holes by the Nile on either bank. Most of the time they stay round about their ⟨lurking-places⟩ and are Reptiles
foretell and
avoid the
rising of the
Nile

---

3 ἀνέφηνεν.   4 συμφυῆ.
5 ⟨ἄλλου⟩ add. Gow.
6 μηκασμός, καὶ διάφορα φθέγματα.
7 τῷ ἄλλῳ.   8 τε καί.

καὶ ἀγαπῶσιν . . .[1], ὡς τὰς οἰκίας τὰς σφετέρας
οἱ ἄνθρωποι· μέλλοντος δὲ τοῦ ποταμοῦ κατὰ τὴν
ὥραν τὴν θέρειον[2] ἀναπλεῖν,[3] πρὸ τριάκοντά που
ἡμερῶν αἱ προειρημέναι ἀσπίδες μετοικίζονται ἐς
τὰ ἀπωτέρω τοῦ Νείλου χωρία, καὶ τοὺς ὄχθους
τοὺς ὑπερέχοντας ἐσέρπουσι, καὶ μέντοι καὶ τὰ
σφῶν αὐτῶν ἔκγονα ἐπάγονται, δῶρον τοῦτο ἴδιον
λαχοῦσαι παρὰ τῆς φύσεως εἰδέναι ποταμοῦ
τοσούτου καὶ οὕτως ἐργατικοῦ τὴν ἀνὰ πᾶν ἔτος
ἐπιδημίαν, καὶ τὴν ἐξ αὐτοῦ κατάληψίν τε καὶ
λύμην φυλάττεσθαι. καὶ αἱ χελῶναι δὲ καὶ οἱ
καρκίνοι καὶ οἱ κροκόδιλοι τὰ ᾠὰ κατὰ τὴν ὥραν
τὴν αὐτὴν μετακομίζουσιν ἐς τὰ ἄβατα τῷ ποταμῷ καὶ ἀνέφικτα· καὶ ἐντεῦθεν ἤδη λογίζονται
οἱ ἐντυγχάνοντες τοῖς τῶν προειρημένων ᾠοῖς ὁ
Νεῖλος ἀνελθὼν ἐς πόσον ἐπαρδεύσει[4] σφίσι τὴν
γῆν.

53. Οἱ ἵπποι οἱ ποτάμιοι τοῦ Νείλου μέν εἰσι
τρόφιμοι· ὅταν δὲ τὰ λήια ἐνακμάζῃ καὶ ὦσιν οἱ
στάχυες ξανθοί, οὐκ ἄρχονται παραχρῆμα κείρειν
αὐτοὺς καὶ ἐσθίειν, ἀλλὰ παραμείβοντες ἔξωθεν
τὸ λήιον στοχάζονται πόσον αὐτοὺς ἐμπλήσει[5]
μέτρον, εἶτα λογισάμενοι τὸ ἀποχρῆσον σφίσιν
ἐμπίπτουσι καὶ ἀναχωροῦσιν ἐπὶ πόδα ἐμπιπλάμε-
νοι, τὸ ῥεῦμα τοῦ ποταμοῦ κατὰ νώτου λαβόντες.
πεφιλοσόφηται δὲ ἄρα τοῦτο αὐτοῖς, ἵνα εἴ τινες
τῶν γεωργῶν ἐπίοιεν ἀμυνούμενοι,[6] οἱ δὲ ἐκ τοῦ
ῥᾴστου ἐς τὸ ὕδωρ καταδραμεῖν ἔχοιεν, τοὺς
πολεμίους ἀντιπροσώπους, ἀλλὰ οὐκ ὄπισθεν οἱ
ἵπποι οὗτοι δοκεύοντες.[7]

[1] Lacuna : ⟨ὑποδρομάς⟩ conj. H.

as attached to them as human beings are to their
own homes. But when in the summertime the river
threatens to overflow, the aforesaid Asps emigrate
some thirty days beforehand to districts further
away from the Nile and creep into bluffs above the
river, and, what is more, bring their young with
them : they have received from Nature this special
gift of being able to foretell the annual visitation of
a river so mighty and so active, and to guard against
being overtaken and destroyed by it. And at the
same season turtles and crabs and crocodiles transfer
their eggs to spots which the river cannot touch or
reach. Hence those who come across the eggs of
the aforesaid creatures calculate to what extent the
Nile will rise and irrigate their land.

53. Hippopotamuses are nurslings of the Nile, and The Hippo-
when the crops are ripe and the ears are yellow they potamus
do not forthwith begin to graze and eat them but
pass along outside the crop and calculate what area will
satisfy them ; and then, having reckoned how much
will be enough, they fall to, and as they fill themselves they withdraw backwards, keeping the river
behind them. Now this move they have cleverly
devised so that, should any farmers attack them in
self-defence, they can run down into the water with
complete ease, on the look out for enemies in front
of them but not looking behind them.

---

² *Anon.*: τὴν ὤ. θερείαν A, τῶν θείων *other* MSS.

³ ἀναπλεῖν, ἀναχθεῖσαι καὶ ὠθούμεναι ὑπό τε πλήθους ὕδατος
καὶ τῶν ἐτησίων ἀνέμων.

⁴ *Reiske* : εἶτα ἀρδεύσει.

⁵ ἐμπλήσειε.

⁶ ἀμυνόμενοι.   ⁷ *Ges* : δοκοῦντες.

AELIAN

54. Ἐν τῇ Μαυρουσίᾳ γῇ αἱ παρδάλεις τοῖς πιθήκοις οὐ κατὰ τὸ καρτερὸν οὐδὲ ὅπως ἂν ἔχωσιν ἀλκῆς τε καὶ ῥώμης ἐπιτίθενται.[1] τὸ δὲ αἴτιον, οὐ χωροῦσιν ὁμόσε, ἀλλὰ ἀποδιδράσκουσιν αὐτὰς καὶ ἐπὶ τὰ δένδρα ἀναθέουσι καὶ ἐκεῖ κάθηνται, τὴν ἐξ ἐκείνων ἐπιβουλὴν φυλαττόμενοι. ἦν δὲ ἄρα ἡ πάρδαλις καὶ τοῦ πιθήκου δολερώτερον. οἵας γοῦν ἐπ᾽ αὐτοῖς παλαμᾶταί τε καὶ ῥάπτει τὰς πάγας. ὅπου πλῆθος πιθήκων κάθηνται, ἐνταῦθα ἐλθοῦσα ἑαυτὴν ὑπέρριψε τῷ δένδρῳ, καὶ κεῖται κατὰ τοῦ δαπέδου ὑπτία, καὶ τὴν μὲν γαστέρα διώγκωσε, παρῆκε δὲ τὰ σκέλη, τὼ δὲ ὀφθαλμὼ κατέμυσε, πιέζει γε μὴν[2] τὸ ἆσθμα, καὶ κεῖται νεκρὰ δή. οἱ δὲ ἄνωθεν τὴν ἐχθίστην ἰδόντες τεθνάναι νομίζουσιν αὐτήν, καὶ ὃ μάλιστα βούλονται, τοῦτο καὶ οἴονται. οὐ μὴν θαρροῦσιν ἤδη, ἀλλὰ πεῖραν καθιᾶσι, καὶ ἔστιν ἡ πεῖρα, ἕνα ἑαυτῶν τὸν δοκοῦντα ἀδεέστατον[3] καταπέμπουσι, βασανίσοντα καὶ κατασκεψόμενον τὸ τῆς παρδάλεως πάθος. ὁ δὲ κάτεισιν οὐ παντελῶς ἀδεής, ἀλλὰ ὀλίγον καταδραμὼν εἶτα ὑπέστρεψεν, τοῦ φόβου ἀναστείλαντος αὐτόν· καὶ κατῆλθε πάλιν, καὶ πλησίον γενόμενος ἀνεχώρησε, καὶ ὑπέστρεψεν αὖθις, καὶ τὼ ὀφθαλμὼ κατεσκέψατο, καὶ τὸ πνεῦμα[4] εἰ μεθίησιν ἐξήτασεν. ἡ δὲ ἀτρεμοῦσα καὶ μάλα ἐγκρατῶς ἐντίθησίν οἱ τὸ κατὰ μικρὰ ἀδεές. προσελθόντος δὲ καὶ παραμένοντος ἀπαθοῦς καὶ οἱ μετέωροι πίθηκοι θαρροῦσιν ἤδη, καὶ καταδραμόντες ἔκ τε ἐκείνου τοῦ δένδρου καὶ τῶν ἄλλων ὅσα πλησίον παραπέφυκεν, ἀθρόοι

---

[1] οὕτως ἐπιτίθενται.   [2] μὴν καὶ συνέχει.
[3] ἀδεέστερον.

354

54. In Mauretania Leopards do not attack Mon- Leopard and Monkeys
keys with force nor with all the strength and power
at their command, the reason being that the Monkeys
do not face them but escape from them and run up
trees and sit there on guard against the designs of
the Leopards. Yet it seems that after all the Leopard is craftier than the Monkey, for such designs
and traps does it contrive for the Monkeys. It comes
to the place where a gathering of Monkeys is seated,
throws itself down beneath a tree, lies on the ground
on its back, inflates its belly, relaxes its legs, closes
both eyes, and even holds its breath, and lies there
like one dead. And the Monkeys looking down upon
their most hated enemy, fancy it to be dead; and
what they most fervently desire, that they believe.
For all that, they do not as yet take courage but make
an experiment, and the experiment is this: they send
down one of their number whom they regard as the
most fearless to test and to scrutinise the state of the
Leopard. So the Monkey descends not altogether
unafraid; but after running down a little way he
turns back, fear causing him to retreat. And a second
time he descends and having approached, withdraws;
and a third time he returns and observes the Leopard's eyes and examines it to see if it is breathing.
But the Leopard, by remaining motionless with the
utmost self-control, inspires a gradual fearlessness in
the Monkey. And since it approaches and remains
close by and takes no harm, the Monkeys up aloft also
now gather courage and run down from that particu-
lar tree and from all others that grow near by, and
assembling in a mass encircle the Leopard and dance

---

⁴ πνεῦμά τε καὶ τὸ ἆσθμα.

γενόμενοι περιέρχονταί τε καὶ περιχορεύουσιν
αὐτήν. εἶτα ἐμπηδήσαντες αὐτῇ καὶ ἐπιβάντες
κατεκυβίστησαν καὶ κατωρχήσαντο κέρτομόν τινα
καὶ πιθήκοις πρέπουσαν ὄρχησιν,[1] καὶ ποικίλως
ἐνυβρίσαντες, ἣν ἔχουσιν ὡς ἐπὶ νεκρᾷ χαρὰν καὶ
ἡδονὴν ἐμαρτύραντο. ἡ δὲ ὑπέμεινε πάντα, εἶτα
ὅταν ἐννοήσῃ κεκμηκέναι ὑπό τε τῆς χορείας
αὐτοὺς καὶ τῆς ὕβρεως, ἀδοκήτως ἀναπηδήσασα
καὶ ἐσθοροῦσα[2] τοὺς μὲν τοῖς ὄνυξι διέξηνε, τοὺς
δὲ τοῖς ὀδοῦσι διεσπάσατο, καὶ τὴν ἐκ τῶν
πολεμίων πανθοινίαν τε καὶ πανδαισίαν ἀφθο-
νώτατα ἔχει. τλημόνως δὲ ἔχειν[3] καὶ καρτερῶς
καὶ γεννικῶς ἡ φύσις κελεύει[4] τὴν πάρδαλιν
ὑπὲρ τοῦ τῶν πολεμίων ἐνυβρισάντων περιγενέσθαι
καρτερικώτατα ἐναθλοῦσαν καὶ μὴ δεομένην εἰπεῖν
τέτλαθι δὴ κραδίη. ὅ γε μὴν τοῦ Λαέρτου ἑαυτὸν
ἐξεκάλυψεν ὀλίγου πρὸ τοῦ καιροῦ, τὴν ἐκ τῶν
παιδισκῶν ὕβριν μὴ φέρων.

55. Ἐν τοῖς Ἰνδοῖς οἱ ἐλέφαντες, ὅταν τι τῶν
δένδρων αὐτόρριζον ἀναγκάζωσιν αὐτοὺς οἱ Ἰνδοὶ
ἐκσπάσαι, οὐ πρότερον ἐμπηδῶσιν[5] οὐδὲ ἐπιχει-
ροῦσι τῷ ἔργῳ πρὶν ἢ διασεῖσαι αὐτὸ καὶ διασκέ-
ψασθαι ἆρά γε[6] ἀνατραπῆναι οἷόν τέ ἐστιν ἢ
παντελῶς ἀδύνατον.

56. Αἱ ἐν Σύροις ἔλαφοι γίνονται μὲν ἐν ὄρεσι
μεγίστοις, Ἀμανῷ τε καὶ Λιβάνῳ καὶ Καρμήλῳ·
ὅταν δὲ βουληθῶσι περαιώσασθαι τὴν θάλατταν,
ἐπὶ τὰς ἠόνας ἀφικνοῦνται ἡ ἀγέλη, καὶ ἀναμέ-

---

[1] Ges : ὀρχηστικήν.
[2] ἐκθοροῦσα.
[3] ἔχει.
[4] κατέχει.

round it. Then they leap upon it and turn somer-
saults on its body and by dancing in triumph a dance
appropriate to monkeys, and by a variety of insults
testify to the joy and delight they feel over the supposed corpse. But the Leopard submits to all this
until it realises that the Monkeys are tired by their
dancing and their insolence, when it leaps up un-
expectedly and springs at them. And some it
lacerates with its claws, others it tears to pieces with
its teeth, and enjoys without stint the ample and
sumptuous banquet provided by its enemies. It is
Nature that bids the Leopard endure with heroic
fortitude, so that it may rise superior to the insults
of its enemies, bearing up with the utmost patience
and finding no need to say ' endure, my heart '
[Hom. *Od.* 20. 18]. Indeed the son of Laertes was
within an ace of revealing himself prematurely
through being unable to tolerate the insults of the
maidservants.

55. In India Elephants, when compelled by the The
natives to pull up some tree, roots and all, do not Elephant
immediately attack it and begin the task, until they
have shaken it and have tested it thoroughly to see
whether in fact it can be overturned, or whether that
is utterly impossible.

56. The Deer of Syria are born on the highest Deer cross-
mountains, on Amanus, on Libanus, and on Carmel. ing the sea
And when they want to cross the sea the herd goes
down to the beaches and waits until the wind drops;

νουσι τοῦ πνεύματος τὴν φθίσιν,[1] καὶ ἡνίκα ἂν
αἴσθωνται πρᾶον αὐτὸ καὶ ἥσυχον καταπνέον,
τηνικαῦτα ἐπιθαρροῦσι τῷ πελάγει. νέουσι δὲ
κατὰ στοῖχον, καὶ ἀλλήλων ἔχονται, τὰ γένεια αἱ
ἑπόμεναι τῶν προηγουμένων τῇ ὀσφύι ἐπερείδου-
σαι· ἡ . . .[2] τελευταία δὲ γενομένη τῇ πρόσθεν
ἐπὶ πάσαις ἑαυτὴν ἐπαναπαύσασα εἶτα οὐραγεῖ.
στέλλονται δὲ ἐπὶ τὴν Κύπρον πόθῳ τῆς πόας
τῆς ἐκεῖ· λέγεται γὰρ εἶναι βαθεῖα καὶ νομὰς
ἀγαθὰς παρέχειν.[3] καὶ λέγουσί γε Κύπριοι εὔγεως
οἰκεῖν χῶρον, καὶ ταῖς Αἰγυπτίων ἀρούραις
τολμῶσιν ἀντικρίνειν τὰς σφετέρας. ἔλαφοι δὲ
καὶ ἕτεραι τήνδε τὴν νῆξιν ἀποδείκνυνται. αἱ γοῦν
Ἠπειρώτιδες ἐς τὴν Κέρκυραν διανήχονται, ἀντί-
πορθμοι δὲ ἀλλήλαις αἵδε εἰσίν.

[1] φύσιν.
[2] Lacuna : ἡ ⟨δὲ ἡγουμένη πρόσθεν, ὅταν κάμῃ,⟩ τελευταία
Jac, comp. Opp. Cyn. 2. 225, Max. Tyr. 12. 3.
[3] ἔχειν.

and as soon as they observe that there is a favourable and gentle breeze, then they brave the open sea. And they swim in single file, holding on to one another, the ones behind supporting their chins on the rumps of those in front . . . *a* takes the last place in the line, and resting itself upon the one next in front of it in the whole troop, brings up the rear. And they make for Cyprus in their longing for the meadows there, for they are said to be deep and to afford excellent pasture. The Cypriots indeed claim that they live in a fertile country, and venture to compare their arable land with that of Egypt. And there are Deer from other countries too which show this same capacity for swimming. For example, the Deer of Epirus swim across to Corcyra : the two countries face each other across a strait.

---

a Some words have been lost; following Jacobs's suggested filling of the lacuna we may translate : ' When the one that has been leading hitherto begins to tire, it drops back to the end of the file, and, *etc.*'

PRINTED IN GREAT BRITAIN BY
RICHARD CLAY AND COMPANY, LTD.,
BUNGAY, SUFFOLK.

THE LOEB CLASSICAL LIBRARY

VOLUMES ALREADY PUBLISHED

Latin Authors

AMMIANUS MARCELLINUS. Translated by J. C. Rolfe. 3 Vols.
(Vols. I. and II. *3rd Imp.*, Vol. III. *2nd Imp. revised.*)

APULEIUS : THE GOLDEN ASS (METAMORPHOSES). W. Adling-
ton (1566). Revised by S. Gaselee. (*7th Imp.*)

S. AUGUSTINE : CITY OF GOD. 7 Vols. Vol. I. G. E.
McCracken.

ST. AUGUSTINE, CONFESSIONS OF. W. Watts (1631). 2 Vols.
(Vol. I. *7th Imp.*, Vol. II. *6th Imp.*)

ST. AUGUSTINE, SELECT LETTERS. J. H. Baxter. (*2nd Imp.*)

AUSONIUS. H. G. Evelyn White. 2 Vols. (*2nd Imp.*)

BEDE. J. E. King. 2 Vols. (*2nd Imp.*)

BOETHIUS : TRACTS and DE CONSOLATIONE PHILOSOPHIAE.
Rev. H. F. Stewart and E. K. Rand. (*6th Imp.*)

CAESAR : ALEXANDRIAN, AFRICAN and SPANISH WARS. A. G.
Way.

CAESAR : CIVIL WARS. A. G. Peskett. (*6th Imp.*)

CAESAR : GALLIC WAR. H. J. Edwards. (*10th Imp.*)

CATO : DE RE RUSTICA ; VARRO : DE RE RUSTICA. H. B. Ash
and W. D. Hooper. (*3rd Imp.*)

CATULLUS. F. W. Cornish ; TIBULLUS. J. B. Postgate ; PER-
VIGILIUM VENERIS. J. W. Mackail. (*13th Imp.*)

CELSUS : DE MEDICINA. W. G. Spencer. 3 Vols. (Vol. I.
3rd Imp. revised, Vols. II. and III. *2nd Imp.*)

CICERO : BRUTUS, and ORATOR. G. L. Hendrickson and H. M.
Hubbell. (*3rd Imp.*)

[CICERO] : AD HERENNIUM. H. Caplan.

CICERO : DE FATO ; PARADOXA STOICORUM ; DE PARTITIONE
ORATORIA. H. Rackham (With De Oratore, Vol. II.)
(*2nd Imp.*)

CICERO : DE FINIBUS. H. Rackham. (*4th Imp. revised.*)

CICERO : DE INVENTIONE, etc. H. M. Hubbell.

CICERO : DE NATURA DEORUM and ACADEMICA. H. Rackham.
(*3rd Imp.*)

CICERO : DE OFFICIIS. Walter Miller. (*7th Imp.*)

CICERO : DE ORATORE. 2 Vols. E. W. Sutton and H. Rackham. (*3rd Imp.*)

CICERO : DE REPUBLICA and DE LEGIBUS ; SOMNIUM SCIPIONIS.
Clinton W. Keyes. (*4th Imp.*)

CICERO : DE SENECTUTE, DE AMICITIA, DE DIVINATIONE.
W. A. Falconer. (*6th Imp.*)

CICERO : IN CATILINAM, PRO FLACCO, PRO MURENA, PRO SULLA.
Louis E. Lord. (*3rd Imp. revised.*)

CICERO : LETTERS TO ATTICUS. E. O. Winstedt. 3 Vols. (Vol. I. *7th Imp.*, Vols. II. and III. *4th Imp.*)

CICERO : LETTERS TO HIS FRIENDS. W. Glynn Williams. 3 Vols. (Vols. I. and II. *3rd Imp.*, Vol. III. *2nd Imp. revised.*)

CICERO : PHILIPPICS. W. C. A. Ker. (*4th Imp. revised.*)

CICERO : PRO ARCHIA, POST REDITUM, DE DOMO, DE HARUS-PICUM RESPONSIS, PRO PLANCIO. N. H. Watts. (*4th Imp.*)

CICERO : PRO CAECINA, PRO LEGE MANILIA, PRO CLUENTIO, PRO RABIRIO. H. Grose Hodge. (*3rd Imp.*)

CICERO : PRO CAELIO, DE PROVINCIIS CONSULARIBUS, PRO BALBO. J. H. Freese and R. Gardner

CICERO : PRO MILONE, IN PISONEM, PRO SCAURO, PRO FONTEIO, PRO RABIRIO POSTUMO, PRO MARCELLO, PRO LIGARIO, PRO REGE DEIOTARO. N. H. Watts. (*3rd Imp.*)

CICERO : PRO QUINCTIO, PRO ROSCIO AMERINO, PRO ROSCIO COMOEDO, CONTRA RULLUM. J. H. Freese. (*3rd Imp.*)

CICERO : PRO SESTIO, IN VATINIUM. J. H. Freese and R. Gardner.

CICERO : TUSCULAN DISPUTATIONS. J. E. King. (*4th Imp.*)

CICERO : VERRINE ORATIONS. L. H. G. Greenwood. 2 Vols. (Vol. I. *3rd Imp.*, Vol. II. *2nd Imp.*)

CLAUDIAN. M. Platnauer. 2 Vols. (*2nd Imp.*)

COLUMELLA : DE RE RUSTICA. DE ARBORIBUS. H. B. Ash, E. S. Forster and E. Heffner. 3 Vols. (Vol. I. *2nd Imp.*)

CURTIUS, Q. : HISTORY OF ALEXANDER. J. C. Rolfe. 2 Vols. (*2nd Imp.*)

FLORUS. E. S. Forster and CORNELIUS NEPOS. J. C. Rolfe. (*2nd Imp.*)

FRONTINUS : STRATAGEMS and AQUEDUCTS. C. E. Bennett and M. B. McElwain. (Vol. I. *3rd Imp.*, Vol. II. *2nd Imp.*)

FRONTO : CORRESPONDENCE. C. R. Haines. 2 Vols. (*3rd Imp.*)

GELLIUS, J. C. Rolfe. 3 Vols. (Vol. I. *3rd Imp.*, Vols. II. and III. *2nd Imp.*)

HORACE : ODES and EPODES. C. E. Bennett. (*14th Imp. revised.*)

HORACE : SATIRES, EPISTLES, ARS POETICA. H. R. Fairclough. (*9th Imp. revised.*)

JEROME : SELECTED LETTERS. F. A. Wright. (*2nd Imp.*)

JUVENAL and PERSIUS. G. G. Ramsay. (*8th Imp.*)

LIVY. B. O. Foster, F. G. Moore, Evan T. Sage, and A. C. Schlesinger. 14 Vols. Vols. I.–XIII. (Vol. I. *5th Imp.*, Vol. V. *4th Imp.*, Vols. II.–IV., VII., IX.–XII. *3rd Imp.*, Vols. VI., VIII., *2nd Imp. revised.*)

LUCAN. J. D. Duff. (*4th Imp.*)

LUCRETIUS. W. H. D. Rouse. (*7th Imp. revised.*)

MARTIAL. W. C. A. Ker. 2 Vols. (Vol. I. *5th Imp.*, Vol. II. *4th Imp. revised.*)

MINOR LATIN POETS : from PUBLILIUS SYRUS to RUTILIUS NAMATIANUS, including GRATTIUS, CALPURNIUS SICULUS, NEMESIANUS, AVIANUS, and others with " Aetna " and the " Phoenix." J. Wight Duff and Arnold M. Duff. (*3rd Imp.*)

OVID : THE ART OF LOVE and OTHER POEMS. J. H. Mozley (*4th Imp.*)

2

Ovid : Fasti. Sir James G. Frazer. (2nd Imp.)
Ovid : Heroides and Amores. Grant Showerman. (6th Imp.)
Ovid : Metamorphoses. F. J. Miller. 2 Vols. (Vol. I. 11th Imp., Vol. II. 9th Imp.)
Ovid : Tristia and Ex Ponto. A. L. Wheeler. (3rd Imp.)
Persius. Cf. Juvenal.
Petronius. M. Heseltine, Seneca Apocolocyntosis. W. H. D. Rouse. (9th Imp. revised.)
Plautus. Paul Nixon. 5 Vols. (Vol. I. 6th Imp., II. 5th Imp., III. 4th Imp., IV. and V. 2nd Imp.)
Pliny : Letters. Melmoth's Translation revised by W. M. L. Hutchinson. 2 Vols. (Vol. I. 7th Imp., Vol. II. 6th Imp.)
Pliny : Natural History. H. Rackham and W. H. S. Jones. 10 Vols. Vols. I.-V. and IX. H. Rackham. Vols. VI. and VII. W. H. S. Jones. (Vols. I.-III. 3rd Imp., Vol. IV. 2nd Imp.)
Propertius. H. E. Butler. (7th Imp.)
Prudentius. H. J. Thomson. 2 Vols.
Quintilian. H. E. Butler. 4 Vols. (Vols. I. and IV. 4th Imp., Vols. II. and III. 3rd Imp.)
Remains of Old Latin. E. H. Warmington. 4 vols. Vol. I. (Ennius and Caecilius.) Vol. II. (Livius, Naevius, Pacuvius, Accius.) Vol. III. (Lucilius and Laws of XII Tables.) Vol. IV. (2nd Imp.) (Archaic Inscriptions.)
Sallust. J. C. Rolfe. (4th Imp. revised.)
Scriptores Historiae Augustae. D. Magie. 3 Vols. (Vol. I. 3rd Imp. revised, Vols. II. and III. 2nd Imp.)
Seneca : Apocolocyntosis. Cf. Petronius.
Seneca : Epistulae Morales. R. M. Gummere. 3 Vols. (Vol. I. 4th Imp., Vols. II. and III. 2nd Imp.)
Seneca : Moral Essays. J. W. Basore. 3 Vols. (Vol. II. 3rd Imp., Vols. I. and III. 2nd Imp. revised.)
Seneca : Tragedies. F. J. Miller. 2 Vols. (Vol. I. 4th Imp., Vol. II. 3rd Imp. revised.)
Sidonius : Poems and Letters. W. B. Anderson. 2 Vols. (Vol. I. 2nd Imp.)
Silius Italicus. J. D. Duff. 2 Vols. (Vol. I. 2nd Imp., Vol. II. 3rd Imp.)
Statius. J. H. Mozley. 2 Vols. (2nd Imp.)
Suetonius. J. C. Rolfe. 2 Vols. (Vol. I. 7th Imp., Vol. II. 6th Imp. revised.)
Tacitus : Dialogus. Sir Wm. Peterson. Agricola and Germania. Maurice Hutton. (6th Imp.)
Tacitus : Histories and Annals. C. H. Moore and J. Jackson. 4 Vols. (Vols. I. and II. 4th Imp., Vols. III. and IV. 3rd Imp.)
Terence. John Sargeaunt. 2 Vols. (7th Imp.)
Tertullian : Apologia and De Spectaculis. T. R. Glover. Minucius Felix. G. H. Rendall. (2nd Imp.)
Valerius Flaccus. J. H. Mozley. (3rd Imp. revised.)
Varro : De Lingua Latina. R. G. Kent. 2 Vols. (2nd Imp. revised.)

3

VELLEIUS PATERCULUS and RES GESTAE DIVI AUGUSTI. F. W. Shipley. (2nd Imp.)
VIRGIL. H. R. Fairclough. 2 Vols. (Vol. I. 19th Imp., Vol. II. 14th Imp. revised.)
VITRUVIUS : DE ARCHITECTURA. F. Granger. 2 Vols. (Vol. I. 3rd Imp., Vol. II. 2nd Imp.)

Greek Authors

ACHILLES TATIUS. S. Gaselee. (2nd Imp.)
AENEAS TACTICUS, ASCLEPIODOTUS and ONASANDER. The Illinois Greek Club. (2nd Imp.)
AESCHINES. C. D. Adams. (3rd Imp.)
AESCHYLUS. H. Weir Smyth. 2 Vols. (Vol. I. 7th Imp., Vol. II. 6th Imp. revised.)
ALCIPHRON, AELIAN, PHILOSTRATUS LETTERS. A. R. Benner and F. H. Fobes.
ANDOCIDES, ANTIPHON, Cf. MINOR ATTIC ORATORS.
APOLLODORUS. Sir James G. Frazer. 2 Vols. (3rd Imp.)
APOLLONIUS RHODIUS. R. C. Seaton. (5th Imp.)
THE APOSTOLIC FATHERS. Kirsopp Lake. 2 Vols. (Vol. I. 8th Imp., Vol. II. 6th Imp.)
APPIAN : ROMAN HISTORY. Horace White. 4 Vols. (Vol. I. 4th Imp., Vols. II.–IV. 3rd Imp.)
ARATUS. Cf. CALLIMACHUS.
ARISTOPHANES. Benjamin Bickley Rogers. 3 Vols. Verse trans. (5th Imp.)
ARISTOTLE : ART OF RHETORIC. J. H. Freese. (3rd Imp.)
ARISTOTLE : ATHENIAN CONSTITUTION, EUDEMIAN ETHICS, VICES AND VIRTUES. H. Rackham. (3rd Imp.)
ARISTOTLE : GENERATION OF ANIMALS. A. L. Peck. (2nd Imp.)
ARISTOTLE : METAPHYSICS. H. Tredennick. 2 Vols. (4th Imp.)
ARISTOTLE : METEOROLOGICA. H. D. P. Lee.
ARISTOTLE : MINOR WORKS. W. S. Hett. On Colours, On Things Heard, On Physiognomies, On Plants, On Marvellous Things Heard, Mechanical Problems, On Indivisible Lines, On Situations and Names of Winds, On Melissus, Xenophanes. and Gorgias. (2nd Imp.)
ARISTOTLE : NICOMACHEAN ETHICS. H. Rackham. (6th Imp. revised.)
ARISTOTLE : OECONOMICA and MAGNA MORALIA. G. C. Armstrong; (with Metaphysics, Vol. II.). (Vol. I. 4th Imp., Vol. II. 3rd Imp.)
ARISTOTLE : ON THE HEAVENS. W. K. C. Guthrie. (3rd Imp. revised.)
ARISTOTLE : On Sophistical Refutations, On Coming to be and Passing Away, On the Cosmos. E. S. Forster and D. J. Furley.
ARISTOTLE : ON THE SOUL, PARVA NATURALIA, ON BREATH. W. S. Hett. (2nd Imp. revised.)
ARISTOTLE : ORGANON, CATEGORIES : On Interpretation, Prior Analytics. H. P. Cooke and H. Tredennick. (3rd Imp.)

4

Aristotle : Parts of Animals. A. L. Peck; Motion and Progression of Animals. E. S. Forster. (4th Imp. revised.)

Aristotle : Physics. Rev. P. Wicksteed and F. M. Cornford. 2 Vols. (Vol. I. 2nd Imp., Vol. II. 3rd Imp.)

Aristotle : Poetics and Longinus. W. Hamilton Fyfe; Demetrius on Style. W. Rhys Roberts. (5th Imp. revised.)

Aristotle : Politics. H. Rackham. (4th Imp. revised.)

Aristotle : Problems. W. S. Hett. 2 Vols. (2nd Imp. revised.)

Aristotle : Rhetorica Ad Alexandrum (with Problems. Vol. II.). H. Rackham.

Arrian : History of Alexander and Indica. Rev. E. Iliffe Robson. 2 Vols. (Vol. I. 3rd Imp., Vol. II. 2nd Imp.)

Athenaeus : Deipnosophistae. C. B. Gulick. 7 Vols. (2nd Imp.)

St. Basil : Letters. R. J. Deferrari. 4 Vols. (2nd Imp.)

Callimachus : Fragments. A. C. Trypanis.

Callimachus, Hymns and Epigrams, and Lycophron. A. W. Mair; Aratus. G. R. Mair. (2nd. Imp.)

Clement of Alexandria. Rev. G. W. Butterworth. (3rd Imp.)

Colluthus. Cf. Oppian.

Daphnis and Chloe. Thornley's Translation revised by J. M. Edmonds; and Parthenius. S. Gaselee. (4th Imp.)

Demosthenes I : Olynthiacs, Philippics and Minor Orations. I.-XVII. and XX. J. H. Vince. (2nd Imp.)

Demosthenes II : De Corona and De Falsa Legatione. C. A. Vince and J. H. Vince. (3rd Imp. revised.)

Demosthenes III : Meidias, Androtion, Aristocrates, Timocrates and Aristogeiton, I and II. J. H. Vince. (2nd Imp.)

Demosthenes IV-VI : Private Orations and In Neaeram. A. T. Murray. (Vol. IV. 3rd Imp., Vols. V. and VI. 2nd Imp.)

Demosthenes VII : Funeral Speech, Erotic Essay, Exordia and Letters. N. W. and N. J. DeWitt.

Dio Cassius : Roman History. E. Cary. 9 Vols. (Vols. I. and II. 3rd Imp., Vols. III.-IX. 2nd Imp.)

Dio Chrysostom. J. W. Cohoon and H. Lamar Crosby. 5 Vols. Vols. I.-IV. 2nd Imp.)

Diodorus Siculus. 12 Vols. Vols. I.-VI. C. H. Oldfather. Vol. VII. C. L. Sherman. Vols. IX. and X. R. M. Geer. Vol. XI. F. Walton. (Vols. I.-IV. 2nd Imp.)

Diogenes Laertius. R. D. Hicks. 2 Vols. (Vol. I. 4th Imp., Vol. II. 3rd Imp.)

Dionysius of Halicarnassus : Roman Antiquities. Spelman's translation revised by E. Cary. 7 Vols. (Vols. I.-V. 2nd Imp.)

Epictetus. W. A. Oldfather. 2 Vols. (Vol. I. 3rd Imp., II. 2nd Imp.)

Euripides. A. S. Way. 4 Vols. (Vol. I. 7th imp., Vol. II. 8th Imp., Vols. III. and IV. 6th Imp.) Verse trans.

Eusebius : Ecclesiastical History. Kirsopp Lake and J. E. L. Oulton. 2 Vols. (Vol. I. 3rd Imp., Vol. II. 5th Imp.)

GALEN : ON THE NATURAL FACULTIES. A. J. Brock. (4th Imp.)
THE GREEK ANTHOLOGY. W. R. Paton. 5 Vols. (Vols. I.–IV. 5th Imp., Vol. V. 3rd Imp.)
GREEK ELEGY AND IAMBUS with the ANACREONTEA. J. M. Edmonds. 2 Vols. (Vol. I. 3rd Imp., Vol. II. 2nd Imp.)
THE GREEK BUCOLIC POETS (THEOCRITUS, BION, MOSCHUS). J. M. Edmonds. (7th Imp. revised.)
GREEK MATHEMATICAL WORKS. Ivor Thomas. 2 Vols. (3rd Imp.)
HERODES. Cf. THEOPHRASTUS : CHARACTERS.
HERODOTUS. A. D. Godley. 4 Vols. (Vol. I. 4th Imp., Vols. II. and III. 5th Imp., Vol. IV. 3rd Imp.)
HESIOD AND THE HOMERIC HYMNS. H. G. Evelyn White. 7th Imp. revised and enlarged.)
HIPPOCRATES and the FRAGMENTS OF HERACLEITUS. W. H. S. Jones and E. T. Withington. 4 Vols. (Vol. I. 4th Imp., Vols. II.–IV. 3rd Imp.)
HOMER : ILIAD. A. T. Murray. 2 Vols. (7th Imp.)
HOMER : ODYSSEY. A. T. Murray. 2 Vols. (8th Imp.)
ISAEUS. E. W. Forster. (3rd Imp.)
ISOCRATES. George Norlin and LaRue Van Hook. 3 Vols. (2nd Imp.)
ST. JOHN DAMASCENE : BARLAAM AND IOASAPH. Rev. G. R. Woodward and Harold Mattingly. (3rd Imp. revised.)
JOSEPHUS. H. St. J. Thackeray and Ralph Marcus. 9 Vols. Vols. I.–VII. (Vol. V. 3rd Imp., Vols. I.–IV., VI. and VII. 2nd Imp.)
JULIAN. Wilmer Cave Wright. 3 Vols. (Vols. I. and II. 3rd Imp., Vol. III. 2nd Imp.)
LUCIAN. A. M. Harmon. 8 Vols. Vols. I.–V. (Vols. I. and II. 4th Imp., Vol. III. 3rd Imp., Vols. IV. and V. 2nd Imp.)
LYCOPHRON. Cf. CALLIMACHUS.
LYRA GRAECA. J. M. Edmonds. 3 Vols. (Vol. I. 5th Imp., Vol. II. revised and enlarged, and III. 4th Imp.)
LYSIAS. W. R. M. Lamb. (2nd Imp.)
MANETHO. W. G. Waddell : PTOLEMY : TETRABIBLOS. F. E. Robbins. (3rd Imp.)
MARCUS AURELIUS. C. R. Haines. (4th Imp. revised.)
MENANDER. F. G. Allinson. (3rd Imp. revised.)
MINOR ATTIC ORATORS (ANTIPHON, ANDOCIDES, LYCURGUS, DEMADES, DINARCHUS, HYPEREIDES). K. J. Maidment and J. O. Burtt. 2 Vols. (Vol. I. 2nd Imp.)
NONNOS : DIONYSIACA. W. H. D. Rouse. 3 Vols. (2nd Imp.)
OPPIAN, COLLUTHUS, TRYPHIODORUS. A. W. Mair. (2nd Imp.)
PAPYRI. NON-LITERARY SELECTIONS. A. S. Hunt and C. C. Edgar. 2 Vols. (Vol. I. 2nd Imp.) LITERARY SELECTIONS. Vol. I. (Poetry). D. L. Page. (3rd Imp.)
PARTHENIUS. Cf. DAPHNIS AND CHLOE.
PAUSANIAS : DESCRIPTION OF GREECE. W. H. S. Jones. 5 Vols. and Companion Vol. arranged by R. E. Wycherley. (Vols. I. and III. 3rd Imp., Vols. II., IV. and V. 2nd Imp.)

PHILO. 10 Vols. Vols. I.-V.; F. H. Colson and Rev. G. H. Whitaker. Vols. VI.–IX.; F. H. Colson. (Vols. I., III., V., and VI. *3rd Imp.*, Vol. IV. *4th Imp.*, Vols. II., VII.–IX. *2nd Imp.*)

PHILO: two supplementary Vols. (*Translation only.*) Ralph Marcus.

PHILOSTRATUS: THE LIFE OF APPOLLONIUS OF TYANA. F. C. Conybeare. 2 Vols. (Vol. I. *4th Imp.*, Vol. II. *3rd Imp.*)

PHILOSTRATUS: IMAGINES; CALLISTRATUS: DESCRIPTIONS. A. Fairbanks. (*2nd Imp.*)

PHILOSTRATUS and EUNAPIUS: LIVES OF THE SOPHISTS. Wilmer Cave Wright. (*2nd Imp.*)

PINDAR. Sir J. E. Sandys. (*8th Imp. revised.*)

PLATO: CHARMIDES, ALCIBIADES, HIPPARCHUS, THE LOVERS, THEAGES, MINOS and EPINOMIS. W. R. M. Lamb. (*2nd Imp.*)

PLATO: CRATYLUS, PARMENIDES, GREATER HIPPIAS, LESSER HIPPIAS. H. N. Fowler. (*4th Imp.*)

PLATO: EUTHYPHRO, APOLOGY, CRITO, PHAEDO, PHAEDRUS. H. N. Fowler. (*11th Imp.*)

PLATO: LACHES, PROTAGORAS, MENO, EUTHYDEMUS. W. R. M. Lamb. (*3rd Imp. revised.*)

PLATO: LAWS. Rev. R. G. Bury. 2 Vols. (*3rd Imp.*)

PLATO: LYSIS, SYMPOSIUM, GORGIAS. W. R. M. Lamb. (*5th Imp. revised.*)

PLATO: REPUBLIC. Paul Shorey. 2 Vols. (Vol. I. *5th Imp.*, Vol. II. *4th Imp.*)

PLATO: STATESMAN, PHILEBUS. H. N. Fowler; ION. W. R. M. Lamb. (*4th Imp.*)

PLATO: THEAETETUS and SOPHIST. H. N. Fowler. (*4th Imp.*)

PLATO: TIMAEUS, CRITIAS, CLITOPHO, MENEXENUS, EPISTULAE. Rev. R. G. Bury. (*3rd Imp.*)

PLUTARCH: MORALIA. 14 Vols. Vols. I.–V. F. C. Babbitt; Vol. VI. W. C. Helmbold; Vol. X. H. N. Fowler. Vol. XII. H. Cherniss and W. C. Helmbold. (Vols. I.–VI. and X. *2nd Imp.*)

PLUTARCH: THE PARALLEL LIVES. B. Perrin. 11 Vols. (Vols. I., II., VI., VII., and XI. *3rd Imp.* Vols. III.–V. and VIII.–X. *2nd Imp.*)

POLYBIUS. W. R. Paton. 6 Vols. (*2nd Imp.*)

PROCOPIUS: HISTORY OF THE WARS. H. B. Dewing. 7 Vols. (Vol. I. *3rd Imp.*, Vols. II.–VII. *2nd Imp.*)

PTOLEMY: TETRABIBLOS. Cf. MANETHO.

QUINTUS SMYRNAEUS. A. S. Way. Verse trans. (*3rd Imp.*)

SEXTUS EMPIRICUS. Rev. R. G. Bury. 4 Vols. (Vol. I. *4th Imp.*, Vols. II. and III. *2nd Imp.*)

SOPHOCLES. F. Storr. 2 Vols. (Vol. I. *10th Imp.* Vol. II. *6th Imp.*) Verse trans.

STRABO: GEOGRAPHY. Horace L. Jones. 8 Vols. (Vols. I., V., and VIII. *3rd Imp.*, Vols. II., III., IV., VI., and VII. *2nd Imp.*)

THEOPHRASTUS: CHARACTERS. J. M. Edmonds. HERODES, etc. A. D. Knox. (*3rd Imp.*)

THEOPHRASTUS : ENQUIRY INTO PLANTS. Sir Arthur Hort, Bart. 2 Vols. (2nd Imp.)
THUCYDIDES. C. F. Smith. 4 Vols. (Vol. I. 5th Imp., Vols. II., III., and IV. 3rd Imp. revised)
TRYPHIODOBUS. Cf. OPPIAN.
XENOPHON : CYROPAEDIA. Walter Miller. 2 Vols. (Vol. I. 4th Imp., Vol. II. 3rd Imp.)
XENOPHON : HELLENICA, ANABASIS, APOLOGY, and SYMPOSIUM. C. L. Brownson and O. J. Todd. 3 Vols. (Vols. I. and III. 3rd Imp., Vol. II. 4th Imp.)
XENOPHON : MEMORABILIA and OECONOMICUS. E. C. Marchant. (3rd Imp.)
XENOPHON : SCRIPTA MINORA. E. C. Marchant. (3rd Imp.)

IN PREPARATION

Greek Authors

AELIAN : ON THE NATURE OF ANIMALS. A. F. Scholfield.
ARISTOTLE : HISTORY OF ANIMALS. A. L. Peck.
PLOTINUS : A. H. Armstrong.

Latin Authors

PHAEDRUS. Ben E. Perry.

DESCRIPTIVE PROSPECTUS ON APPLICATION

London   WILLIAM HEINEMANN LTD
Cambridge, Mass.   HARVARD UNIVERSITY PRESS